Analysis and Deformulation of Polymeric Materials

Paints, Plastics, Adhesives, and Inks

TOPICS IN APPLIED CHEMISTRY

Series Editors: Alan R. Katritzky, FRS
Kenan Professor of Chemistry
University of Florida, Gainesville, Florida

Gebran J. Sabongi
Laboratory Manager, Encapsulation Technology Center
3M Company, St. Paul, Minnesota

Current volumes in the series:

ANALYSIS AND DEFORMULATION OF POLYMERIC MATERIALS
Paints, Plastics, Adhesives, and Inks
Jan W. Gooch

CHEMISTRY AND APPLICATIONS OF LEUCO DYES
Edited by Ramaiah Muthyala

FROM CHEMICAL TOPOLOGY TO THREE-DIMENSIONAL GEOMETRY
Edited by Alexandru T. Balaban

LEAD-BASED PAINT HANDBOOK
Jan W. Gooch

ORGANOFLUORINE CHEMISTRY
Principles and Commercial Applications
Edited by R. E. Banks, B. E. Smart, and J. C. Tatlow

PHOSPHATE FIBERS
Edward J. Griffith

POLY(ETHYLENE GLYCOL) CHEMISTRY
Biotechnical and Biomedical Applications
Edited by J. Milton Harris

RADIATION CURING
Science and Technology
Edited by S. Peter Pappas

RESORCINOL
Its Uses and Derivatives
Hans Dressler

TARGET SITES FOR HERBICIDE ACTION
Edited by Ralph C. Kirkwood

A Continuation Order Plan is available for this series. A continuation order will bring delivery of each new volume immediately upon publication. Volumes are billed only upon actual shipment. For further information please contact the publisher.

Analysis and Deformulation of Polymeric Materials

Paints, Plastics, Adhesives, and Inks

Jan W. Gooch
Polymers and Coatings Consultant
Atlanta, Georgia

Plenum Press • New York and London

Library of Congress Cataloging-in-Publication Data

On file

ISBN 0-306-45541-2

A Division of Plenum Publishing Corporation
233 Spring Street, New York, N. Y. 10013

http://www.plenum.com

10 9 8 7 6 5 4 3 2 1

Printed in the United States of America

Preface

This book is designed for the chemist, formulator, student, teacher, forensic scientist, or others who wish to investigate the composition of polymeric materials. The information within these pages is intended to arm the reader with the necessary working knowledge to analyze, characterize, and deformulate materials.

The structure of the Contents is intended to assist the reader in quickly locating the subject of interest and proceed to it with a minimum of expended time and effort. The Contents provides an outline of major topics and relevant materials characterized for the reader's convenience. An introduction to analysis and deformulation is provided in Chapter 1 to acquaint the reader with analytical methods and their applications. Extensive references are provided as additional sources of information. All tables are located in the Appendix, beginning on p. 235.

GUIDE FOR USE

This is a practical book structured to efficiently use the reader's time with a minimum effort of searching for entries and information by following these brief instructions:

1. Search the Contents and/or Index for a subject within the text.
2. Analysis/deformulation principles are discussed at the outset to familiarize the reader with analysis methods and instruments; followed by formulations, materials, and analysis of paint, plastics, adhesives, and inks; and finally reformulation methods to test the results of analysis.
3. Materials and a wide assortment of formulations are discussed within the text by chapter/section number.
4. Materials are referred to by various names (trivial, trade, and scientific), and these are listed in tables and cross-referenced to aid the reader.

ACKNOWLEDGMENTS

I wish to thank the following people for their contributions to this book: Lisa Detter-Hoskin; Garth Freeman; John Sparrow; Joseph Schork; Gary Poehlein, Kash Mittal; John Muzzy; Paul Hawley; Ad Hofland; Tor Aasrum; James Johnson; Linda, Sonja, Luther, and Lottie Gooch.

Contents

5. Paint Materials

List of Figures

CHAPTER 1

CHAPTER 2

CHAPTER 3

CHAPTER 6

CHAPTER 9

CHAPTER 12

CHAPTER 15

1

Deformulation Principles

1.1. INTRODUCTION

You have a manufactured product or an unknown formulated material, and you want to know its composition. How do you go about it without spending an enormous amount of time and money? This book is designed to answer those questions in great detail.

Just identifying a solid or liquid substance can be a challenging experience, and accurately analyzing a multicomponent formulation can be an exhausting one. In liquid or solid forms, a paint can resemble an adhesive, ink, or plastic material. Therefore, we will explore extensively how to distinguish types of formulations and how to efficiently, economically, and, hopefully, painlessly deformulate it.

Formulations can be mixtures of materials of widely varying concentrations and forms. To investigate any formulated plastic, paint, adhesive, or ink material, the investigator must have a plan to deformulate or reverse engineer, then analyze each separated component. A typical formulation requires very specific isolation of a mixture of chemical compounds before an identification of individual components can be attempted. The state and chemical nature of materials vary widely, and require a host of analytical tools. Historically, the strategy for analysis has varied as widely. Strategy is provided for using proven methods to untangle and characterize multicomponents from a single formulation.

The structure of this book as outlined in the Contents consists of a logical scheme to allow the reader to identify a particular area of interest. The basic scheme consists of formulations, materials used in the formulation, and followed by methods of deformulation.

The reader is referred to texts on qualitative and quantitative chemistry principles and techniques for precise laboratory methods.

There is a "deformulation" chapter following each paint, plastics, adhesives, and inks materials chapter. Many of the deformulation principles are similar. For this reason, the information is usually discussed once and referred to in other deformulation chapters to eliminate repetition of the material.

Standard materials found in formulations are well characterized, and the results are presented in each case. The reader will find these characterizations invaluable when comparing experimental results for purposes of identification.

1.2. CHARACTERIZATION OF MATERIALS

Though materials come in different forms such as solids and liquids, methods for accurate analysis are available. Successful analysis depends on isolation of individual components and a proper selection of tools for investigation.

The typical properties of materials and methods of analysis are listed in Table 1.1 (see Appendix, p. 235). Types of analysis are discussed in Chapters 2 (surface analysis) and 3 (bulk analysis) together with corresponding analytical instruments. No investigation can be performed without the proper tools, and materials such as polymers and pigments require corresponding instrumentation for identification and characterization such as infrared spectroscopy and X-ray diffraction. The methods and equipment for surface and bulk analysis are discussed in Chapters 2 and 3. The emphasis is on information that is valuable to the user without going into great detail about theory or hardware. The user will need to identify a competent operator of equipment (or laboratory) to acquire the necessary analytical data.

It is seldom necessary to use all of the tools in Table 1.1 to identify components in a formulation, but analysis by more than one method is recommended for confirmation. In other words, what degree of confidence is required?

A standard or control specimen of a material is always recommended for comparison to the specimen under study.

1.3. FORMULATION AND DEFORMULATION

A paint, plastic, adhesive, or ink is actually a mixture of materials to create a formulation. Almost all formulations are types of dispersions including emulsions and suspensions, and separation of the phases is the first step of deformulation. The formulation is the useful form of materials to perform a task which is often a commercial product. Physical measurements can be performed on a formulation such as weight per gallon. However, the formulation must be treated as a mixture and subdivided into its individual components. Only then can analysis of each material begin. The general scheme for analysis of formulations is illustrated in Fig. 1.1 showing methods of identifying each component.

The first concern relates to whether the formulated materials are in solid or liquid form. If the specimen is a liquid, then solids are separated using gravity or increased gravity called *centrifugation*. Separation of solids from fluids is described by Stokes's law (Weast, 1978): When a small sphere (or particle) falls under the action of gravity through a viscous medium, it ultimately acquires a constant velocity V (cm/sec),

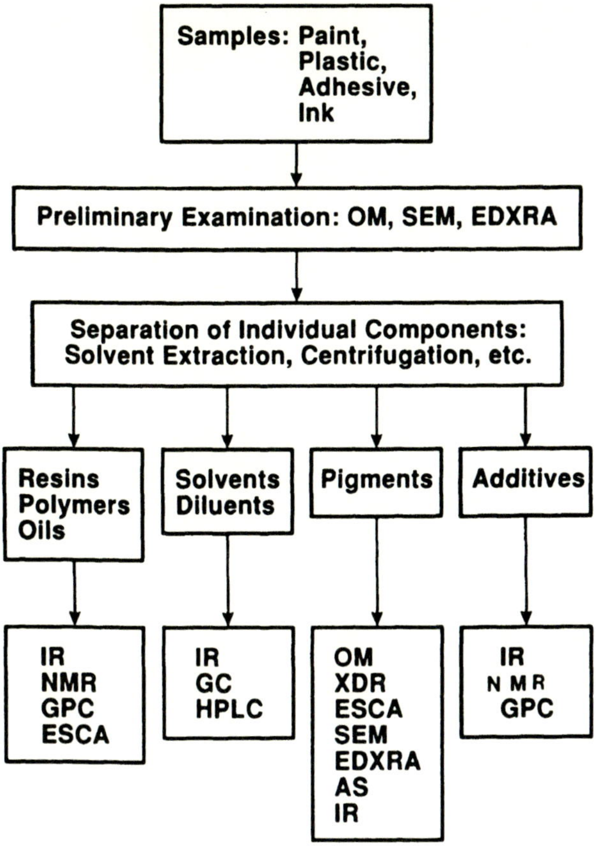

Figure 1.1. Basic deformulation scheme for paint, plastics, adhesives, and inks.

$$V = [2ga^2(d_1 - d_2)]/9\eta$$

where a (cm) is the radius of the sphere, d_1 and d_2 (g/cm^3) the densities of the sphere and the medium, respectively, η (dyn-sec/cm^2, or poise) the viscosity, and g (cm/sec^2) the gravity.

From Stokes's law, the greater the differences in density of the particle and the medium, the greater is the rate of separation. Also, the closer the particle resembles a perfect sphere, the greater is the rate of sedimentation and separation. A liquid formulation is subjected to several orders of gravity by spinning in a mechanical centrifuge. Earth's gravity causes particles to naturally fall through fluids such as water and air, but mechanical centrifugation greatly accelerates the motion of the particle. Mechanical centrifugation can reduce the time for separation to a couple of hours compared to years at natural gravity conditions. Centrifugal force is defined as

$$F = (mv^2)/R$$

where F (dyn) is force, m (g) is mass, v (cm/sec) is velocity, and R (cm) is radius of rotation. From this equation, increasing velocity dramatically increases force by the square of the velocity. Many dispersions never separate under natural gravity, or filtration.

A liquid specimen is centrifuged or filtered to separate major components such as resin/solvent fraction and pigments which can be further separated. A laboratory centrifugation separation is illustrated in Fig. 1.2. A photograph of a Fisher Marathon centrifuge is shown in Fig. 1.3. Centrifugation of components is an efficient method of separating emulsions and suspensions as all of the components separate in individual layers by density. Decreasing the temperature of a liquid suspension can sometimes aid the separation, and can reduce the vapor pressure of a volatile solvent like acetone. Temperature control is important because heat is generated during centrifugation. A centrifuge with temperature control is shown in Fig. 1.3 with a fixed angle rotor and centrifuge tube. No filtering is required when using centrifugation. However, dissolved resins and polymers in solvents do not

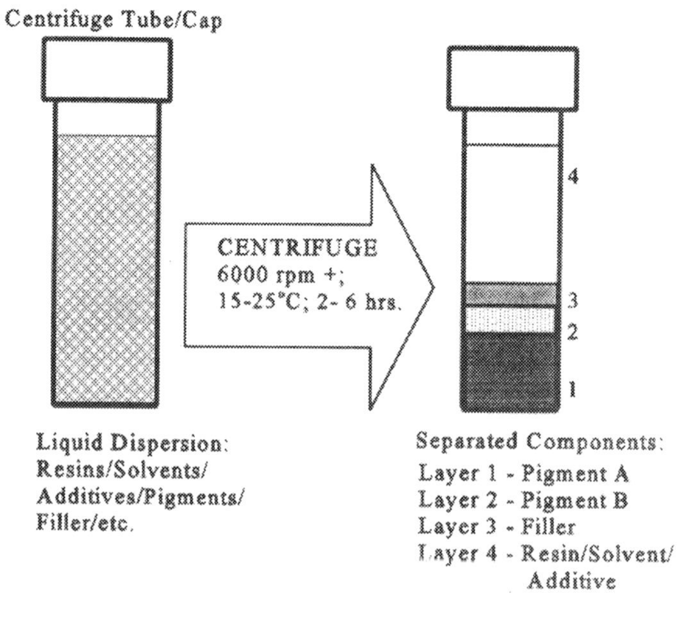

Figure 1.2. Separation of dispersed components from formulations.

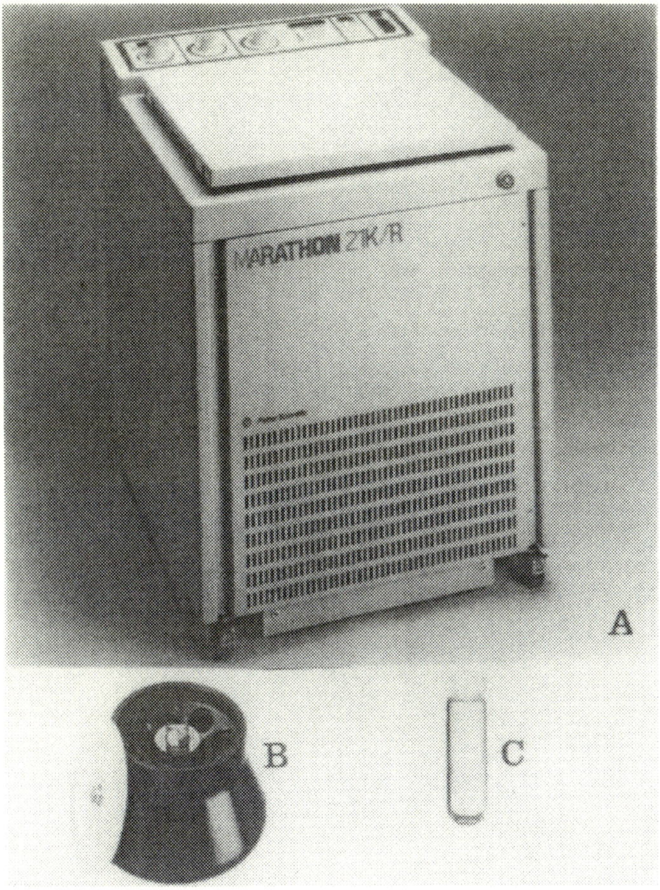

Figure 1.3. Photograph of Fisher Marathon Model 21K/R General-Purpose Refrigerated Centrifuge, maximum speed 13,300 rpm, temperature range –20 to –40°C (A) Centrifuge; (B) eight place fixed angle rotor; and (C) Nalgene polypropylene copolymer centrifuge tubes with screw caps. Reprinted with permission of Fisher Scientific Company.

separate by centrifugation. Following separation, each component can be individually examined and identified.

A solid formulation such as a paint chip or a plastic part must be analyzed as a mixture of components, using surface reflectance methods with microscopic resolution.

In the following pages, formulations are investigated with many examples and step-by-step procedures. Formulations of popular and widely used products are presented to give the reader an understanding of how a product is formulated for the consumer market.

2

Surface Analysis

2.1. LIGHT MICROSCOPY (LM)

2.1.1. Fundamentals

Light microscopy (Hemsley, 1984; McCrone, 1974) is useful for studying the pigments for color, particle size and distribution, and concentration in films. Although light microscopy is useful for studying polymer surfaces (Hemsley, 1984), its use for the study of surfaces has decreased considerably since the commercial introduction of scanning electron microscopes (SEM). These instruments will resolve detail one-tenth as large (20 nm = 0.02 μm) as that resolved by the light microscope, and the in-focus depth of field of the SEM is 100–300 times that of the light microscope. A Leica Strata Lab Monocular Microscope in shown in Fig. 2.1.

There are other advantages of the SEM, including ease of sample preparation, elemental analysis by energy-dispersive X-ray analyzer, and, usually, excellent specimen contrast. The light microscope is still important because the cost of an SEM is 10 to 50 times that of an adequate light microscope. In addition, there are many routine surface examinations easily performed by light optics that do not justify use of the SEM. There are at least a few surface characterization problems for which the SEM cannot be used: surfaces of materials unstable under high vacuum or high-energy electron bombardment, samples too bulky for the SEM sample compartment, and samples requiring manipulation on the surface during examination and vertical resolution of detail below 250 μm. Also, the natural color of the specimen (e.g., paint pigment) is observed with the light microscope whereas it cannot be determined in the electron microscope.

It is wise to examine a specimen with an optical microscope before proceeding to other methods of examination. A simple visual inspection may provide the necessary information for identification.

Often, of course, both the light microscope and the SEM are used to examine paint materials. The stereobinocular microscope is needed if only to quickly decide

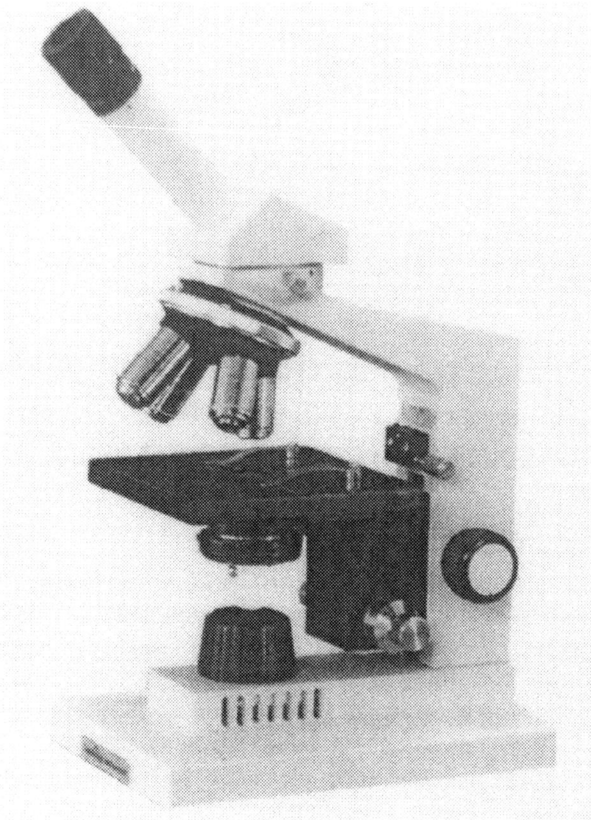

Figure 2.1. Photograph of Leica Strate Lab Monocular Microscope. Reprinted with permission of Leica Instruments Co.

what areas to study or to examine the pertinent areas in terms of the total sample including color. Even SEM examination should begin at low magnification and never be increased more than necessary.

There are accessories for the light microscope that greatly enhance its ability to resolve detail, differentiate different compositions, or increase contrast. Any microscopist who has attempted to observe thin coatings on paper, e.g., ink lines, with the SEM soon goes back to the light microscope. The Nomarski interference contrast system on a reflected light microscope gives excellent rendition of surface detail for metals, ceramics, polymers, or biological tissue. The SEM is 10 times better than the light microscope in horizontal resolution but 20 times worse in vertical resolution.

Characterization of a surface refers to topography, elemental composition, and solid-state structure. All three are usually studied by what is often termed *morpho-*

logical analysis, i.e., shape characteristics. Surface geometry or topography is obviously a matter of morphology. The light microscopist may have to enhance contrast of transparent, colorless surfaces like paper or ceramics by a surface treatment (e.g., an evaporated-metal coating).

Elemental composition determination is often possible by study of morphology although it perhaps can be made easier by surface etching, staining, or examination by polarized light.

When micromorphological studies fail, the investigator then proceeds to the electron microscope for topography, to the electron beam probe (EBP), electron spectroscopy chemical analysis (ESCA), or the scanning electron microscope (SEM) with energy-dispersive X-ray analysis (EDXRA) for elemental analysis.

• Topography. The topography of a surface greatly affects wear, friction, reflectivity, catalysis, and a host of other properties. Many techniques are used to study surfaces, but most begin with visual examination supplemented by increasing magnification of the light microscope. Straightforward microscopy may be supplemented by either sample-preparation techniques or use of specialized microscope accessories.

There are two general methods of observing surfaces, dark-field and bright-field. Each of these, however, can be obtained with transmitted light from a substage condenser and with reflected light from above the preparation. For bright-field top lighting, the microscope objective itself must act as condenser for the illuminating beam, or dark-field transmitted light. The condenser numerical aperture (NA) must exceed the NA of the objective, and a central cone of the condenser illuminating beam, equal in angle to the maximum objective angular aperture, must be opaque.

The stereobinocular microscope is an arrangement of two separate compound microscopes, one for each eye, looking at the same area of an object. A Leica SZ6 Series Stereoscope is shown in Fig. 2.2. Because each eye views the object from a different angle, separated by about 14°, a stereoimage is obtained. The physical difficulty of orienting two high-power objectives close enough together for both to observe the same object limits the NA to about 0.15 and the magnification to about 200×.

The erect image is an advantage, and the solution to most surface problems starts with the stereomicroscope. There is ample working distance between the objective and the preparation, and the illumination is flexible. Many stereos permit transmitted illumination and some permit bright-field top lighting. At worst, one can shine a light down one bodytube and observe the bright-light image with the second bodytube.

The resolution of a stereobinocular microscope is only 2 μm, 20 times larger than the limit of a mono-objective microscope. Unfortunately, increased resolution is paid for by a smaller working distance and a smaller depth of field. It becomes more difficult, as a result, to reflect light from a surface, using side spotlights, as

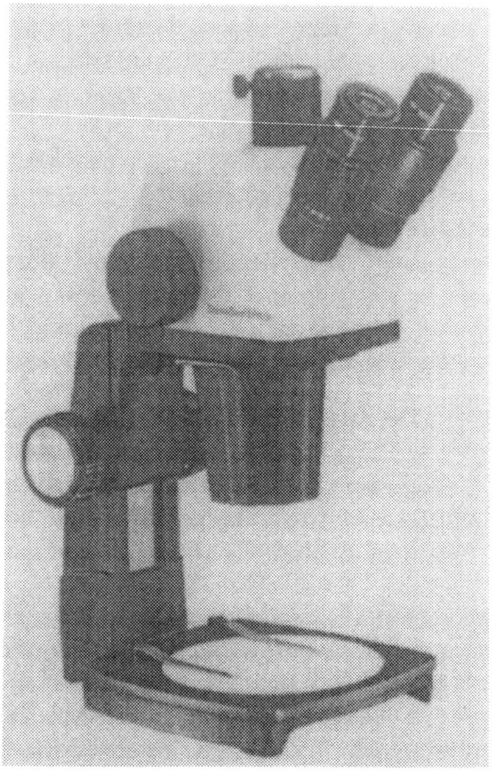

Figure 2.2. Photograph of Leica SZ6 Series Stereoscope. Reprinted with permission of Leica Instruments Co.

the objective NA increases. The angle between the light rays and the surface must decrease rapidly as the NA increases and the working distance decreases. The surface should be uncovered, i.e., no cover slip. All objectives having NA > 0.25 should be corrected for uncovered preparations.

The annular mirror is a dark-field system: scratches on a polished metal surface, for example, appear white on a dark field. The central mirror, on the other hand, is a bright-field system, and scratches on a polished metal appear dark on a bright field.

When surface detail is not readily visible because contrast is low, phase contrast is a useful means of enhancing contrast. Phase contrast enhances optical path differences and, as surface detail generally involves differences in optical path (differences in height), these differences are more apparent to the eye by phase contrast.

It is an advantage to be able to generate black-and-white or color photomicrographs of the specimen through a microscope. All major microscope manufacturers offer such equipment.

The following is a discussion on sample treatment procedures used to enhance contrast. There is one kind of surface difficult to study and virtually impossible to photograph by light microscopy. This is the surface of any transparent, colored, multicomponent substance, e.g., paper, particle-filled polymers, and pigments. So much light penetrates the surface only to be refracted and reflected back to the observer that the surface itself is lost in glare.

This problem is solved, however, by evaporating a thin film of metal onto the surface. The metal (usually aluminum, chromium, or gold) may be evaporated under vacuum in straight lines at any angle to the surface, from grazing to normal incidence. An angle of about 30° is often used; under these conditions, the heights of surface elevations can be calculated from shadow lengths.

Transparent film replicas of opaque surfaces are studied by transmission light microscopy. This leads to the possibility of using transmission phase contrast or interferometry and the best possible optics. In addition to these obvious advantages, replication is almost the only way to study contoured surfaces. The position of the particles relative to the surface geometry is also preserved by replication.

A direct way of examining a surface profile (i.e., coating or film) is to make a cross section and turn the surface up on edge for microscopical study. This usually involves mounting the piece in a cured polymeric resin mount, then grinding and polishing down to the desired section.

An interesting variation of this sectioning procedure is to make the section at an angle other than normal to the surface. This has the effect of magnifying the heights of elevations.

- Chemical composition and solid-state structure.

- Morphological analysis. Characterization of a surface includes not only topography but also chemical composition and solid-state structure. An experienced microscopist can identify many microscopic objects in the same way all of us identify macroscopic objects, that is, by shape, size, surface detail, color, luster, and the like. Descriptive terms (McCrone, 1974) found useful for surfaces include: angular, cemented, cracked, cratered, dimpled, laminar, orange-peel, pitted, porous, reticulated, smooth, striated, and valleyed. The nature of the surface helps to identify that substance.

Measurements of reflectance on polished surfaces can be used to calculate the refractive indices of transparent substances and to give specific reflectance data for opaque substances. The methods are discussed in detail by Cameron (1961). Reflectance and microhardness data are tabulated by Bowie and Taylor (1958) in a system for mineral identification.

- Staining surfaces. According to McCrone in Kane and Larrabee (1974), staining a surface, either chemically or optically, helps to differentiate different

Figure 2.3. Photomicrograph of paint specimen.

parts of a composite surface and to identify the various phases. A variety of stains are available for diverse surfaces. Mineral sections are etched with hydrofluoric acid and then stained with $Na_3CO\ (NO_2)_6$ to differentiate quartz (unetched), feldspars (etched but unstained), and potassium feldspars (etched and stained yellow). Isings (1961) selectively stains unsaturated elastomers with osmium tetroxide.

2.1.2. Equipment

Examples of Leica mono- and stereomicroscopes are given in Figs. 2.1 and 2.2. A photomicrograph of a paint specimen is shown in Fig. 2.3. The optical microscope has a depth of view which is apparent from this image, but this paint specimen will be viewed with an electron microscope and the surface will appear flatter.

2.1.3. Applications

Light microscopy is useful for observing solid forms of paint, plastics, adhesives, and inks and especially for pigments, fibers, or other solid particles. The resin or polymer portion of the material is not resolvable with light microscopy, with the exception of crystallites in polyethylene. However, there are many important observations that can be made using light microscopy:

1. The interface at an adhesive bond showing good adhesion, contamination, etc.
2. Pigments, fibers, and other particles of all types and colors
3. Erosion, deterioration, inclusions, and contaminants
4. Fractures, cracks and pinholes (Roulin-Moloney, 1989)
5. Refractive index (Hemsley, 1984)

2.2. ELECTRON MICROSCOPY (EM)

2.2.1. Fundamentals

Electron microscopy is useful for studying the pigments, particle size and distribution, and surfaces where very high resolution is required.

There is hardly a field in materials science where the physical nature of the surface is not an important feature. For example, in fatigue fracture, cracks nucleate at the surfaces of materials and the rate at which they nucleate is greatly influenced by the detailed topography of the surfaces. In the field of thin-film devices, the manufacturing tendency has been to reduce the size of electronic components. Surface-to-volume ratios are now exceedingly high. Young (1971) points out that we are not far from the point where we can anticipate devices employing single layers of atoms. However, the device industry, which presently employs films in the 10- to 100-Å range, suffers very high failure rates because of surface imperfections, stacking-fault intersections, voids in the films, thermally induced pits, and multiple steps. As a result of these deficiencies, large resources have been employed to control the imperfections by close control of processing variables. In other areas, elaborate polishing, cleaning, and smoothing techniques have been developed in an effort to eliminate the variability associated with surfaces. However, none of these efforts can improve on a detailed knowledge of the actual surface topography.

• Transmission electron microscopy (TEM). The purpose of this discussion is to describe how transmission electron microscopy has been, or can be, applied to the study of paint surfaces. The transmission microscope (Kane and Larrabee, 1974) is similar to the ordinary optical microscope in that it simultaneously illuminates the whole specimen area and employs Gaussian optics to generate the image. This is the only type of electron microscopic instrument to be considered here. A comparative review of the capability of all kinds of topographic measurers has been given by Young (1971), and the flying-spot and other types of instruments are treated in detail by Johari (1974). However, it is worth pointing out briefly the advantages and disadvantages of the transmission microscope with respect to the scanning microscope, its most serious competitor, at least in terms of numbers. Unlike the transmission microscope, the scanner illuminates only one spot on the specimen at a time and forms its image sequentially. The transmission microscope (as is generally true of types that employ Gaussian optics) has greater resolving power than an equivalent scanner, and it spreads the illumination over the whole specimen rather than concentrating it in one high-density spot. As a consequence, the scanner must employ a much smaller beam current than the transmission microscope and, in my experience, causes much less overall specimen damage than the transmission microscope in highly susceptible materials such as polymers. On the other hand, the transmission microscope, working with metals and regular accelerating voltages (100–150 kV), and equipped with a good decontamination device, can operate virtually ad infinitum without serious deterioration of the area under observation. The same is hardly likely in the case of a scanning instrument, unless it also is equipped with a good decontamination device.

Flying-spot instruments permit point-by-point analysis of surface properties. At first sight, it would appear that transmission microscopes, illuminating the whole sample, would not be capable of such application. In general, this is so. However, a new transmission microscope, the EMMA 4, has been developed with combined transmission microscope and probe capability by the introduction of a "minilens" in the illumination system (Cooke and Duncumb, 1969; Jacobs, 1971). This instrument should be considered a special case of microprobe analysis, also treated in this volume (Hutchins, 1974). EMMA 4 has demonstrated considerable power in a number of applications and could easily be applied to surfaces, but it will not be further considered here because the primary emphasis is on the topography of paint.

A great advantage of the scanning instrument is its ability to deal with bulk specimens. Unfortunately, nonconducting samples have to be given a light coating of metal, typically gold; otherwise, charging effects will seriously impair the resolution of the image. Transmission microscopes are not subject to this limitation and the techniques to be described here apply universally to all materials. Such a statement is, of course, "theoretical" because numerous practical problems beset the preparation of all kinds of materials for observation in the transmission microscope.

In the transmission microscope, the electrons that form the image must pass through the specimen; thus, the specimen thickness is limited to a few thousand angstroms, or to a few micrometers for a high-voltage instrument. If one is to study the surfaces of solids, two approaches are possible. In one approach, a replica of the surface can be made—for example, a carbon replica can be made by vacuum-depositing a 100- to 1000-Å film on the surface—and be carefully removed by some etching technique and then mounted in the microscope. The image obtained from such a replica does represent the surface topography, but it is frequently subject to distortion and artifacts and is often difficult to interpret. Moreover, the process of replication seriously cuts down the resolution ultimately obtainable with the instrument.

In the other approach, it is necessary to plate a suitable material onto the surface of interest and then to section a slice normal to that surface. The section is then mounted for observation in the microscope and it permits one to observe the surface in profile. The resolving power of the instrument can be fully exploited by this method (the profile method) and it has the additional advantage of revealing the surface topography in relation to the underlying structure of the material.

The scope of this theme is too broad to permit detailed description of any kind of instrument or of the theory by which it is employed. Many excellent books have been written on the microscope itself (Klemperer, 1953; Thomas, 1962; Haine and Cosslett, 1961; Heidenreich, 1964; Grivet, 1965; Hirsh *et al.*, 1965; Amelinckx, 1964, 1970; Hall, 1966; Wyckoff, 1949), on methods of preparing specimens (Wyckoff, 1949; Kay, 1961; Thomas, 1971), and on the theory of contrast (Heidenreich, 1964; Hirsh *et al.*, 1965; Amelinckx, 1964, 1970), and here I provide only a very brief description of contrast principles and specimen-preparation methods and applications where replication and sectioning techniques have been successfully employed to study surfaces, with the aim of illustrating the scope of the instrument, the resolution obtained, and the limitations of the methods.

• Contrast theory. The problem now is to interpret the electron images obtained by the two approaches available for studying surfaces: the replication and profile methods. Because the electrons pass through the samples, the images formed from them are going to be strongly affected by the interaction of the electrons with the material of the sample. The atomic spacings of most materials and the wavelengths of the electrons obtained from the accelerating voltages employed are suitable for diffraction effects to occur.

Many different types of inelastic scattering occur (Hirsh *et al.*, 1965; Amelinckx, 1964, 1970), including plasma losses, photon interactions, and bremsstrahlung radiation. The net effect is that some of the incident electrons are deflected from the collimated, axially parallel beam focused on the specimen by the illumination system. These deflected beams are focused at different points in the back focal plane of the objective lens. To obtain contrast in the image, an objective aperture is inserted in the back focal plane to block the scattered beams and to permit only the direct beam to form an image in the projection lens system of the microscope. This image is called the bright-field image and its details are determined by the extent to which scattering has occurred in different regions of the specimen. Alternatively, one can form a dark-field image by shifting the objective aperture laterally so as to block the direct beam and to permit only one of the scattered beams to pass into the image system of the microscope. The different information contained in the bright- and dark-field images can be employed to determine many details about the imperfections contained within the specimen or at its surface.

Although this method of obtaining contrast is quite general, the scattering processes involved are going to vary widely for different materials, and it is convenient to discriminate between those that occur in the two approaches employable for studying surfaces. In the replication method, most replicas are essentially amorphous. The diffraction of electrons from replicas is therefore going to differ from the type that occurs in profile sections which are more likely to be crystalline. In replicas, the diffraction patterns (i.e., the distribution of electron intensity in the back focal plane) are hazy with a fairly high intensity scattered at a Bragg angle corresponding to the most populous interatomic spacing. As the structure is generally uniform, intensity distributions in the electron images are also uniform unless the thickness of the replica varies. Heidenreich (1964) worked out in detail the contrast to be expected from such specimens.

It usually happens that the materials used for replication, such as carbon, are so transparent to electrons that small thickness variations produce no observable contrast. It is usual, therefore, to enhance contrast by shadowing the replica with a heavy metal, which produces marked variations in contrast. In addition, the shadows help to bring out height differences in the specimen and open the way to obtain quantitative information about the surface topography via stereomicrometry.

For profile specimens, the ordered nature of the crystals will give rise to marked elastic scattering of the incident beam. If the specimen is monocrystalline, the diffraction pattern will be a spot pattern, readily identifiable by the techniques described in much more detail elsewhere (Hirsh *et al.*, 1965). As the theory of electron diffraction is well understood, detailed quantitative information can be obtained from the specimen by tilting it in seriatim to different orientations and exciting a variety of Bragg reflections (Heidenreich, 1964; Grivet, 1965). This information can be obtained about both the crystallography of the specimen and the defects within it.

• Techniques. Replication techniques have been developed to a considerable degree of sophistication, comprising both one- and two-stage methods, and make use of a wide variety of replicating materials, depending on the application (Kay, 1961). Plastic replicas have a serious resolution limitation in that the molecule of the plastic itself may be larger than the resolving power of the instrument; the aggregate of the replica can interfere, then, with the fine details of the surface of interest. Consequently, shadowed carbon replicas, having much better resolution, are used almost exclusively in the most exacting work.

• Transmission scanning electron microscopy (TSEM). Although most commercial SEMs are used to study surface features, signals transmitted through thin samples can be collected by a suitable detector placed below the sample, and thus SEM can be used in the transmission mode (TSEM). Comparison of the TSEM with a conventional transmission electron microscope (TEM) shows that the two microscopes are equivalent, so that data obtained from the two microscopes are equivalent, and thus data obtained from a TEM theoretically can also be obtained from a TSEM (Jones and Boyde, 1970; Zeitler, 1971).

Specially built TSEMs with a field-emission source and an ion-pumped vacuum system have been used to obtain point resolutions of 5 Å and to resolve atoms of uranium (Crewe, 1970).

• Scanning electron microscopy (SEM). A detailed examination of material is vital to any investigation relating to the processing properties and behavior of materials. Characterization includes information relating to topographical features, morphology, habit and distribution, identification of differences based on chemistry, crystal structure, physical properties, and subsurface features.

Before the advent of the SEM (Johari, 1971), several tools such as the optical microscope, the transmission electron microscope, the electron microprobe analyzer, and X-ray fluorescence were employed to accomplish partial characterization; this information was then combined for a fuller description of materials. Each of these tools has proficiency in one particular aspect and complements the information obtainable with other instruments. These bits of information are limited because of the inherent limitations of each method such as the invariably cumbersome specimen preparation, specialized techniques of observation, and interpretation of the results.

In comparison with other tools, the SEM serves to bridge the gap between the optical microscope and the transmission microscope, although the TSEM approaches the resolution and magnification obtainable with the TEM. The SEM has a magnification of 3 to 100,000×, a resolution of about 200–250 Å, and a depth of field at least 300 times or more that of the light microscope which results in the three-dimensional high-quality photographs of coating and pigments. Because of the large depth of focus and large working distance, the SEM permits direct examination of rough conductive samples at all magnifications without special preparation. All surfaces have to be coated with a thin conductive layer of, e.g., carbon, gold, or palladium. All electron microscopy instruments are strictly topological viewing tools (i.e., only the immediate surface is visible).

The SEM has so many material-characterization capabilities that it is often considered the ideal tool for material characterization (Johari, 1971; Howell and Boyde, 1972; Boyde, 1970).

2.2.2. Equipment

The Hitachi scanning electron microscope is shown in Fig. 2.4. SEMs are available in different sizes, but usually in a desk-size console depending on the capabilities. Micrographs can be conveniently generated in black and white and/or color. Also, EDXRA spectrograms are usually available from the same SEM instrument. Both capabilities can be used together and SEM images can be high-

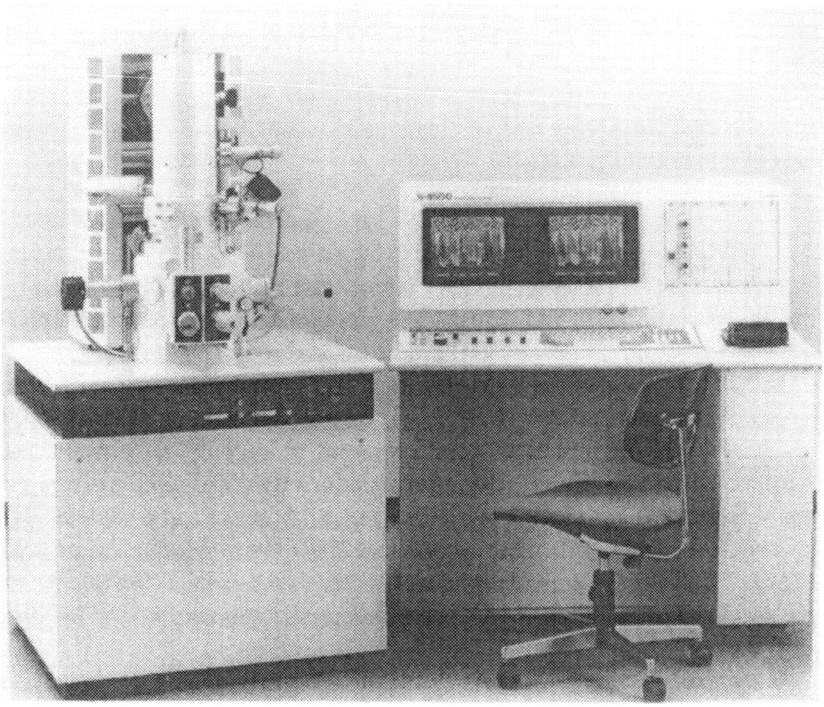

Figure 2.4. Photograph of Hitachi S-4500 Scanning Electron Microscope. Reprinted with permission of Hitachi Instruments Co.

lighted for the presence of elements (usually to a minimum atomic number of 5) which is very impressive in colors.

2.2.3. Applications

Using a combination of SEM and EDXRA, a specimen (e.g., paint chip) can be examined to vividly show pigment particles and their elemental composition. The identification of the pigments can be estimated and if required, compared to other specimens. This technique is often used to match paint fragments from automobile accidents. The same technique can be applied for plastic or adhesives. In Fig. 2.5, a SEM micrograph of a paint specimen, note the flat appearance of the image, and the high resolution of individual particles.

Inks are particularly observable with SEM and EDXRA as the solid specimens always are thin films of printed materials.

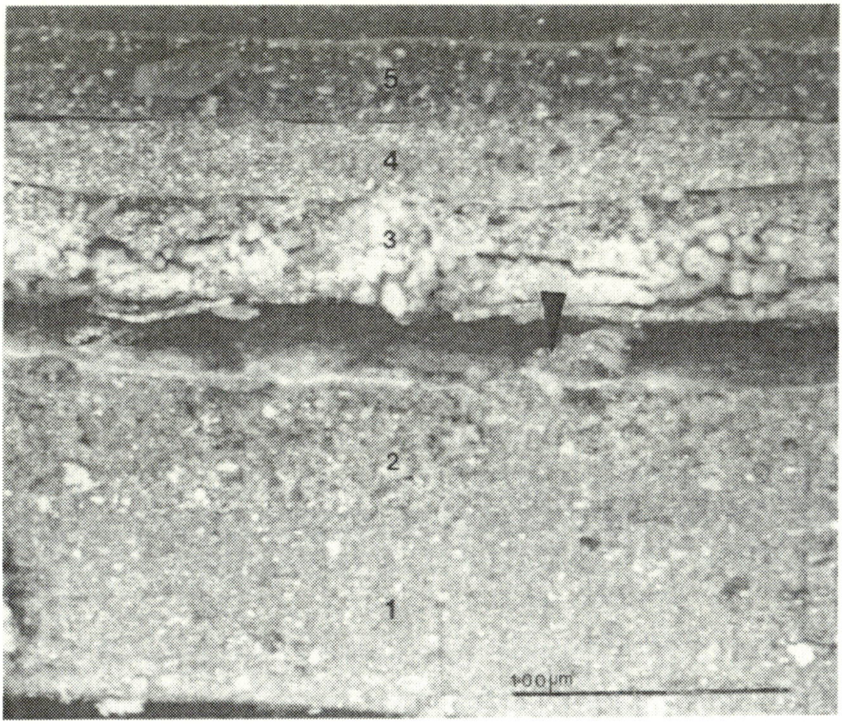

Figure 2.5. SEM micrograph of multilayered lead paint chip. (Arrowhead indicates mica particle analyzed in Fig. 2.6.)

2.3. Energy-Dispersive X-Ray Analysis (EDXRA)

2.3.1. Fundamentals

Use of X-ray spectroscopy (Gilfrich, 1974; Johari and Samuda, 1974) tremendously enhances the analytical value of the SEM in material characterization by providing chemical analysis of the sample along with surface topology.

A brief description of the two X-ray detection methods is warranted before comparing them. In the wavelength diffractometer (WD) method, a crystal of a known spacing d separates X rays according to Bragg's law, $n\lambda = 2d \sin\theta$, so that at a diffraction angle θ (collection of 2θ), X rays of specific wavelengths are detected. To cover the whole range, the diffractometers are usually equipped with many crystals. Even then, considerable time is needed to obtain an overall spectrum of all elements present. The resolution of the crystal in separating X rays of different wavelengths is very good (on the order of 10 eV), but the efficiency is very poor.

To improve the collection efficiency, curve-crystal fully focusing diffractometers are used.

For nondispersive (ED) spectrometers, the energy of an incoming X-ray photon is converted into an electric pulse in a lithium-drifted silicon crystal. A bias voltage applied to the crystal collects this charge, which is proportional to the energy of the X ray. This pulse is amplified, converted to a voltage pulse, and fed into a multichannel analyzer. The analyzer sorts out the pulses according to their energy and stores them in the memory of the correct channel. The resulting spectrum can be displayed on a cathode-ray tube (CRT), plotted on a chart, or printed out numerically.

Characteristic X rays emitted under the effect of the electron beam provide information about the nature and amount of elements present in the volume excited by the primary beam. EDXRA attachments, consisting of a lithium-drifted silicon crystal, a multichannel analyzer, and necessary electronics, are finding increasing use on many SEM models. This method is capable of detecting elements with

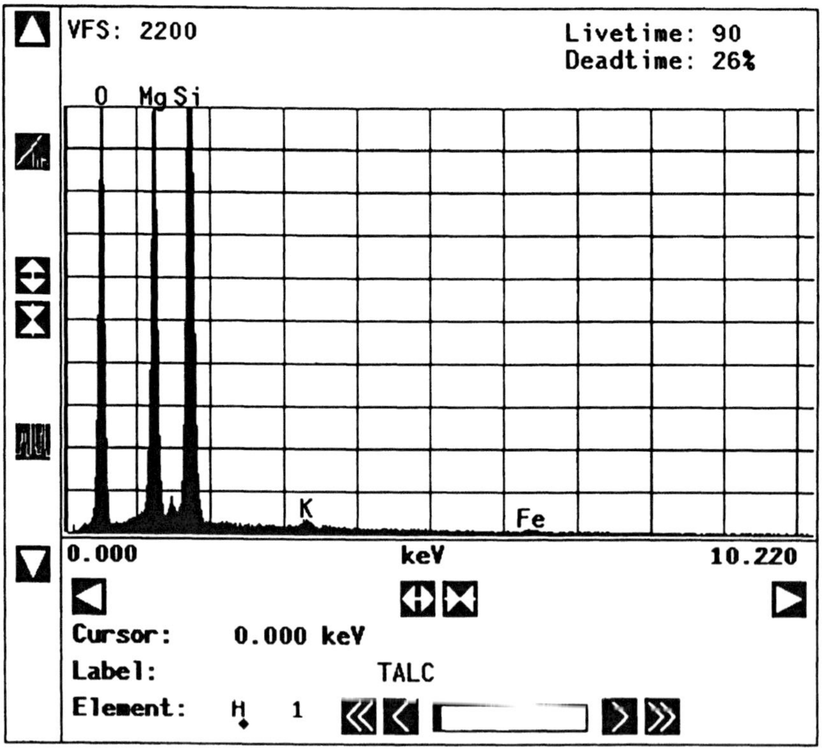

Figure 2.6. EDXRA spectrogram of talc mica particle shown in SEM micrograph of Fig. 2.5.

atomic number down to 9 (fluorine) in the SEM and 5 (boron) in the TSEM with a detectability limit of 0.5% by volume. A spectrogram of elements is generated and can be presented on a CRT, printed graphically for a permanent record, or stored on magnetic disk. In a spectrogram, the *x–y* plot consists of wavelength versus intensity, and the area under the peaks is indicative of the amount present. Wavelength diffractometers, used with electron beam probe microanalyzers, are also available as an accessory on the SEM.

The disadvantage of EDXRA is the lack of quantitative data which are available from electron probe microanalysis. The data are semiquantitative, but very quickly generated.

2.3.2. Equipment

The EDXRA equipment is contained in a typical SEM (see Section 2.3).

2.3.3. Applications

The application of EDXRA accompanies SEM (see discussion on SEM). A specimen can be quickly scanned for elemental composition before investing time in more complicated and quantitative methods. An EDXRA spectrogram of a paint specimen is shown in Fig. 2.6.

2.4. ELECTRON PROBE MICROANALYSIS (EPM)

2.4.1. Fundamentals

Electron probe microanalysis (Hutchins, 1974) is an analytical technique that may be used to determine the chemical composition of a solid specimen weighing as little as 10^{-11} g and having a volume as small as 1 μm^3. The primary advantage of electron probe microanalysis over other analytical methods is the possibility of obtaining a quantitative analysis of a specimen.

The selected area of the specimen is bombarded with a beam of electrons (Duncumb, 1969). The accelerating voltage of the electrons (typically 10–30 kV) determines the depth of penetration into the specimen. The degree of beam focusing determines the diameter of the analyzed volume. The electron bombardment of the specimen causes the emission of an X-ray spectrum that consists of characteristic X-ray lines of elements present in the bombarded volume. The chemical analysis is accomplished by the dispersion of this X-ray spectrum and the quantitative measurement of the wavelength and intensity of each characteristic line. The wavelengths present identify the emitting elements, and the line intensities are related to the concentration of the corresponding elements.

The four major instrument subsystems are:

1. An electron optical system of high stability is needed to produce a focused beam of electrons on the specimen. The electron energy should be variable in steps from 5 to 30 keV.
2. A specimen airlock, a stage with *xyz* motion, and an optical microscope must be incorporated into the instrument so that the desired area of the specimen can be positioned under the electron beam.
3. An energy or wavelength spectrometer is required to disperse the X rays so that the characteristic lines can be assigned to specific elements.
4. Readout and recording electronics are needed to display and record the characteristic X-ray intensities as a function of energy, wavelength, and/or specimen position.

There are two basic types of analyses, and both may be either qualitative or quantitative.

1. A spot analysis consists of an analysis for all detectable elements on one spot of a much larger specimen. This analysis may be representative of the entire specimen or it may be an analysis of an unusual region.
2. A distribution analysis determines the distribution of one or more elements as a function of position on the specimen. A distribution analysis is used to detect compositional gradients on a specimen surface; the average composition of the specimen is often known from a bulk analysis performed by other methods.

A qualitative spot analysis can be completed quickly by scanning the spectrometer through the portion of the X-ray spectrum detectable with the instrument. A strip chart recording of X-ray intensity versus wavelength or an oscilloscope trace of X-ray intensity versus energy is obtained. Peaks are assigned to emitting elements with the aid of tables.

2.4.2. Equipment

The Acton MS64EBP Electron Beam Microanalyzer is shown in Fig. 2.7. This instrument is manufactured by Cameca, Inc., Stamford, Connecticut. The optical stereoviewer is shown near the base of the instrument.

2.4.3. Applications

The electron probe is a valuable tool for obtaining quantitative elemental data from specimens. The technique requires more time than does EDXRA examination, and it is useful to first scan the specimen with EDXRA to determine the presence of the major elements. The detection limit is lower than for EDXRA, but must be determined for each instrument. An electron probe spectrogram of a paint specimen is shown in Fig. 2.8.

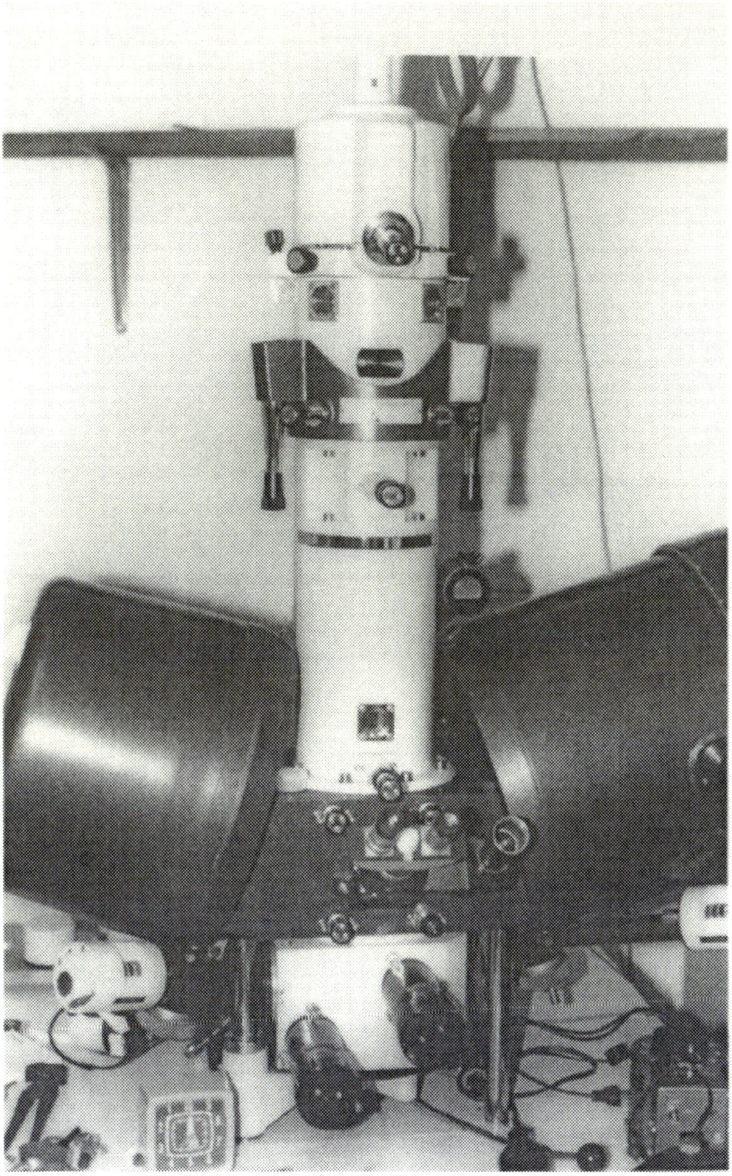

Figure 2.7. Photograph of Acton MS64EBP Electron Beam Microanalyzer. Reprinted with permission of Cameca, Inc.

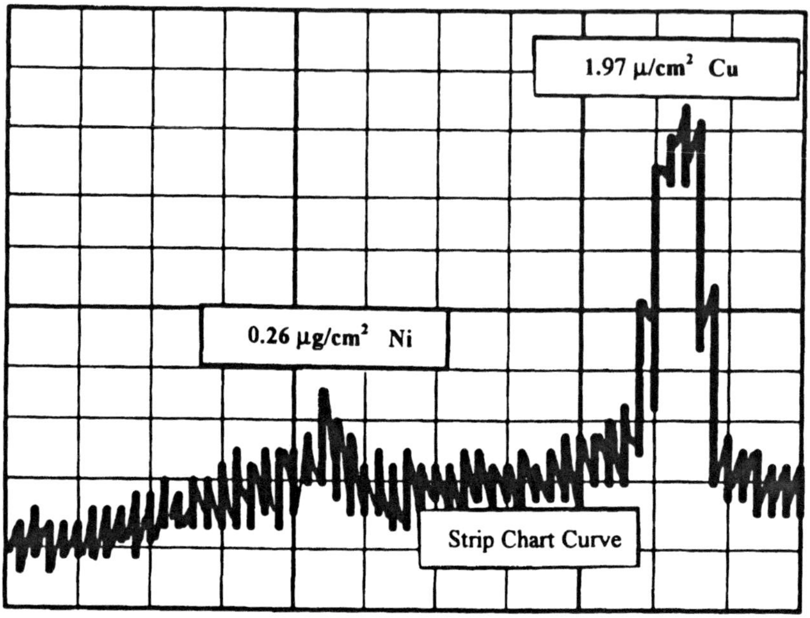

Figure 2.8. Electron beam microanalyzer spectrogram of chemically deposited nickel and copper on high-purity aluminum foil. (From Hutchins, 1974.)

2.5. AUGER SPECTROSCOPY (AES)

2.5.1. Fundamentals

This technique is most powerful, providing analysis of the first few atom layers (10 Å or less) on the surface of the sample (Chang, 1971).

Auger spectroscopy explores the electronic energy levels in atoms and solids. The term "Auger process" has come to denote any electron deexcitation in which the deexcitation energy is transferred to a second electron, the "Auger electron." Because of the discrete nature of most electronic energy levels, the Auger process can be analyzed by measuring the energy distribution of Auger electrons. Low-energy Auger electrons (<1 keV) can escape from only the first several atom layers of a surface because they are strongly absorbed by even a monolayer of atoms. This gives Auger spectroscopy its high surface sensitivity.

Auger electron analysis with the SEM requires appropriate energy-analyzing equipment, but more importantly a much better vacuum system than is currently available in most instruments. Because AES analyzes only the first few surface layers (10 Å or less on the surface of the sample), the samples must be free of any surface film. The required energy-analyzing equipment is available as standard commercial items. Standard Auger spectra are available for all elements for various

Figure 2.9. Photograph of Perkin–Elmer Auger Electron Spectrometer. Reprinted with permission of Perkin–Elmer Corp.

Auger transitions. Overlapping spectra from two elements may create some problems of elemental separation, but with high-resolution energy-analyzing equipment, procedures similar to those used with X rays can be employed to obtain elemental separation.

Specimens examined in the SEM mode must be coated with a conductive layer similar to the process in conventional SEM instruments. Specimen charging occurs if not coated. See MacDonald (1971) and Chang (1971) for excellent review articles on AES. A reference for Auger spectra is L. A. Davis *et al.*, *Handbook of Auger Electron Spectra Microscopy,* Perkin–Elmer Corporation, 6509 Flying Cloud Drive, Eden Prairie, MN 55344.

2.5.2. Equipment

A Perkin–Elmer Auger spectroscope is shown in Fig. 2.9.

2.5.3. Applications

The AES method is very useful for thorough, and low detection limit, elemental identification and especially for layers immediately under the surface. The technique is slower than SEM and the instrument is more expensive. An AES spectrogram of alumina is shown in Fig. 2.10.

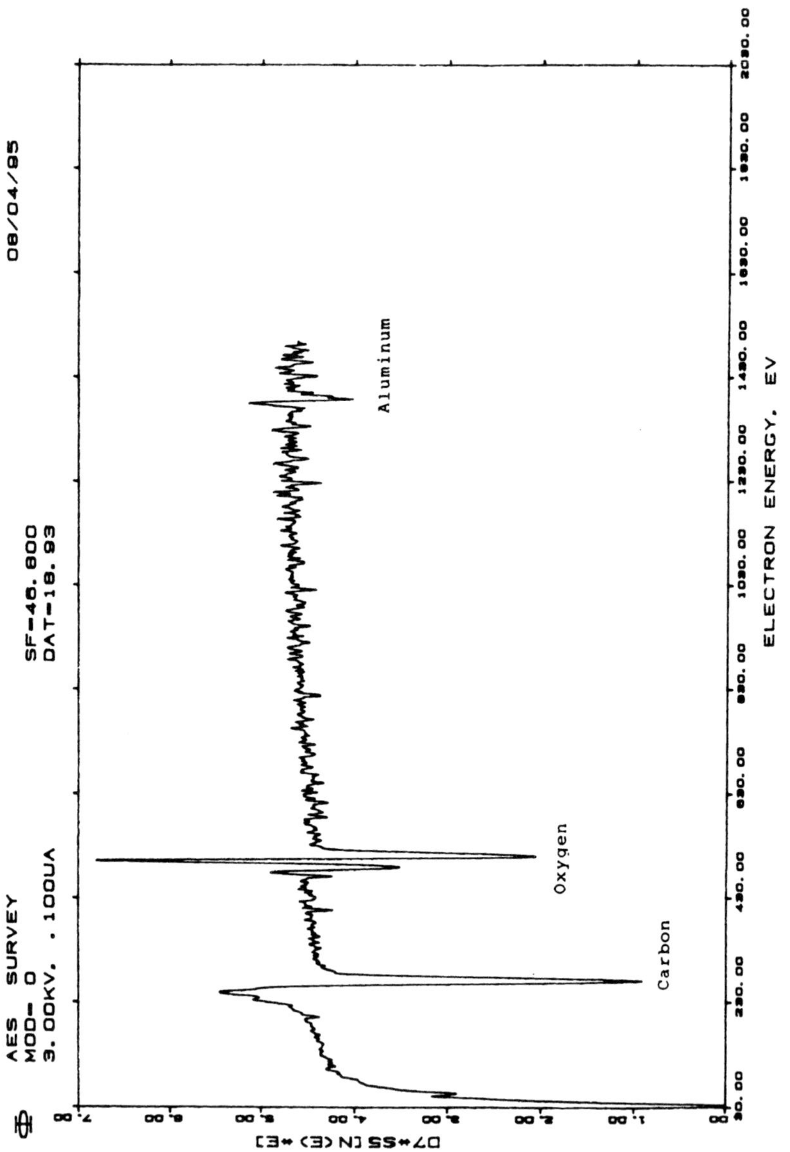

Figure 2.10. AES spectrum of alumina, Al_2O_3.

2.6. SCANNING ION MASS SPECTROSCOPY (SIMS)

2.6.1. Fundamentals

A mass spectrometer is an apparatus that produces a supply of gaseous ions from a sample, separates the ions in either space or time according to their mass-to-charge ratios, and provides an output record or display indicating the intensity of the separated ions.

Mass spectrometry is a term describing an analysis whereby matter is affected by means of ionization of the matter followed by separation of the ions according to their mass-to-charge ratio and recording of a measure of the numbers of the various ions.

2.6.2. Equipment

A leading SIMS instrument is the Perkin–Elmer PHI 7200 TOF-SIMS shown in Fig. 2.11 and manufactured by:

Perkin–Elmer Corporation
Physical Electronics Division
6509 Flying Cloud Drive
Eden Prairie, MN 55344

Figure 2.11. Photograph of Perkin–Elmer Scanning Ion Mass Spectrometer. Reprinted with permission of Perkin–Elmer Corp.

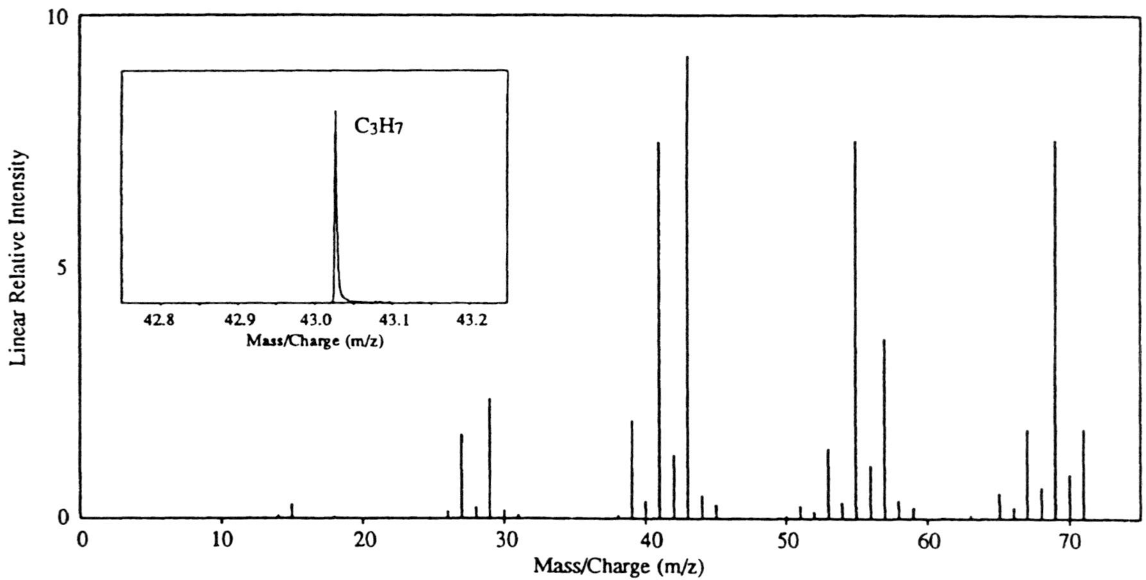

Figure 2.12. TOF-SIMS spectrogram of polypropylene specimen.

This instrument has been successfully used for analysis of polymers, biomaterials, adhesives, and insulators.

2.6.3. Applications

Analysis of polymeric materials is a good application of SIMS, where metallic elements are not often observed. A SIMS spectrogram of a polymer specimen is shown in Fig. 2.12. Use of the instrument is time consuming and most of the data derived can be generated with electron spectroscopy chemical analysis (ESCA).

2.7. ELECTRON SPECTROSCOPY CHEMICAL ANALYSIS (ESCA)

2.7.1. Fundamentals

ESCA is useful for the determination of chemical composition of materials (Barr, 1994). It is a microanalytical surface method. Micrometer-size areas on a surface can be focused and explored with ESCA. Historically, ESCA was developed from the photoelectron sciences. The term ESCA was coined by Professor Kai Siegbahn *et al.* (1969) in Uppsala, Sweden.

Figure 2.13. Photograph of Surface Science Laboratories, Model SSX-100 Small Spot Electron Spectroscopy Chemical Analysis Spectrometer. Reprinted with permission of Surface Science Laboratories.

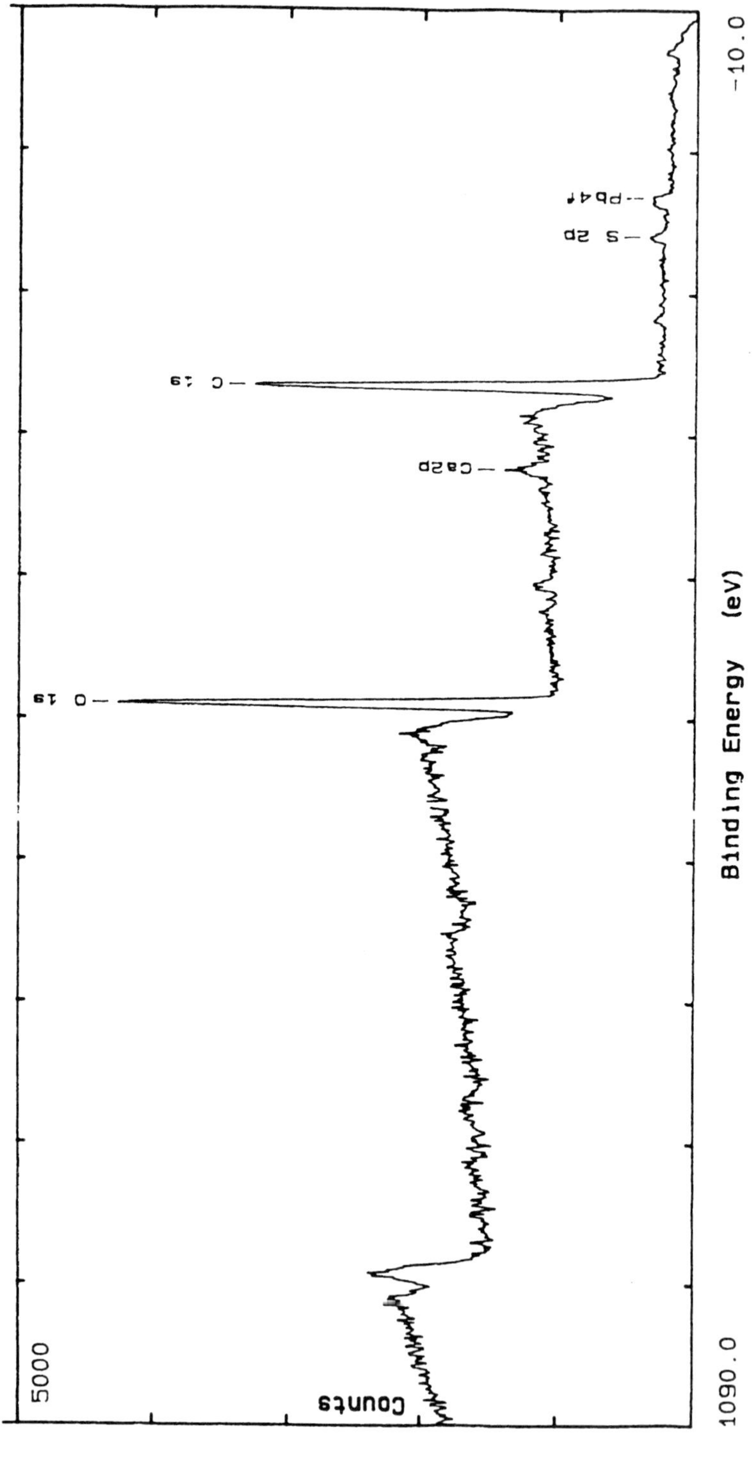

Figure 2.14. ESCA spectrogram of paint pigment, lead carbonate, and calcium sulfate.

The advantage of ESCA lies in its ability to provide detailed chemical information about the surface–near surface regions of solid materials. The principal feature of ESCA that contains the chemical information is the "chemical shift," a term employed to designate the changes in "binding energy" apparently induced in many core-level, photoelectron lines as a result of changes in the chemical environment of the material. The binding energy is then correlated to a spectrogram of "binding energy versus counts," enabling the identification of chemical groups which are useful for identifying the element or compound.

2.7.2. Equipment

A Surface Sciences Instruments ESCA instrument is shown in Fig. 2.13.

2.7.3. Applications

The ESCA method is very useful for chemical analysis of solid materials, especially small specimens. Metallic and nonmetallic elements can be detected, and the data are semiquantitative. An ESCA spectrogram of a polymer specimen is shown in Fig. 2.14.

ESCA offers a unique means for detecting a wide range of elements and groups at low detection limits, but particularly important for elements and chemical groups found in resins, polymers, and pigments. The fine resolution of examination makes it a valuable tool for investigating a mixture of resins, polymers, pigments, and other particles.

2.8. INFRARED SPECTROSCOPY (IR) FOR SURFACE ANALYSIS

2.8.1. Fundamentals

The following fundamental information can be found in Willard *et al.* (1974). The infrared region of the electromagnetic spectrum extends from the red end of the visible spectrum to the microwaves; that is, the region includes radiation at wavelengths between 0.7 and 500 μm, or, in wave numbers, between 14,000 and 20 cm^{-1}. The spectral range of greatest use is the mid-infrared region, which covers the frequency range from 200 to 4000 cm^{-1} (50 to 2.5 μm). Infrared spectroscopy involves the twisting, bending, rotating, and vibrational motions of atoms in a molecule. On interaction with infrared radiation, portions of the incident radiation are absorbed at particular wavelengths. The multiplicity of vibrations occurring simultaneously produces a highly complex absorption spectrum, which is uniquely characteristic of the functional groups comprising the molecule and of the overall configuration of the atoms as well. Suggested review articles on the fundamentals of infrared spectroscopy are Bellamy (1958), Colthup *et al.* (1964), Gianturco (1965), Herberg (1945), and Nakanishi (1962).

An extensive discussion of IR analysis is contained in Chapter 3, so only IR analysis that pertains to surface investigations will be discussed here.

When a three-atom system is part of a larger molecule, it is possible to have bending or deformation vibrations. These are vibrations that imply movement of atoms out from the bonding axis. Four types can be distinguished:

1. Deformation or scissoring. The two atoms connected to a central atom move toward and away from each other with deformation of the valence angle.
2. Rocking or in-plane bending. The structural unit swings back and forth in the symmetry plane of the molecule.
3. Wagging or out-of-plane bending. The structural unit swings back and forth in a plane perpendicular to the molecule's symmetry plane.
4. Twisting. The structural unit rotates back and forth around the bond that joins it to the rest of the molecule.

Splitting of bending vibrations caused by in-plane and out-of-plane vibrations is found with larger groups joined by a central atom. An example is the doublet produced by the *gem*-dimethyl group. Bending motions produce absorption at lower frequencies than fundamental stretching modes.

Molecules composed of several atoms vibrate not only according to the frequencies of the bonds, but also at overtones of these frequencies. When one tone vibrates, the rest of the molecule is involved. The harmonic (overtone) vibrations possess a frequency that represents approximately integral multiples of the fundamental frequency. A combination band is the sum of, or the difference between, the frequencies of two or more fundamental or harmonic vibrations. The uniqueness of an infrared spectrum arises largely from these bands which are characteristic of the whole molecule. The intensities of overtone and combination bands are usually about 1/100th of those of fundamental bands.

The intensity of an infrared absorption band is proportional to the square of the rate of change of dipole moment with respect to the displacement of the atoms. In some cases, the magnitude of the change in dipole moment may be quite small, producing only weak absorption bands, as in the relatively nonpolar C=N group. By contrast, the large permanent dipole moment of the C=O group causes strong absorption bands, which is often the most distinctive feature of an infrared spectrum. If no dipole moment is created, as in the C=C bond (when located symmetrically in the molecule) undergoing stretching vibration, then no radiation is absorbed and the vibrational mode is said to be infrared inactive. Fortunately, an infrared inactive mode will usually give a strong Raman signal.

As defined by quantum laws, the vibrations are not random events but can occur only at specific frequencies governed by the atomic masses and strengths of the chemical bonds. Mathematically, this can be expressed as

$$\bar{\nu} = \frac{1}{2\pi c}\sqrt{\frac{k}{\mu}}$$

where ν is the frequency of the vibration, *c* is the velocity of light, k is the force constant, and μ is the reduced mass of the atoms involved. The frequency is greater the smaller the mass of the vibrating nuclei and the greater the force restoring the nuclei to the equilibrium position. Motions involving hydrogen atoms are found at much higher frequencies than are motions involving heavier atoms. For multiple bond linkage, the first constants of double and triple bonds are roughly two and three times those of the single bonds, and the absorption position becomes approximately two and three times higher in frequency. Interaction with neighbors may alter these values, as will resonating structures, hydrogen bonds, and ring strain.

Example. Calculate the fundamental frequency expected in the infrared absorption spectrum for the C–O stretching frequency. The value of the force constant is 5.0×10^5 dyn cm^{-1}.

$$\bar{\nu} = \frac{1}{(2)(3.14)(3 \times 10^{10})}\sqrt{\frac{(5 \times 10^5)(12 + 16)(6.023 \times 10^{23})}{(12)(16)}} = 1110\ cm^{-1}$$

• Microscopic infrared spectroscopy. The microscopic infrared photometer is the perfect tool for analysis of surfaces for the purpose of chemical identification of organic materials. This is the Fourier transform (FT) infrared spectroscopy method, but with a microscopic attachment. The instrument is extremely useful for identifying microscopic particles up to large pieces. The Perkin–Elmer System 2000 FT-IR Microscope instrument is shown in Fig. 2.15, and the optical operation is diagrammed in Fig. 2.16.

In conventional FT-IR microscopes, typically infrared optics have been added to standard optical microscopes. The mechanical coupling of the two subsystems and the switching between the viewing modes present sources of inaccuracies and interfere with conventional infrared study of samples. Cassegrain optical assemblies mounted into a frame with a precision optical microscope give the advantage of rapid switching. Additional features are fixed-stereo, zoom-stereo, and video viewing options; a vernier-calibrated sample x,y,z stage; and multiple illumination positions. It can be seen that this recent development in IR analysis has produced the ultimate IR instrument for surface analysis of solid materials.

Very small samples including paint and plastic chips and organic fibers can be analyzed by this method with minimal sample preparation. Also, the analysis can be conducted in the reflectance or transmission mode if the sample is transparent or translucent.

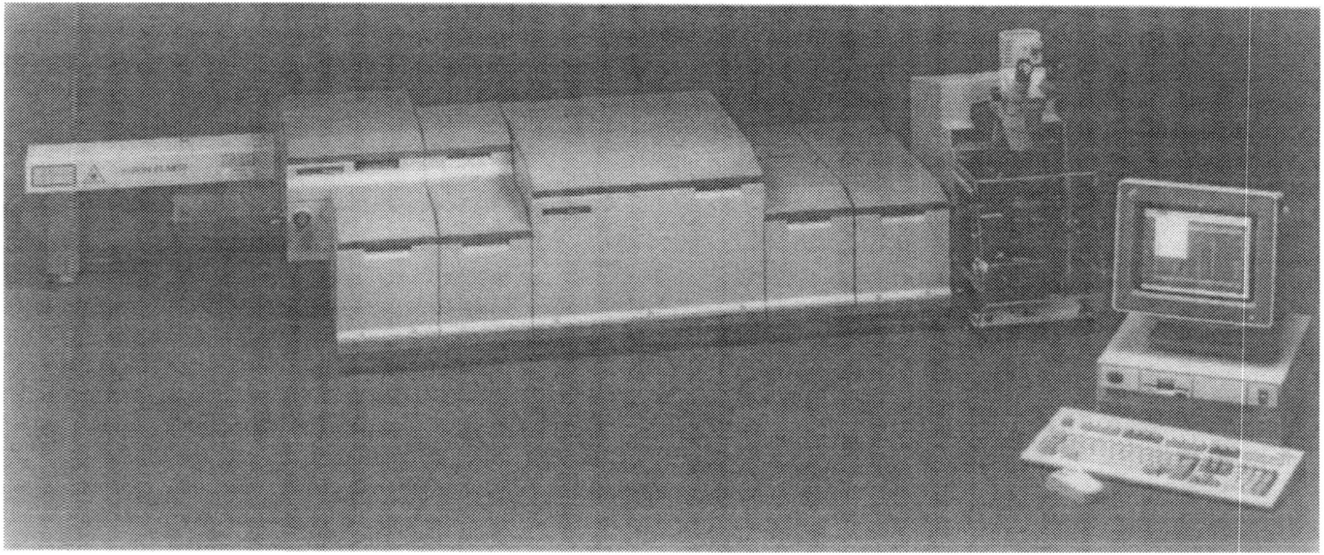

Figure 2.15. Photograph of Perkin–Elmer FT-IR System 2000, microscopic Cassegrain optical assemblies. Reprinted with permission of Perkin–Elmer Corp.

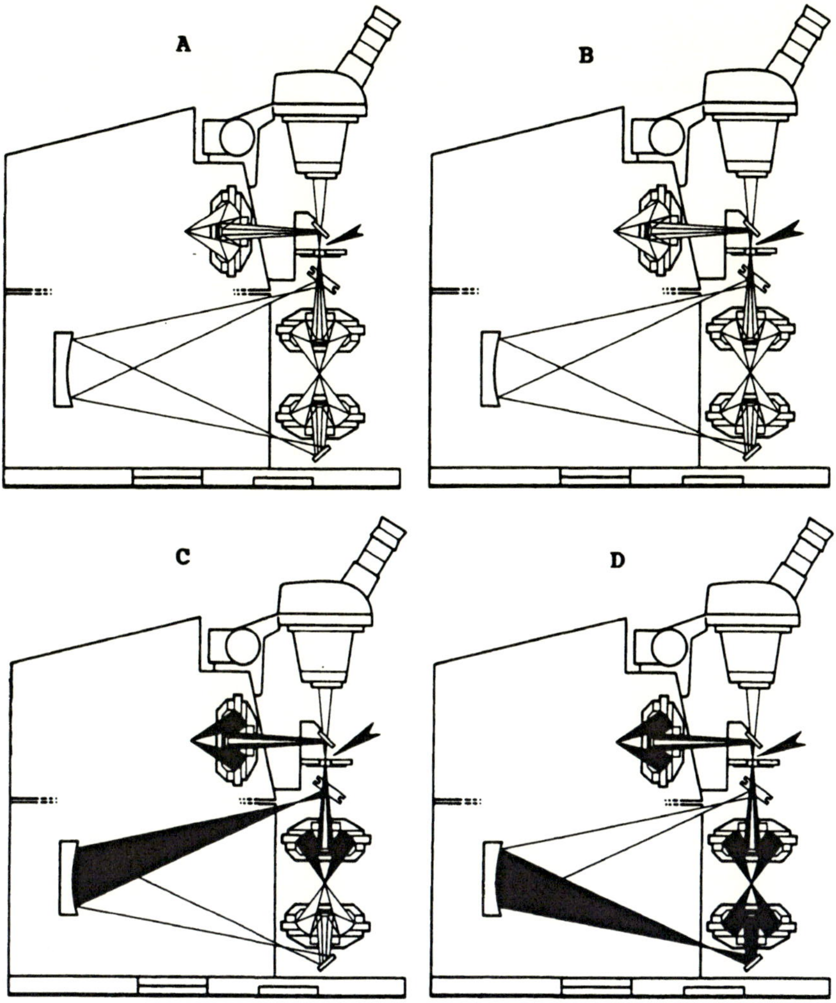

Figure 2.16. Perkin–Elmer FT-IR Microscope. (A) Optical path–sample preparation, (B) optical path–sample viewing, (C) optical path–reflectance infrared, and (D) optical path–transmittance infrared. (Arrowhead indicates sample position.) Reprinted with permission of Perkin–Elmer Corp.

• Attenuated total reflectance (ATR). The scope and versatility of infrared spectroscopy as a qualitative analytical tool have been increased substantially by the attenuated total reflectance, also known as internal reflectance technique (Harrick, 1967; Wilkes, 1972). When a beam of radiation enters a plate (or prism), it will be reflected internally if the angle of incidence at the interface between sample and plate is greater than the critical angle (which is a function of refractive index). On internal reflection, all of the energy is reflected. However, the beam appears to penetrate slightly (from a fraction of a wavelength up to several wavelengths) beyond the reflecting surface, and then return. When a material is placed in contact with the reflecting surface, the beam will lose energy at those wavelengths where the material absorbs due to an interaction with the penetrating beam. This attenuated radiation, when measured and plotted as a function of wavelength, will give rise to an absorption spectrum characteristic of the material which resembles an infrared spectrum obtained in the normal manner.

Most ATR work is done by means of an accessory readily inserted in, and removed from, the sampling space of a conventional infrared spectrophotometer.

• Correlation of infrared spectra with molecular structure. The infrared spectrum of a compound is essentially the superposition of absorption bands of specific functional groups, yet subtle interactions with the surrounding atoms of the molecule impose the stamp of individuality on the spectrum of each compound. Table 2.1 lists chemical groups and their infrared absorption frequencies. For qualitative analysis, one of the best features of an infrared spectrum is that the absorption or the lack of absorption in specific frequency regions can be correlated with specific stretching and bending motions and, in some cases, with the relationship of these groups to the remainder of the molecule. Thus, by interpretation of the spectrum, it is possible to state that certain functional groups are present in the material and that certain others are absent. With this datum, the possibilities for the unknown can sometimes be narrowed so sharply that comparison with a library of pure spectra permits identification.

a. Near-infrared region. In the near-infrared region, which meets the visible region at about 12,500 cm^{-1} (0.8 μm) and extends to about 4000 cm^{-1} (2.5 μm), are found many absorption bands resulting from harmonic overtones of fundamental bands and combination bands often associated with hydrogen atoms. Among these are the first overtones of the O–H and N–H stretching vibrations near 7140 cm^{-1} (1.4 μm) and 6667 cm^{-1} (1.5 μm), respectively, combination bands resulting from C–H stretching, and deformation vibrations of alkyl groups at 4548 cm^{-1} (2.6 μm). Thicker sample layers (0.5–10 mm) compensate for lessened molar absorptivities. The region is accessible with quartz optics, and this is coupled with greater sensitivity of near-infrared detectors and more intense light sources. The near-infrared region is often used for quantitative work.

Water has been analyzed in glycerol, hydrazine, Freon, organic films, acetone, and fuming nitric acid. Absorption bands at 2.76, 1.90, and 1.40 μm are used depending on the concentration of the test substance. Where interferences from other absorption bands are severe or where very low concentrations of water are being studied, the water can be extracted with glycerol or ethylene glycol.

Near-infrared spectrometry is a valuable tool for analyzing mixtures of aromatic amines. Primary aromatic amines are characterized by two relatively intense absorption bands near 1.97 and 1.49 μm. The band at 1.97 μm is a combination of N–H bending and stretching modes and the one at 1.49 μm is the first overtone of the symmetric N–H stretching vibration. Secondary amines exhibit an overtone band but do not absorb appreciably in the combination region. Secondary amines exhibit an overtone band but do not absorb appreciably in the combination region. These differences in absorption provide the basis for rapid, quantitative analytical methods. The analyses are normally carried out on 1% solutions in CCl_4, using 10-cm cells. Background corrections can be obtained at 1.575 and 1.915 μm. Tertiary amines do not exhibit appreciable absorption at either wavelength. The overtone and combination bands of aliphatic amines are shifted to about 1.525 and 2.000 μm, respectively. Interference from the first overtone of the O–H stretching vibration at 1.40 μm is easily avoided with the high resolution available with near-infrared instruments.

b. Mid-infrared region. Many useful correlations have been found in the mid-infrared region. This region is divided into the "group frequency" region, i.e., 4000 to 1300 cm^{-1} (2.5 to 8 μm), and the "fingerprint" region, 1300 to 650 cm^{-1} (8.0 to 15.4 μm). In the group frequency region the principal absorption bands may be assigned to vibration units consisting of only two atoms of a molecule, i.e., units that are more or less dependent only on the functional group responsible for the absorption and not on the complete molecular structure. Structural influences do reveal themselves, however, as significant shifts from one compound to another. In the deviation of information from an infrared spectrum, prominent bands in this region are noted and assigned first.

In the interval from 4000 to 2500 cm^{-1} (2.5 to 4.0 μm), the absorption is characteristic of hydrogen stretching vibrations with elements of mass 19 or less. When coupled with heavier masses, the frequencies overlap the triple-bond region. The intermediate frequency range, 2500 to 1540 cm^{-1} (4.0 to 6.5 μm), is often termed the unsaturated region. Triple bonds, and very little else, appear from 2500 to 2000 cm^{-1} (4.0 to 5.0 μm). Double-bond frequencies fall in the region from 2000 to 1540 cm^{-1} (5.0 to 6.5 μm). By judicious application of accumulated empirical data, it is possible to distinguish among C=O, C=C, C=N, N=O, and S=O bands. The major factors in the spectra between 1300 and 650 cm^{-1} (7.7 to 15.4 μm) are single-band stretching frequencies and bending vibrations (skeletal frequencies) of polyatomic systems which involve motions of bonds linking a substituent group of the remainder of the molecule. This is the fingerprint region. Multiplicity is too great for assured individual identification, but collectively the absorption bands aid in identification.

c. Far-Infrared Region. The region between 667 and 10 cm^{-1} (15 to 1000 μm) contains the bending vibrations of carbon, nitrogen, oxygen, and fluoride with atoms heavier than mass 19, and additional bending motions in cyclic or unsaturated systems. The low-frequency molecular vibrations found in the far-infrared are particularly sensitive to changes in the overall structure of the molecule. When studying the conformation of the molecule as a whole, the far-infrared bands differ often in a predictable manner for different isometric forms of the same basic compound. The far-infrared frequencies of organometallic compounds are often sensitive to the metal ion or atom, and this, too, can be used advantageously in the study of coordination bonds. Moreover, this region is particularly well suited to the study of organometallic or inorganic compounds whose atoms are heavy and whose bonds are inclined to be weak (Ferraro, 1968).

d. Molecular Structure Analysis. After the presence of a particular fundamental stretching frequency has been established, closer examination of the shape and exact position of an absorption band often yields additional information. The shape of an absorption band around 3000 cm^{-1} (3.3 μm) gives a rough idea of the CH group present. Alkyl groups have their C–H stretching frequencies lower than 3000 cm^{-1}, whereas for alkenes and aromatics they are slightly higher than 3000 cm^{-1}. The CH_3 group gives rise to an asymmetric stretching mode at 2960 cm^{-1} (3.38 μm) and a symmetric mode at 2870 cm^{-1} (3.48 μm). For $-CH_2-$ these bands occur at 2930 cm^{-1} (3.42 μm) and 2850 cm^{-1} (3.51 μm).

Next, one should examine regions where characteristic vibrations from bending motions occur. For alkanes, bands at 1460 cm^{-1} (6.85 μm) and 1380 cm^{-1} (7.25 μm) are indicative of a terminal methyl group attached to carbon exhibiting in-plane bending motions; if the latter band is split into a doublet at about 1397 and 1370 cm^{-1} (7.16 and 7.30 μm), geminal methyls are indicated. The symmetrical in-plane bending is shifted to lower frequencies when the methyl group is adjacent to >C=O (1360–1350 cm^{-1}), –S– (1325 cm^{-1}), and silicon (1250 cm^{-1}). The in-plane scissor motion of $-CH_2-$ at 1470 cm^{-1} (6.80 μm) indicates the presence of that group. Four or more methylene groups in a linear arrangement gives rise to a weak rocking motion at about 720 cm^{-1} (13.9 μm).

The substitution pattern of an aromatic ring can be deduced from a series of weak but very useful bands in the region 2000 to 1670 cm^{-1} (5 to 6 μm) coupled with the position of the strong bands between 900 and 650 cm^{-1} (11.1 and 15.4 μm) which are related to the out-of-plane bending vibrations. Ring stretching modes are observed near 1600, 1570, and 1500 cm^{-1} (6.25, 6.37, and 6.67 μm). These characteristic absorption patterns are also observed with substituted pyridines and polycyclic benzenoid aromatics.

The presence of an unsaturated C=C linkage introduces the stretching frequency at 1650 cm^{-1} (6.07 μm), which may be weak or nonexistent if symmetrically located in the molecule. Mono- and trisubstituted olefins give rise to more intense bands than *cis* or *trans*-distributed olefins. Substitution by a nitrogen or oxygen functional group greatly increases the intensity of the C=C absorption band. Conjugation with an aromatic nucleus causes a slight shift to lower frequency, but with a second C=C or C=O, the shift to lower frequency is 40 to 60 cm^{-1} with a substantial increase in intensity. The out-of-plane bending vibrations of the hydrogens on a C=C linkage are very valuable. A vinyl group gives rise to two bands at about 990 cm^{-1} (10.1 μm) and 910 cm^{-1} (11.0 μm). The $=CH_2$ (vinylidene) band appears near 895 cm^{-1} (11.2 μm) and is a very prominent feature of the spectrum. *Cis*- and *trans*-disubstituted olefins absorb near 685–730 cm^{-1} (13.7–14.6 μm) and 965 cm^{-1} (10.4 μm), respectively. The single hydrogen in a trisubstituted olefin appears near 820 cm^{-1} (12.2 μm).

In alkynes the ethynyl hydrogen appears as a needle-sharp and intense band at 3300 cm^{-1} (3.0 μm). The absorption band for –C=C– is located approximately in the range from 2100 to 2140 cm^{-1} (4.76–4.67 μm) when terminal, but in the region from 2260 to 2190 cm^{-1} (4.42–4.56 μm) when nonterminal. The intensity of the latter type band decreases as the symmetry of the molecule increases; it is best identified by Raman spectroscopy. When the acetylene linkage is conjugated with a carbonyl group, however, the absorption becomes very intense.

For ethers, the one important band appears near 1100 cm^{-1} (9.09 μm) and is related to the antisymmetric stretching mode of the –C–O–C– links. It is quite strong and may dominate the spectrum of a simple ether.

For alcohols, the most useful absorption is that related to the stretching of the O–H bond. In the free or unassociated state, it appears as a weak but sharp band at about 3600 cm^{-1} (2.78 μm). Hydrogen bonding will greatly increase the intensity of the band and move it to lower frequencies and, if the hydrogen bonding is especially strong, the band becomes quite broad. Intermolecular hydrogen bonding is concentration dependent, whereas intramolecular hydrogen bonding is not concentration dependent. Measurements in solution under different concentrations are invaluable. The spectrum of an acid is quite distinctive in shape and breadth in the high-frequency region. The distinction between the several types of alcohol is often possible on the basis of the C–O stretching absorption bands.

The carbonyl group is not difficult to recognize; it is often the strongest band in the spectrum. Its exact position in the region, extending from about 1825 to 1575 cm^{-1} (5.48 to 6.35 μm), is dependent on the double-bond character of the carbonyl group. Anhydrides usually show a double absorption band.

Aldehydes are distinguished from ketones by additional C–H stretching frequency of the CHO group at about 2720 cm^{-1} (3.68 μm). In esters, two bands related to C–O stretching and bending are recognizable, between 1300 and 1040 cm^{-1} (7.7 and 9.6 μm), in addition to the carbonyl band. The carboxyl group shows bands arising from the superposition of C=O, C–O, C–OH, and O–H vibrations. Of five characteristic bands, three (2700, 1300, and 943 cm^{-1}; 3.7, 7.7, and 10.6 μm) are associated with vibrations of the carboxyl OH. They disappear when the carboxylate ion is formed. When the acid exists in the dimeric form, the O–H stretching band; at 2700 cm^{-1} disappears, but the absorption band at 943 cm^{-1} related to OH out-of-plane bending of the dimer remains.

Of particular interest in a primary amine (or amide) are the N–H stretching vibrations at about 3500 and 3400 cm^{-1} (2.86 and 2.94 μm), the in-plane bending of N–H at 1610 cm^{-1} (6.2 μm), and the out-of-plane bending of $-NH_2$ at about 830 cm^{-1} (12.0 μm), which is broad for primary amines. By contrast, a secondary amine exhibits a single band in the high-frequency region at about 3350 cm^{-1} (2.98 μm). The high-frequency bands broaden and shift about 100 cm^{-1} to lower frequency when involved in hydrogen bonding. When the amine salt is formed, these bands are markedly broadened and lie between 3030 and 2500 cm^{-1} (3.3 and 4.0 μm) resembling the COOH bands in this region.

The nitro group is characterized by two equally strong absorption bands at about 1560 and 1350 cm^{-1} (6.41 and 7.40 μm), the asymmetric and symmetric stretching frequencies. In an N-oxide, only a single very intense band is present in the region from 1300 to 1200 cm^{-1} (7.70 to 8.33 μm). In addition, there are C–N stretching and various bending vibrations whose positions should be checked. Quite analogous bands are observed for bonds between S and O; all are intense. Stretching frequencies of SO_2 appear around 1400–1310 and 1230–1120 cm^{-1} (7.14–7.63 and 8.13–8.93 μm); for S=O at 1200–1040 cm^{-1} (8.33–9.62 μm); and for S–O around 900–700 cm^{-1} (11.11–14.28 μm).

- Compound identification. In many cases the interpretation of the infrared spectrum on the basis of characteristic frequencies will not be sufficient to permit positive identification of a total unknown, but perhaps the type of class of compound can be deduced. One must resist the tendency to over interpret a spectrum, that is, to attempt to interpret and assign all of the observed absorption bands, particularly those of moderate and weak intensity in the fingerprint region. Once the category is established, the spectrum of the unknown is compared with spectra of appropriate known compounds for an exact spectral match. If the exact compound happens not to be in the file, particular structure variations within the category may assist in suggesting possible answers and eliminating others. Several collections of spectra are available commercially (ASTM-Wyandotte Index, 1963; Nyquist and Kagel, 1971; Aldrich, 1995; Sadtler Research Laboratories, 1963; *Infrared Spectroscopy—Its Use in the Coatings Industry*, 1969).

2.8.2. Equipment

A microscopic infrared spectroscope is shown in Fig. 2.15 and the many different modes of operation in Fig. 2.16.

2.8.3. Applications

The ATR method is useful for reflecting IR energy off the surface of a specimen and generating a spectrum to identify the material, if possible. Organic materials are usually identifiable with ATR or other IR methods, but not all pigments are identifiable with IR.

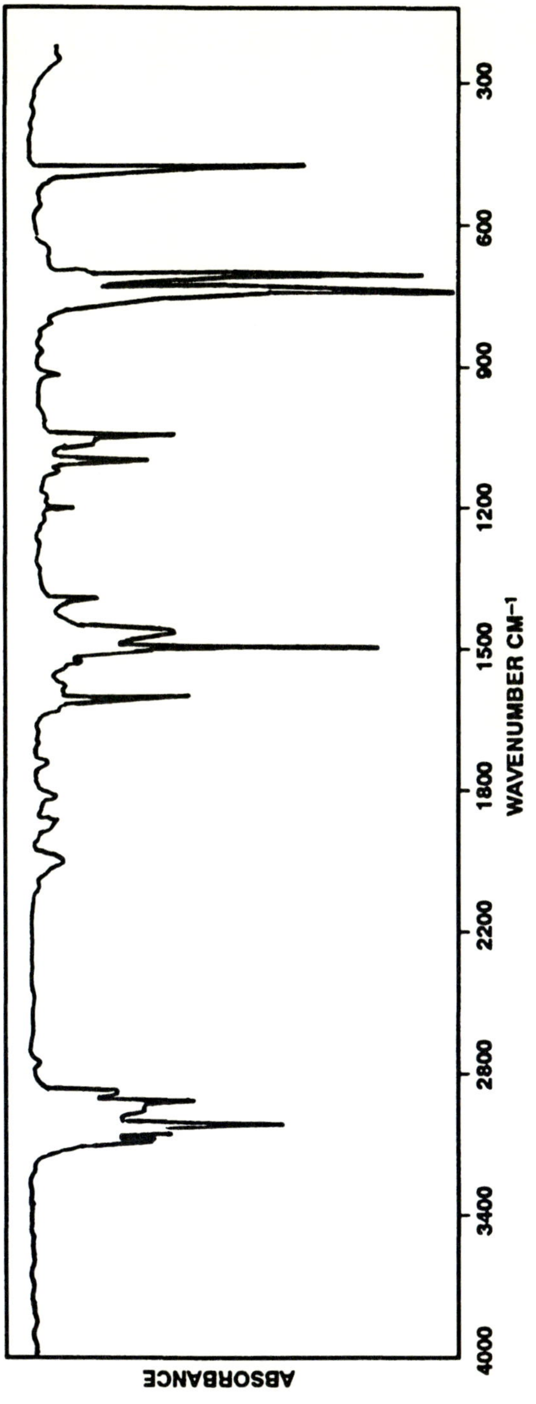

Figure 2.17. Infrared spectrum of toluene.

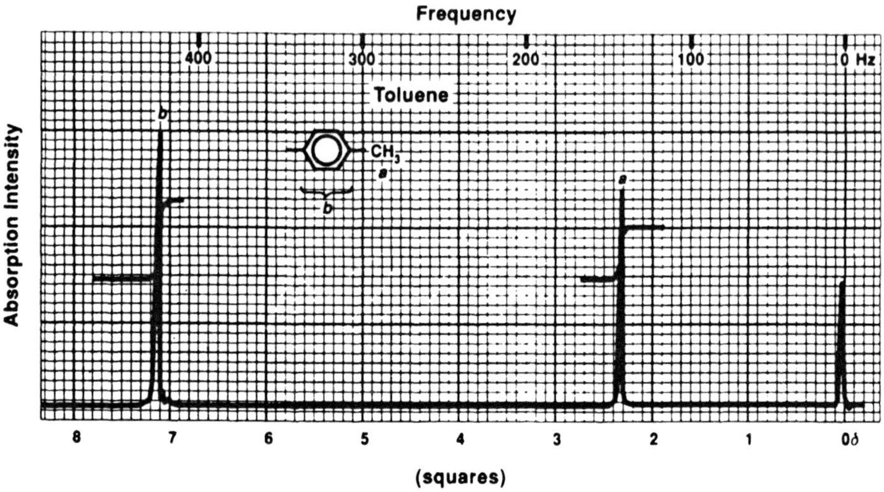

Figure 2.18. ^{1}H-NMR spectrum of toluene.

The microscopic FTIR is the most useful tool for identifying a wide range of specimen sizes, and particularly useful for simultaneously analyzing a mixture of materials without physical separation. The technique often avoids the laborious task of dissolving a resin or polymer in solvent and filtering and/or centrifuging particles. It is the only type of instrument that can analyze individual microscopic particles. The FTIR spectrum of toluene is shown in Fig. 2.17 (the ^{1}H-NMR spectrum of toluene is presented in Fig. 2.18). The absorbance peaks indicate $-CH_3$ and C_6H_5- of toluene. Interpretation of IR spectra is discussed further in Chapter 3.

2.9. SURFACE ENERGY AND CONTACT ANGLE MEASUREMENT

2.9.1. Fundamentals

A surface has a surface energy, and it is representative of a chemical structure, even if only superficially. For example, Teflon has a very low surface energy (< 20 dyn/cm) and is difficult to wet, paint, and so forth. This is because the surface of the wetting agent must be lower than the substrate, and few substances possess a surface energy lower than Teflon's. The measurement of a test liquid on a substrate is shown in Fig. 2.19. The contact angles of a series of liquids are measured and a plot of "cos θ versus surface energy (dyn/cm)" is generated. The extrapolation of the curve to cos θ = 1 is the corresponding surface energy (dyn/cm) of the test substrate (see Fig. 2.21). The instrument for measuring contact angle is a goniometer.

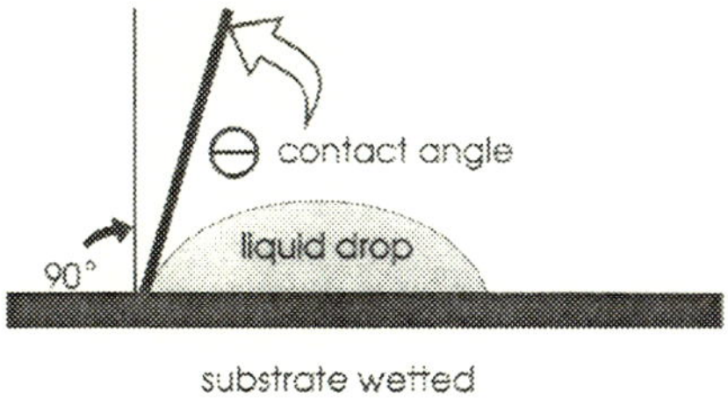

Figure 2.19. Measurement of contact angle of a solid material using a goniometer.

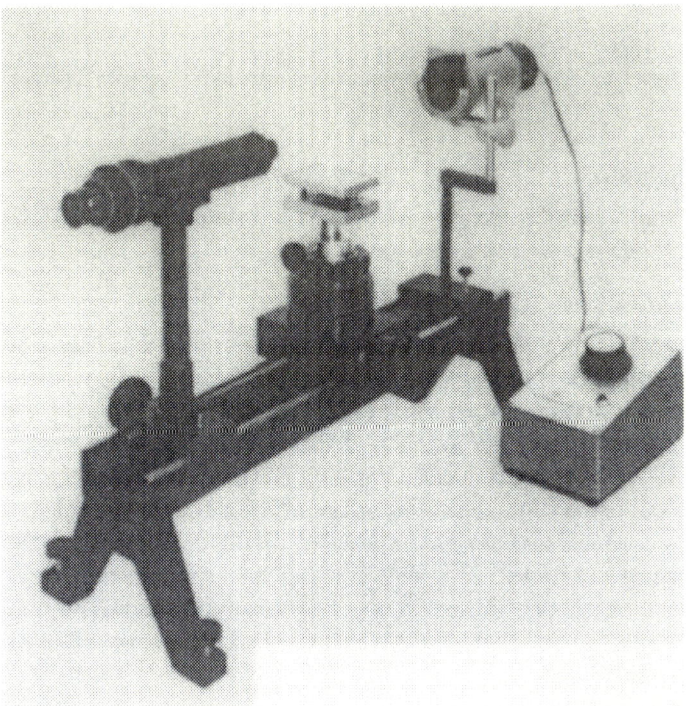

Figure 2.20. Photograph of Ramé–Hart NRL Contact Angle Goniometer. White arrow indicates position of specimen. Reprinted with permission of Ramé–Hart, Inc.

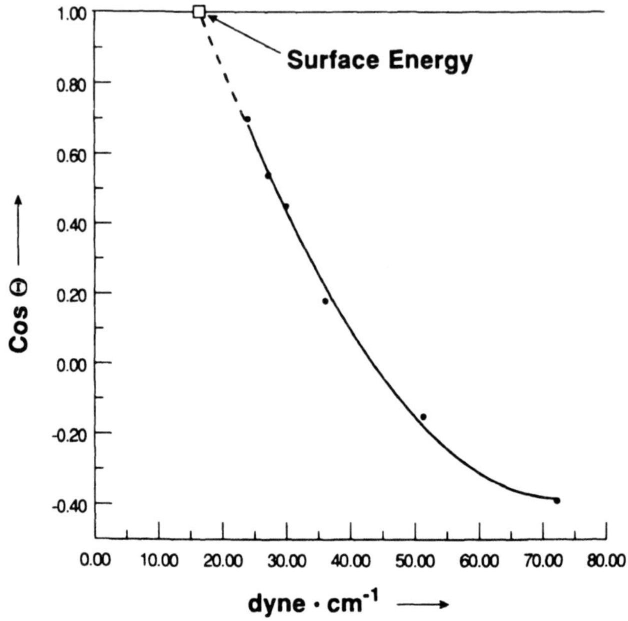

Figure 2.21. Surface energy determination of polytetrafluoroethylene (Teflon).

2.9.2. Equipment

The Ramé–Hart Contact Angle Goniometer is shown in Fig. 2.20. The position of the specimen is indicated by the arrowhead.

2.9.3. Applications

An example of a contact angle measurement is shown in Fig. 2.19. cosθ is plotted against known surface energies of control liquids, and an extrapolation is made to cosθ = 1 which is the surface energy (or surface tension) of the specimen. The low surface energy of Teflon is determined in Fig. 2.21. Most polymers (Shafrin, 1977) demonstrate a surface energy greater than 20 dyn/cm. The surface energy is a function of the chemical nature of the substrate and often, important clues to the chemical structure can be found by first determining the surface energy. Surface energy determination is not expensive, the measurement is very sensitive, and the goniometer is not difficult to use. For example, trace quantities of a silicon adhesion agent may reside on the surface of a substrate and are difficult to detect except by contact angle.

3

Bulk Analysis

3.1. ATOMIC SPECTROSCOPY (AS)

3.1.1. Fundamentals

Atomic spectroscopy is actually not one technique but three (Willard *et al.*, 1974): atomic absorption, atomic emission, and atomic fluorescence. Of these, atomic absorption (AA) and atomic emission are the most widely used. Our discussion will deal with them and an affiliated technique, inductively coupled plasma (ICP)-mass spectrometry.

- Atomic absorption. Atomic absorption (Willard *et al.*, 1974) is the process that occurs when a ground-state atom absorbs energy in the form of light of a specific wavelength and is elevated to an excited state. The amount of light energy absorbed at this wavelength will increase as the number of atoms of the selected element in the light path increases. The relationship between the amount of light absorbed and the concentration of an analyte present in known standards can be used to determine unknown concentrations by measuring the amount of light they absorb. Instrument readouts can be calibrated to display concentrations directly.

The basic instrumentation for atomic absorption requires a primary light source, an atom source, a monochromator to isolate the specific wavelength of light to be used, a detector to measure the light accurately, electronics to treat the signal, and a data display or a logging device to show the results. The atom source used must produce free analyte atoms from the sample. The source of energy for free atom production is heat, the most common source being an air–acetylene or nitrous oxide–acetylene flame. The sample is introduced as an aerosol into the flame. The flame burner head is aligned so that the light beam passes through the flame, where the light is absorbed.

- Graphite furnace atomic absorption. The major limitation of atomic absorption using flame sampling (flame AA) is that the burner–nebulizer system is a relatively inefficient sampling device. Only a small fraction of the sample reaches the flame, and the atomized sample passes quickly through the light path. An improved sampling device would atomize the entire sample and retain the atomized sample in the light path for an extended period to enhance the sensitivity of the technique. Electrothermal vaporization using a graphite furnace provides those features.

With graphite furnace atomic absorption (GFAA), the flame is replaced by an electrically heated graphite tube. A sample is introduced directly into the tube, which is then heated in a programmed series of steps to remove the solvent and major matrix components and then to atomize the remaining sample. All of the analyte is atomized, and the atoms are retained within the tube (and the light path, which passes through the tube) for an extended period. As a result, sensitivity and detection limits are significantly improved.

Graphite furnace analysis times are longer than those for flame sampling, and fewer elements can be determined using GFAA. However, the enhanced sensitivity of GFAA and the ability of GFAA to analyze very small samples and directly analyze certain types of solid samples significantly expand the capabilities of atomic absorption.

- Atomic emission. Atomic emission spectroscopy (Willard *et al.*, 1976; Dean and Raines, 1974) is a process in which the light emitted by excited atoms or ions is measured. The emission occurs when sufficient thermal or electrical energy is available to excite a free atom or ion to an unstable energy state. Light is emitted when the atom or ion returns to a more stable configuration or the ground state. The wavelengths of light emitted are specific to the elements that are present in the sample.

The basic instrument used for atomic emission is very similar to that used for atomic absorption with the difference that no primary light source is used for atomic emission. One of the more critical components for atomic emission instruments is the atomization source (Grove, 1971) because it must also provide sufficient energy to excite the atoms and atomize them.

The earliest energy sources for excitation were simple flames, but these often lacked sufficient thermal energy to be a truly effective source. Later, electrothermal sources such as arc/spark systems were used, particularly when analyzing solid samples. These sources are useful for doing qualitative and quantitative work with solid samples, but are expensive, difficult to use, and have limited applications.

Because of the limitations of the early sources, atomic emission initially did not enjoy the universal popularity of atomic absorption. This changed dramatically with the development of the inductively coupled plasma (ICP) as a source for atomic emission. The ICP eliminates many of the problems associated with past emission sources and has caused a dramatic increase in the utility and use of emission spectroscopy.

• Inductively coupled plasma (ICP). The ICP (Berlin, 1970) is an argon plasma maintained by the interaction of a radio frequency (RF) field and ionized argon gas. The ICP is reported to reach temperatures as high as 10,000 K, with the sample experiencing useful temperatures between 5500 and 8000 K. These temperatures allow complete atomization of elements, minimizing chemical interference effects.

The plasma is formed by a tangential stream of argon gas flowing between two quartz tubes. RF power is applied through the coil, and an oscillating magnetic field is formed. The plasma is created when the argon is made conductive by exposing it to an electrical discharge which creates seed electrons and ions. Inside the induced magnetic field, the charged particles (electrons and ions) are forced to flow in a closed annular path. As they meet resistance to their flow, heating takes place and additional ionization occurs. The process occurs almost instantaneously, and the plasma expands to its full dimensions.

As viewed from the top, the plasma has a circular, "doughnut" shape. The sample is injected as an aerosol through the center of the doughnut. This characteristic of the ICP confines the sample to a narrow region and provides an optically thin emission source and a chemically inert atmosphere. This results in a wide dynamic range and minimal chemical interactions in an analysis. Argon is also used as a carrier gas for the sample.

• ICP-mass spectroscopy. As its name implies, ICP-mass spectrometry (ICP-MS) is the synergistic combination of an inductively coupled plasma with a quadrupole mass spectrometer (Birks, 1959). ICP-MS uses the ability of the argon ICP to efficiently generate singly charged ions from the elemental species within a sample. These ions are then directed into a quadrupole mass spectrometer.

The function of the mass spectrometer is similar to that of the monochromator in an AA or ICP emission system. However, rather than separating light according to its wavelength, the mass spectrometer separates the ions introduced from the ICP according to their mass-to-charge ratio. Ions of the selected mass/charge are directed to a detector which counts the number of ions present. Because of the similarity of the sample introduction and data handling techniques, using an ICP-MS is very much like using an ICP emission spectrometer.

ICP-MS combines the multielement capabilities and broad linear working range of ICE emission with the exceptional detection limits of GFAA. It is also one of the few analytical techniques that permit the quantitation of elemental isotopic concentrations and ratios.

• Selection of the proper atomic spectroscopy technique. With the availability of a variety of atomic spectroscopy techniques such as flame atomic absorption, graphite furnace atomic absorption, ICP emission, and ICE-mass spectrometry, laboratory managers must decide which technique is best suited for the analytical problems of their laboratory. Because atomic spectroscopy techniques complement each other so well, it may not always be clear which technique is optimal for a particular laboratory. A clear understanding of the analytical problem in the laboratory and the capabilities provided by the different techniques is necessary. Important criteria for selecting an analytical technique include detection limits, analytical working range, sample throughput, cost, interferences, ease of use, and the availability of proven methodology. These criteria are discussed below for flame AA, GFAA, ICE emission, and ICE-MS.

• Atomic spectroscopy detection limits. The detection limits achievable for individual elements represent a significant criterion of the usefulness of an analytical technique for a given analytical problem. Without adequate detection limit capabilities, lengthy analytical concentration procedures may be required prior to analysis.

Generally, the best detection limits are attained using ICE-MS or GFAA. For mercury and those elements that form hydrides, the cold vapor mercury or hydride generation techniques offer exceptional detection limits.

Most manufacturers (e.g., Perkin–Elmer) define detection limits very conservatively with either a 95 or 98% confidence level, depending on established conventions for the analytical technique. This means that if a concentration at the detection limit were measured many times, it could be distinguished from a zero or baseline reading in 95% (or 98%) of the determinations.

Figure 3.1. Photograph of Perkin–Elmer 3100 Atomic Absorption Spectrometer. Reprinted with permission of Perkin–Elmer Corp.

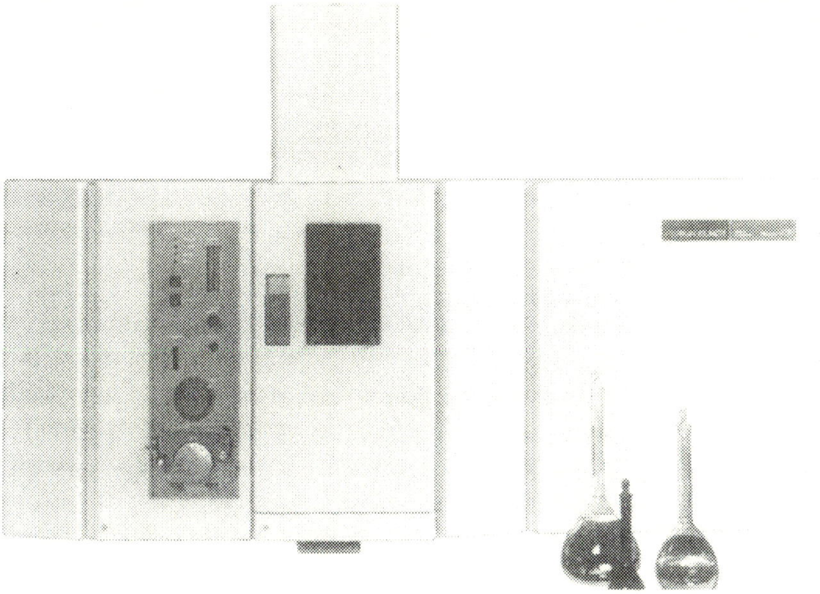

Figure 3.2. Photograph of Perkin–Elmer Plasma 400 ICI Emission Spectrometer. Reprinted with permission of Perkin–Elmer Corp.

3.1.2. Equipment

Figures 3.1 and 3.2 show a Perkin–Elmer 3100 Atomic Absorption Spectrometer and a Perkin–Elmer Plasma 400 ICI Emission Spectrometer.

3.1.3 Applications

Atomic spectroscopy has many uses for analysis of materials, and especially for inorganic pigments that contain metals. Trace concentrations are measurable using these methods.

3.2. INFRARED SPECTROSCOPY (IR) FOR BULK ANALYSIS

3.2.1. Fundamentals

Much of the following information is taken from Willard *et al.* (1974). The infrared region of the electromagnetic spectrum extends from the red end of the visible spectrum to the microwaves; that is, the region includes radiation at wavelengths between 0.7 and 500 μm, or, in wave numbers, between 14,000 and 20 cm^{-1}. The spectral range of greatest use is the mid-infrared region, which covers the frequency range from 200 to 4000 cm^{-1} (50 to 2.5 μm). Infrared spectroscopy involves the twisting, bending, rotating, and vibrational motions of atoms in a molecule. On interaction with infrared radiation, portions of the incident radiation are absorbed at particular wavelengths. The multiplicity of vibrations occurring simultaneously produces a highly complex absorption spectrum, which is uniquely characteristic of the functional groups comprising the molecule and of the overall configuration of the atoms as well. Suggested review articles on the fundamentals of infrared spectroscopy are Bellamy (1958), Colthup *et al.* (1964), Gianturco (1965), Herberg (1945), and Nakanishi (1962).

• Molecular vibrations. Atoms or atomic groups in molecules are in continuous motion with respect to each other. The possible vibrational modes in a polyatomic molecule can be visualized from a mechanical model of the system. Atomic masses are represented by balls, their weight being proportional to the corresponding atomic weight. The atomic masses are arranged in accordance with the actual space geometry of the molecule. Mechanical springs, with forces that are proportional to the bonding forces of the chemical links, connect and keep the balls in positions of balance. If the model is suspended in space and struck by a blow, the balls will appear to undergo random chaotic motions. However, if the vibrating model is observed with a stroboscopic light of variable frequency, certain light frequencies will be found at which the balls appear to remain stationary. These represent the specific vibrational frequencies for these motions.

For infrared absorption to occur, two major conditions must be fulfilled. First, the energy of the radiation must coincide with the energy difference between the excited and ground states of the molecule. Radiant energy will then be absorbed by the molecule, increasing its natural vibration. Second, the vibration must entail a change in the electrical dipole moment, a restriction that distinguishes infrared from Raman spectroscopy.

Stretching vibrations involve changes in the frequency of the vibration of bonded atoms along the bond axis. In a symmetrical group such as methylene, there are identical vibrational frequencies. For example, the asymmetric vibration occurs in the plane of the paper and also in the plane at right angles to the paper. In space these two are indistinguishable and said to be one doubly degenerate vibration. In the symmetric stretching mode there will be no change in the dipole moment as the two hydrogen atoms move equal distances in opposite directions from the carbon atom, and the vibration will be infrared inactive. If there is a change in the dipole moment, the centers of highest positive charge (hydrogen) and negative charge (carbon) will move in such a way that the electrical center of the group is displaced from the carbon atom. These vibrations will be observed in the infrared spectrum of the methylene group.

3.2.2. Equipment

It is convenient to divide the infrared region into three segments with the dividing points based on instrumental capabilities. Different radiation sources, optical systems, and detectors are needed for the different regions. The standard infrared spectrophotometer is an instrument covering the range from 4000 to 650 cm^{-1} (2.5 to 15.4 μm).

Grating instruments offer higher resolution that permits separation of closely spaced absorption bands, more accurate measurements of band positions and intensities, and higher scanning speeds for a given resolution and noise level. Modern spectrophotometers generally have attachments that permit speed suppression, scale expansion, repetitive scanning, and automatic control of slit, period, and gain. Accessories such as beam condensers, reflectance units, polarizers, and micro cells can usually be added to extend versatility or accuracy.

Temperature and relative humidity in the room housing the instrument must be controlled.

• Spectrometers. Most infrared spectrophotometers are double-beam instruments in which two equivalent beams of radiant energy are taken from the source. By means of a combined rotating mirror and light interrupter, the source is flicked alternately between the reference and sample paths. In the optical-null system, the detector responds only when the intensity of the two beams is unequal. Any imbalance is corrected for by a light attenuator (an optical wedge or shutter comb) moving in or out of the reference beam to restore balance. The recording pen is coupled to the light attenuator. Although very popular, the optical-null system has serious faults. Near zero transmittance of the sample, the reference-beam attenuator will move in to stop practically all light in the reference beam. Both beams are then blocked, no energy is passed, and the spectrometer has no way of determining how close it is to the correct transmittance value. The instrument will go dead. However, in the mid-infrared region, the electrical beam-radioing method is not an easy means of avoiding the deficiencies of the optical-null system. To a large extent it is trading optical and mechanical problems for electronic problems.

Monochromators employing prisms for dispersion utilize a Littrow 60° prism-plane mirror mount. Mid-infrared instruments employ a sodium chloride prism for the region from 4000 to 650 cm^{-1} (2.5 to 15.4 μm), with a potassium bromide or cesium iodide prism and optics for the extension of the useful spectrum to 400 cm^{-1} (25 μm) or 270 cm^{-1} (37 μm), respectively. Quartz monochromators, designed for the ultraviolet–visible region, extend their coverage into the near-infrared (to 2500 cm^{-1} or 4 μm).

To cover the wide wavelength range, several gratings with different ruling densities and associated higher-order filters are necessary. This requires some complex sensing and switching mechanisms for automating the scan with acceptable accuracy. Because of the nature of the blackbody emission curve, a slit programming mechanism must be employed to give near-constant energy and resolution as a function of wavelength. The principal limitation is energy. Resolution and signal-to-noise ratio are limited primarily by the emission of the blackbody source and the noise-equivalent power of the detector. Two gratings are often mounted back to back so that each need be used only in the first order; the gratings are changed to 2000 cm^{-1} (5.0 μm) in mid-infrared spectrometers. Grating instruments incorporate a sine-bar mechanism to drive the grating mount when a wavelength readout is desired, and a cosecant-bar drive when wave numbers are desired. Undesired overlapping can be eliminated with a fore-prism or by suitable filters.

The filters are inserted near a slit or slit image when the required size of the filter is not excessive. The circular variable filter is simple in construction. It is frequently necessary to use gratings as reflectance filters when working in the far-infrared so as to remove unwanted second and higher orders from the light incident on the far-infrared grating. For this purpose, small plane gratings are used which are blazed for the wavelength of the unwanted radiation. The grating acts as mirror reflecting the wanted light into the instrument and diffracting the shorter wavelengths out of the beam; grating "looks" like a good mirror to wavelengths longer than the groove spacing.

- Interferometric (Fourier transform) spectrometer (Low, 1970). The basic configuration of the interferometer portion of a Fourier transform spectrometer includes two plane mirrors at a right angle to each other and a beam splitter at 45° to the mirrors. Modulated light from the source is collimated and passes to the beam splitter which divides it into two equal beams for the two mirrors. An equal thickness of support material (without the semireflection coating), called the compensator, is placed in one arm of the interferometer to equalize the optical path length in both arms. When these mirrors are positioned so that the optical path lengths of the reflected and transmitted beams are equal, the two beams will be in phase when they return to the beam splitter and will constructively interfere. Displacing the movable mirror by one-quarter wavelength will bring the two beams 180° out of phase and they will destructively interfere. Continuing the movement of the mirror in either direction will cause the field to oscillate from light to dark for each quarter-wavelength movement of the mirror, corresponding to $\lambda/2$ changes. When the interferometer is illuminated by monochromatic light of wavelength λ, and the mirror is moved with a velocity v, the signal from the detector has a frequency $f = 2v/\lambda$. A plot of signal versus mirror distance is a pure cosine wave. With polychromatic light, the output signal is the sum of all the cosine waves, which is the Fourier transform of the spectrum. Each frequency is given an intensity modulation, f, which is proportional both to the frequency of the incident radiation and to the speed of the moving mirror. For example, with a constant mirror velocity of 0.5 mm/sec, radiation of 1000 cm^{-1} (10 μm and a frequency of 3×10^{14} Hz) will produce a detector signal of 50 Hz. For 5-μm radiation, the signal is 100 Hz, and so on. An appropriate inverse transformation of the interferogram will give the desired spectrum. Rather than dispersing polychromatic radiation as would a conventional dispersive spectrometer, the Fourier transform spectrometer performs a frequency transformation. Data reduction requires digital computer techniques and analog conversion devices.

To make any sense out of the intensity measurement, the displacement of the movable mirror has to be known precisely. With a constant velocity of mirror motion, the mirror should move as far and as smoothly as possible. If the velocity is precise, an electronically timed coordinate can be generated for the interferogram. Severe mechanical problems limit this approach. The interferometer itself, however, can be used to generate its own time scale. In addition to processing the incoming spectral radiation, a line from a laser source is used to produce a discrete signal which is time-locked to the mirror motion and hence to the interferogram. This is the fringe-reference system and is analogous to the frequency/field lock in NMR. The mirror position can be determined by measuring the laser line interferogram, counting the fringes as the mirror moves from the starting position—denoted by a burst of light from an incandescent source.

Dispersion or filtering is not required, so that energy-wasting slits are not needed, and this is a major advantage. With energy at a premium in the far-infrared, the superior light-gathering power of the interferometric spectrometer is a welcome asset for this spectral region.

In the near- and mid-infrared, germanium coated on a transparent salt, such as NaCl, KBr, or CsI, is a common beam splitter material. In far-infrared spectrometers, the beam splitter is a thin film of Mylar whose thickness must be chosen for the spectral region of interest. For example, a Mylar film 0.25 mil thick can effectively cover the range from 60 to 375 cm^{-1}.

Resolution is related to the maximum extent of mirror movement so that a 1-cm movement results in 1-cm^{-1} resolution and a 2-cm movement yields 0.1-cm^{-1} resolution. Resolution can also be doubled by doubling the measurement times, or resolution can be traded for rapid response. Because the detector of the interferometer "sees" all resolution elements throughout the entire scan time, the signal-to-noise ratio, S/N, is proportional to T, where T is the measurement time. For example, when examining a spectrum composed of 2000 resolution elements with an observation time of 1 sec per element assumed for the desired S/N, the interferometric measurement is complete in 1 sec. Improving the S/N by a factor of 2 would require only 4 sec to complete the measurement. Comparable times for a dispersive spectrometer are 33 and 72 min, respectively. Repetitive signal-averaged scans are very feasible with an interferometer.

- Sampling handling. Infrared instrumentation has reached a remarkable degree of standardization as far as the sample compartment of various spectrometers is concerned. Sample handling itself, however, presents a number of problems in the infrared region. No rugged window material for cuvettes exists that is transparent and also inert over this region. The alkali halides are widely used, particularly sodium chloride, which is transparent at wavelengths as long as 16 μm (625 cm^{-1}). Cell windows are easily fogged by exposure to moisture and require frequent repolishing. Silver chloride is often used for moist samples, or aqueous solutions, but it is soft, easily deformed, and darkens on exposure to visible light. Teflon has only C–C and C–F absorption band. For frequencies under 600 cm^{-1}, a polyethylene cell is useful. Crystals of high refractive index produce strong, persistent fringes.

• Liquids and solutions. Samples that are liquid at room temperature are usually scanned in their neat form, or in solution. The sample concentration and path length should be chosen so that the transmittance lies between 15 and 70%. For neat liquids this will represent a very thin layer, about 0.001–0.05 mm in thickness. For solutions, concentrations of 10% and cell lengths of 0.1 mm are most practical. Unfortunately, not all substances can be dissolved in a reasonable concentration in a solvent that is nonabsorbing in regions of interest. When possible, the spectrum is obtained in a 10% solution of CCl_4 in a 0.1-mm cell in the region 4000 to 1333 cm^{-1} (2.5 to 7.5 μm), and in a 10% solution of CS_2 in the region 1333 to 650 cm^{-1} (7.5 to 15.4 μm). To obtain solution spectra of polar materials that are insoluble in CCl_4 or CS_2, chloroform, methylene chloride, acetonitrile, and acetone are useful solvents. Sensitivity can be gained by going to longer path lengths if a suitably transparent solvent can be found. In a double-beam spectrophotometer a reference cell of the same path length as the sample cell is filled with pure solvent and placed in the reference beam. Moderate solvent absorption, now common to both beams, will not be observed in the recorded spectrum. However, solvent transmittance should never fall under 10%.

The possible influence of a solvent on the spectrum of a solute must not be overlooked. Particular care should be exercised in the selection of a solvent for compounds that are susceptible to hydrogen-bonding effects. Hydrogen bonding through an –OH or –NH– group alters the characteristic vibrational frequency of that group; the stronger the hydrogen bonding, the greater is the lowering of the fundamental frequency. To differentiate between inter- and intramolecular hydrogen bonding, a series of spectra at different dilutions, yet having the same number of absorbing molecules in the beam, must be obtained. If, as the dilution increases, the hydrogen-bonded absorption band decreases while the unbonded absorption band increases, the bonding is intermolecular. Intramolecular bonding shows no comparable dilution effect.

Infrared solution cells are constructed with windows sealed and separated by thin gaskets of copper and lead that have been wetted with mercury. The whole assembly is securely clamped together. As the mercury penetrates the metal, the gasket expands, producing a tight seal. The cell is provided with tapered fittings to accept the needles of hypodermic syringes for filling. In demountable cells, the sample and spacer are placed on one window, covered with another window, and the entire sandwich is clamped together.

- Films. Spectra of liquids not soluble in a suitable solvent are best obtained from capillary films. A large drop of the neat liquid is placed between two rock-salt plates which are then squeezed together and mounted in the spectrometer in a suitable holder. Plates need not have high polish, but must be flat to avoid distortion of the spectrum.

For polymers, resins, and amorphous solids, the sample is dissolved in any reasonably volatile solvent, the solution poured onto a rock-salt plate, and the solvent evaporated by gentle heating. If the solid is noncrystalline, a thin homogeneous film is deposited on the plate which then can be mounted and scanned directly. Sometimes polymers can be "hot pressed" onto plates.

- Mulls. Solids can be reduced to particles, and examined as a thin paste or mull by grinding the pulverized solid (about 9 mg) in a greasy liquid medium. The suspension is pressed into an annular groove in a demountable cell. Multiple reflections and reflections off the particles are lessened by grinding the particles to a size an order of magnitude less than the analytical wavelength and surrounding the particles by a medium whose refractive index more closely matches theirs than does air. Liquid media include mineral oil or Nujol, hexachlorobutadiene, perfluorokerosene, and chlorofluorocarbon gases (fluorc lubes). The latter are used when the absorption by the mineral oil masks the presence of C–H bands. For qualitative analysis the mull technique is rapid and convenient, but quantitative data are difficult to obtain; even halides may be used, particularly CsI or CsBr for measurements at longer wavelengths. Good dispersion of the sample in the matrix is critical; moisture must be absent. Freeze-drying the sample is often a necessary preliminary step.

KBr wafers can be formed, without evacuation, in a Mini-Press^{R}. Two highly polished bolts are turned against each other in a steel cylinder. Pressure is applied with wrenches for about 1 min to 75 to 100 mg of powder, the bolts are removed, and the cylinder is installed in its slide holder in any spectrophotometer.

Quantitative analyses can be performed as a measurement can be made of the weight ratio of sample to internal standard added in each disk or wafer.

The appearance and intensity of an ATR spectrum will depend on the difference of the indices of refraction between the reflection crystal and the rarer medium containing the absorber, and on the internal angle of incidence. Thus, a reflection crystal of relatively high index of refraction should be used. Two materials found to perform most satisfactorily for the majority of liquid and solid samples are KRS-5 and AgCl. KRS-5 is a tough and durable material with excellent transmission properties. Its index of refraction is high enough to permit well-defined spectra of nearly all organic materials, although it is soluble in basic solutions.

AgCl is recommended for aqueous samples because of its insolubility and lower refractive index. An overall angle of incidence should be selected that is far enough from the average critical angle of sample versus reflector so that the change of the critical angle through the region of changing index of refraction (the absorption band) has a minimum effect on the shape of the ATR band. Unfortunately, when the index of refraction of the crystal is considerably greater than that of the sample so that little distortion occurs, the total absorption is reduced. With multiple reflection equipment, however, ample absorption can be obtained at angles well away from the critical angle when an internal standard is incorporated.

- Pellet technique. The pellet technique involves mixing the fine ground sample (1–100 μg) and potassium bromide powder, and pressing the mixture in an evacuable die at sufficient pressure (60,000–100,000 psi) to produce a transparent disk. Grinding-mixing is conveniently done in a vibrating ball-mill (Wig-L-Bug).

- Infrared probe. Resembling a specific ion electrode, the infrared probe contains a sensitive element that is dipped into the sample. To operate it, the user selects the proper wavelength by rotating a calibrated, circular variable filter, then adjusts the gain and slits to bring the meter to 100%. Next, the probe is lowered into the sample. The meter indicates the absorbance. This value can be converted into concentration by reference to a previously prepared calibration curve. To detect the presence or absence of a particular functional group, one scans through the portion of the spectrum where the absorption bands characteristic of that group appear.

The infrared probe utilizes attenuated total reflection to obtain the absorption information. The probe crystal is made from a chemically inert material such as germanium or synthetic sapphire. The reflecting surfaces are masked so that the same area is covered by sample each time an analysis is made. A single-beam optical system is employed, chopped at 45 Hz. Because the air path is less than 5 cm, as opposed to well over 1 m in conventional infrared spectrophotometers, absorption related to atmospheric water vapor and carbon dioxide is insignificant.

- Quantitative analysis. The application of infrared spectroscopy as a quantitative analytical tool varies widely from one laboratory to another. However, the use of high-resolution grating instruments materially increases the scope and reliability of quantitative infrared work. Quantitative infrared analysis is based on Beer's law; apparent deviations arise from either chemical or instrumental effects. In many cases, the presence of scattered radiation makes the direct application of Beer's law inaccurate, especially at high values of absorbance. As the energy available in the useful portion of the infrared is usually quite small, it is necessary to use rather wide slit widths in the monochromator. This causes a considerable change in the apparent value of the molar absorptivity; therefore, molar absorptivity should be determined empirically.

The baseline method involves selection of an absorption band of the substance under analysis that does not fall too close to the bands of other matrix components. The value of the incident radiant energy P_0 is obtained by drawing a straight line tangent to the spectral absorption curve at the position of the sample's absorption band. The transmittance P is measured at the point of maximum absorption. The value of log (P_0/P) is then plotted against concentration.

Many possible errors are eliminated by the baseline method. The same cell is used for all determinations. All measurements are made at points on the spectrum that are sharply defined by the spectrum itself; thus, there is no dependence on wavelength settings. Use of such ratios eliminates changes in instrument sensitivity, source intensity, or changes in adjustment of the optical system.

Pellets from the disk technique can be employed in quantitative measurements. Uniform pellets of similar weight are essential, however, for quantitative analysis. Known weights of KBr are taken, plus a known quantity of the test substance from which absorbance data a calibration curve can be constructed. The disks are weighed and their thickness measured at several points on the surface with a dial micrometer. The disadvantage of measuring pellet thickness can be overcome by using an internal standard. Potassium thiocyanate makes an excellent internal standard. It should be preground, dried, and then reground, at a concentration of 0.2% by weight with dry KBr. The final mix is stored over phosphorous pentoxide. A standard calibration curve is made by mixing about 10% by weight of the test substance with the KBr–KSCN mixture and then grinding ratio of the thiocyanate absorption at 2125 cm^{-1} (4.70 μm) to a chosen absorption of the test substance is plotted against percent concentration of the sample.

For quantitative measurements, the single-beam system has some fundamental characteristics that can result in greater sensitivity and better accuracy than the double-beam systems. All other things being equal, a single-beam instrument will automatically have a greater signal-to-noise ratio. There is a factor of 2 advantage in looking at one beam all the time rather than two beams half the time. Electronic switching gives another factor of 2 advantage. Thus, in any analytical situation where background noise is appreciable, the single-beam spectrometer should be superior.

- Correlation of infrared spectra with molecular structure.

Example. An IR spectrum shows characteristic absorption peaks (for toluene's, see Fig. 2.17). From Table 2.1 chemical bonds and absorption frequencies—the peaks indicate a monosubstitute aromatic ring structure, namely, $-CH_3$ and

C_6H_5–, which is toluene. The NMR spectrum of toluene seen in Fig. 2.18 confirms this conclusion.

3.3. X-RAY DIFFRACTION (XRD)

3.3.1. Fundamentals

Every atom in a crystal scatters an X-ray beam (Bertin, 1970) incident on it in all directions. Because even the smallest crystal contains a very large number of atoms, the chance that these scattered waves would constructively interfere would be almost zero except for the fact that the atoms in crystals are arranged in a regular, repetitive manner. The condition for diffraction of a beam of X rays from a crystal is given by the Bragg equation (Birks, 1959, 1963; Bunn, 1961; Clark, 1955). Atoms located exactly on the crystal planes contribute maximally to the intensity of the diffracted beam; atoms exactly halfway between the planes exert maximum destructive interference and those at some intermediate location interfere constructively or destructively depending on their exact location but with less than their maximum effect. Furthermore, the scattering power of an atom for X rays depends on the number of electrons it possesses. Thus, the position of the diffraction beams from a crystal depends only on the size and shape of the repetitive unit of a crystal and the wavelength of the incident X-ray beam whereas the intensities of the diffracted beams depend also on the type of atoms in the crystal and their location in the fundamental repetitive unit, the unit cell (Henke *et al.*, 1970, Liebhafsky *et al.*, 1960; Liebhafsky, 1964). No two substances will have absolutely identical diffraction patterns when one considers both the direction and intensity of all diffracted beams (Robertson, 1953; Sproull, 1946); however, some similar, complex organic compounds may have almost identical patterns. The diffraction pattern is thus a "fingerprint" of a crystalline compound and the crystalline components of a mixture can be identified individually.

• Reciprocal lattice concept. Diffraction phenomena can be interpreted most conveniently with the aid of the reciprocal lattice concept. A plane can be represented by a line drawn normal to the plane; the spatial orientation of this line describes the orientation of the plane. Furthermore, the length of the line can be fixed in an inverse proportion to the interplanar spacing of the plane that it represents.

When a normal is drawn to each plane in a crystal and the normals are drawn from a common origin, the terminal points of these normals constitute a lattice array. This is called the reciprocal lattice (Birks, 1953; Bragg, 1933) because the distance of each point from the origin is reciprocal to the interplanar spacing of the planes that it represents. There exists in an individual cell of a crystalline structure, near the origin, the traces of several planes in a unit cell of a crystal, namely, the (100),

(001), (101), and (102) planes. The normals to these planes, also indicated, are called the reciprocal lattice vectors, α_{hkl}, and are defined by

$$\alpha_{hkl} = \frac{\lambda}{d_{hkl}}$$

In three dimensions, the lattice array is described by three reciprocal lattice vectors whose magnitudes are given by

$$a^* = \alpha_{100} = \frac{\lambda}{d_{100}}$$

$$b^* = \alpha_{101} = \frac{\lambda}{d_{101}}$$

$$c^* = \frac{\alpha}{001} = \frac{\lambda}{d_{001}}$$

and whose directions are defined by three interaxial angles α^*, β^*, γ^*.

Writing the Bragg equation in a form that relates the glancing angle θ most clearly to the other parameters, we have

$$\sin\theta_{hkl} = \frac{\lambda / d_{hkl}}{2}$$

The numerator can be taken as one side of a right triangle with θ as another angle and the denominator its hypotenuse. The diameter of a circle represents the direction of the incident X-ray beam. A line through the origin of the circle and forming the angle θ with the incident beam, represents a crystallographic plane that satisfies the Bragg diffraction condition. A line forming the angle θ with the crystal plane and 2θ with the incident beam, represents the diffracted beam's direction. Another line is the reciprocal lattice vector to the reciprocal lattice point P_{hkl} lying on the circumference of a circle. The vector α_{hkl} originates at the point on a circle where the direct beam leaves the circle. The Bragg equation is satisfied when and only when a reciprocal lattice point lies on the "sphere of reflection," a sphere formed by rotating the circle on the diameter.

Thus, the crystal in a diffraction experiment can be pictured at the center of a sphere of unit radius, and the reciprocal lattice of this crystal is centered at the point where the direct beam leaves the sphere. Because the orientation of the reciprocal lattice bears a fixed relation to that of the crystal, if the crystal is rotated, the reciprocal lattice can be pictured as rotating also. When a reciprocal lattice point intersects the sphere, a reflection emanates from the crystal at the sphere's center and passes through the intersecting reciprocal lattice point.

• Diffraction patterns. If the X-ray beam is monochromatic, there will be only a limited number of angles at which diffraction of the beam can occur. The actual angles are determined by the wavelength of the X rays and the spacing between the various planes of the crystal. In the *rotating crystal method*, monochromatic X radiation is incident on a single crystal which is rotated about one of its axes.

In a modification of the single-crystal method, known as the Weissenberg method, the photographic film is moved continuously during the exposure parallel to the axis of rotation of the crystal. All reflections are blocked out except those that occur in a single layer line. This results in a film that is somewhat easier to decipher than a simple rotation photograph. Still other techniques are used; one, the precession method, results in a photograph that gives an undistorted view of a plane in the reciprocal lattice of the crystal.

In the *powder method*, the crystal is replaced by a large collection of very small crystals, randomly oriented, and a continuous cone of diffracted rays is produced. There are some important differences, however, with respect to the rotating crystal method. The cones obtained with a single crystal are not continuous because the diffracted beams occur only at certain points along the cone, whereas the cones with the powder method are continuous. Furthermore, although the cones obtained with rotating single crystals are uniformly spaced about the zero level, the cones produced in the powder method are determined by the spacings of prominent planes and are not uniformly spaced. Because of the random orientation of the crystallites, the reciprocal lattice points generate a sphere of radius α_{hkl} about the origin of the reciprocal lattice. A number of these spheres intersect the sphere of reflection.

• Camera design. Cameras are usually constructed so that the film diameter has one of the three values 57.3, 114.6, or 143.2 mm. The reason for this can be understood by considering the calculations involved. If the distance between corresponding arcs of the same cone of diffracted rays is measured and called S, then

$$4\theta_{\text{rad}} = \frac{S}{R}$$

where θ_{rad} is the Bragg angle measured in radians and R is the radius of the film in the camera. The angle, θ_{deg}, measured in degrees, is then

$$\theta_{\text{deg}} = \frac{57.295s}{4R}$$

where 57.295 equals the value of a radian in degrees. Therefore, when the camera diameter ($2R$) is equal to 57.3 mm, $2\theta_{deg}$ may be found by measuring S in millimeters. When the diameter is 114.59 mm, $2\theta_{deg} = S/2$, and when the diameter is 143.2 mm, $\theta_{deg} = 2(S/10)$.

Once the angle θ has been calculated, the equation can be used to find the interplanar spacing, using values of wavelength λ. Sets of tables are available that give the interplanar spacing for the angle 2θ for the types of radiation most commonly used.

- X-ray powder data file. For most purposes, the identification of a powder pigment specimen is desired; its diffraction pattern is compared with diagrams of known substances until a match is obtained. This method requires that a library of standard films be available. An X-ray data card for sodium chloride is shown in Fig. 3.3. Alternatively, *d* values calculated from the diffraction diagram of the unknown substance are compared with the *d* values of over 5000 entries, which are listed on plain cards, Keysort cards, and IBM cards in the X-ray powder data file (Switzer, 1948). An index volume is available with the file. The cataloging scheme (American Society of Testing Materials, 1955) used to classify different cards lists the three most intense reflections in the upper left corner of each card. The cards are then arranged in sequence of decreasing *d* values of the most intense reflections, based on 100 for the most intense reflection observed.

To use the file to identify a sample containing one component, the *d* value for the darkest line of the unknown is looked up first in the index. Because more than one listing containing the first *d* value probably exists, the *d* values of the next two darkest lines are then matched against the values listed. Finally, the various cards involved are compared. A correct match requires that all of the lines on the card and film agree. It is also good practice to derive the unit cell from the observed interplanar spacings and to compare it with that listed in the card.

If the unknown contains a mixture, each component must be identified individually. This is done by treating the list of *d* values as if they belonged to a single component. After a suitable match for one component is obtained, all of the lines of the identified component are omitted from further consideration. The intensities of the remaining lines are rescaled by setting the strongest intensity equal to 100 and repeating the entire procedure.

Reexamination of the cards in the file is a continuing process so as to eliminate errors and remove deficiencies. Replacement cards for substances bear a star in the upper right corner.

X-ray diffraction furnishes a rapid, accurate method for the identification of the crystalline phases present in a material. Sometimes it is the only method available for determining which of the possible polymorphic forms of a substance are present, for example, carbon in graphite or in diamond. Differentiation among various oxides such as FeO, Fe_2O_3, and Fe_3O_4, or between materials present in such

d	282	1.99	1.63	3.258	NaCl
I / I1	100	55	15	13	SODIUM CHLORIDE HALITE

Rad. Cu λ 1.5405 Filter Dia.
Cut off I / I1
Ref. Swanson and Fuyat, NBS Circular 539, Vol. II, 41 (1953)

Sys. Cubic S,G,O ^5_H — Fm3m
a_a S,6402 b_0 c_0 A C
α β γ Z 4 Dx 2.164
Ref. Ibid.

εα ηωβ 1.542 εγ Sign
2V D mp Color
Ref. Ibid.

An ACS reagent grade sample recrystallized twice from hydrochloric acid.
X-ray pattern at 26°C.

Replaces 1-0993, 1-0994, 2-0815

dÅ	I / I1	hkl	dÅ	I / I1	hkl
3.258	13	111			
2.821	100	200			
1.994	55	220			
1.701	2	311			
1.626	15	222			
1.410	6	400			
1.294	1	331			
1.261	11	420			
1.1515	7	422			
1.0855	1	511			
0.9969	2	440			
0.9533	1	531			
0.9401	3	600			
0.8917	4	620			
0.8601	1	533			
0.8503	3	622			
0.8141	2	444			

Figure 3.3. X-ray data card for sodium chloride. (Source: American Society for Testing Materials.)

mixtures as KBr + NaCl, KCl + NaBr, or all four is easily accomplished with X-ray diffraction. On the contrary, chemical analysis would show only the ions present and not the actual state of combination. The presence of various hydrates is another possibility.

- Quantitative analysis. X-ray diffraction is adaptable to quantitative applications because the intensities of the diffraction peaks of a given compound in a mixture are proportional to the fraction of the material in the mixture. However, direct comparison of the intensity of a diffraction peak in the pattern obtained from a mixture is fraught with difficulties. Corrections are frequently necessary for the differences in absorption coefficients between the compound being determined and the matrix. Preferred orientations must be avoided. Internal standards help but do not overcome the difficulties entirely.

Structural Applications. A discussion of the complete structural determination for a crystalline substance is beyond the scope of this book.

Microradiographic methods are based on absorption and the contrast in the images is the result of differences in absorption coefficients from point to point. X-ray diffraction topography depends for image contrast on point-to-point changes in the direction or the intensity of beams diffracted by planes in the crystal.

3.3.2. Equipment

A Ragaku X-Ray Diffractometer is shown in Fig. 3.4.

3.3.3. Applications

The greatest application for X-ray diffraction is for the identification of inorganic pigments, fillers, and fibers. X-ray spectra can identify the degree of crystallinity, type of crystalline structure, and, usually, the identification of a crystalline material if there are no serious interferences. In the case of particles that may be found in plastics or paint, a microprobe can isolate an individual particle for examination.

Only crystalline materials produce a response to X-ray diffraction. However, it is important to know if a substance is crystalline, amorphous, or a combination of the two. For example, carbon fibers and graphite have a very similar appearance, but carbon fibers are totally amorphous and graphite fibers are totally crystalline. Placing a gram or so of each in a sample holder and subjecting them to X radiation will quickly determine which is which, i.e., no peaks for the carbon fibers.

Polymers have crystallinity also, i.e., over 95% HDPE polyethylene consists of orthorhombic crystals. Polymers that possess crystallinity usually are only semicrystalline, but a well-calibrated X-ray diffractometer is the best method to measure the degree of crystallinity in a polymer and make correlations to density and other properties.

Figure 3.4. Photograph of Rigaku X-Ray Diffractometer. Reprinted with permission of Rigaku. Inc.

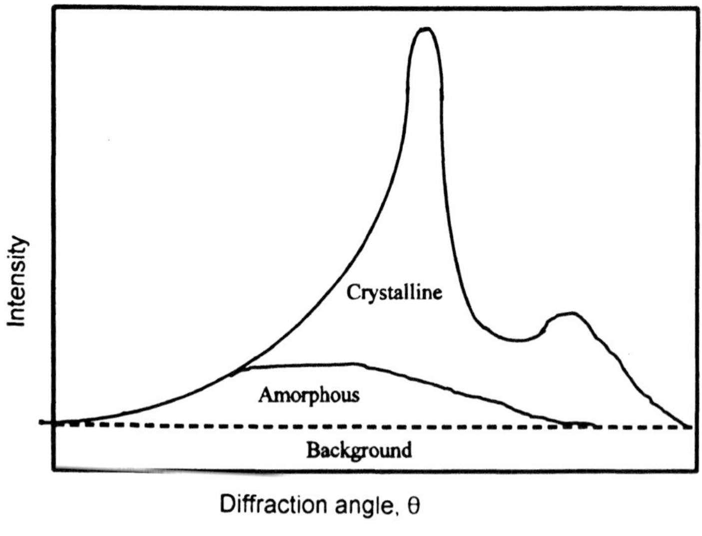

Figure 3.5. X-ray diffraction spectrum of lead pigment specimen.

When particles occur in polymers and other materials, it is necessary to isolate them by dissolving the polymer and filter or centrifuge the sediments. However, the X-ray microprobe is the easiest method as the sample only has to be cut or prepared to reveal a fresh surface. Surface preparation time is minimal and time is always valuable.

An X-ray diffraction spectrum of a lead pigment specimen is shown in Fig. 3.5.

3.4. GEL PERMEATION (GPC), HIGH-PRESSURE LIQUID (HPLC), AND GAS CHROMATOGRAPHY (GC)

3.4.1. Fundamentals

Molecules can be fractionated according to their constitution, configuration, or molecular weight by chromatographic methods. Adsorption chromatography is rarely used. Elution chromatography and gel permeation (size exclusion) chromatography are more often used.

Chromatography, as discussed in this book, consists of a chromatography column, a carrier gas or liquid, a detector, and an injection port. The specimen is introduced into the injection port with a calibrated syringe, and the carrier gas or liquid travels through the column while reacting with the packing material in the column. The interaction between the sample and the column packing material

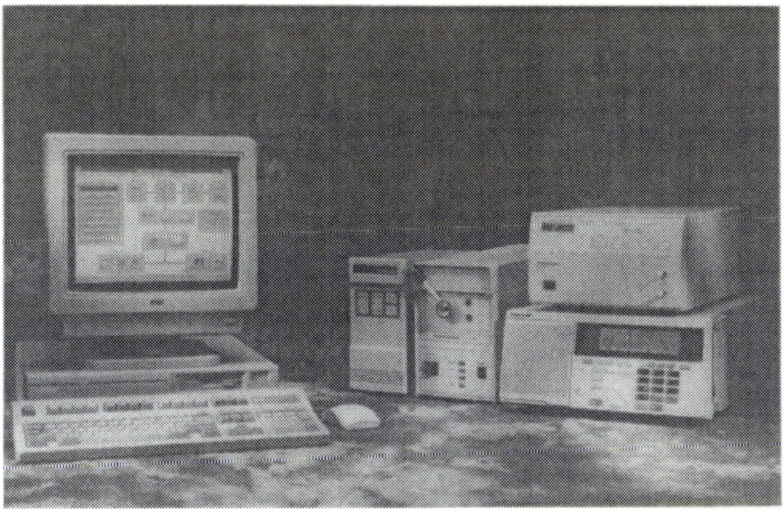

Figure 3.6. Photograph of Perkin–Elmer Gel Permeation Chromatograph. Reprinted with permission of Perkin–Elmer Corp.

causes a change in the rate of travel of the sample through the column (separation of sizes of molecules, separation by chemical species, etc.).

3.4.2. Equipment

Perkin–Elmer Gel Permeation Chromatograph (GPC), Integral 4000 High Performance Liquid Chromatograph (LC), and AutoSystem XL Gas Chromatograph (GC) are pictured in Figs. 3.6, 3.7, and 3.8, respectively.

3.4.3. Applications

- Gel permeation. GPC measures molecular weight and immediately reveals a high-molecular-weight material in the presence of a material of much lower molecular weight, e.g., a solvent (Collins *et al.*, 1973; Elias, 1977). GPC is most valuable for the following uses:

1. Measurement of molecular weight of soluble polymers, resins, and rosins
2. Measurement of molecular weight distribution

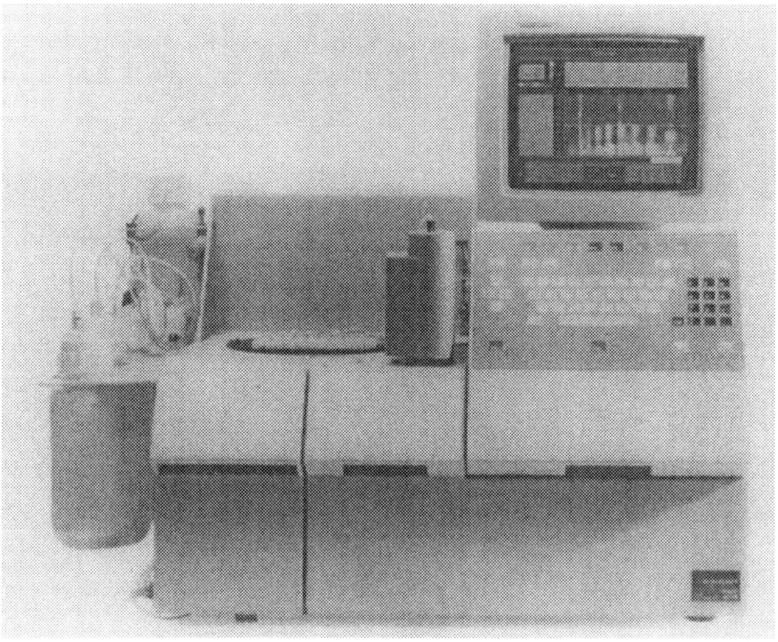

Figure 3.7. Photograph of Perkin–Elmer Integral 4000 High Performance Liquid Chromatograph. Reprinted with permission of Perkin–Elmer Corp.

Figure 3.8. Photograph of Perkin–Elmer AutoSystem XL Gas Chromatograph. Reprinted with permission of Perkin–Elmer Corp.

3. Determination of a low-molecular-weight species such as a solvent

GPC is a separation technique based on differences in molecular size, and use is made of the one-to-one relationship between size and mass for linear polymers of a single chemical type in making this determination. GPC is a liquid–liquid chromatographic separation in which columns are packed with porous gel particles, the pore sizes being of the same order of magnitude as the sizes of dissolved polymer molecules.

GPC can compare the molecular weight and distribution of materials which is useful for determining sources as materials often differ with supplier. Samples with molecular weights as low as 100 can be resolved with the proper column, but GPC is most useful for polymers and resins with masses above 1000 g/mole. A polymeric or resin sample of material to be analyzed is dissolved in carrier solvent or liquid and transported through a column such as Styrogel (cross-linked polystyrene column). The highest-molecular-weight fractions elute through the column first and

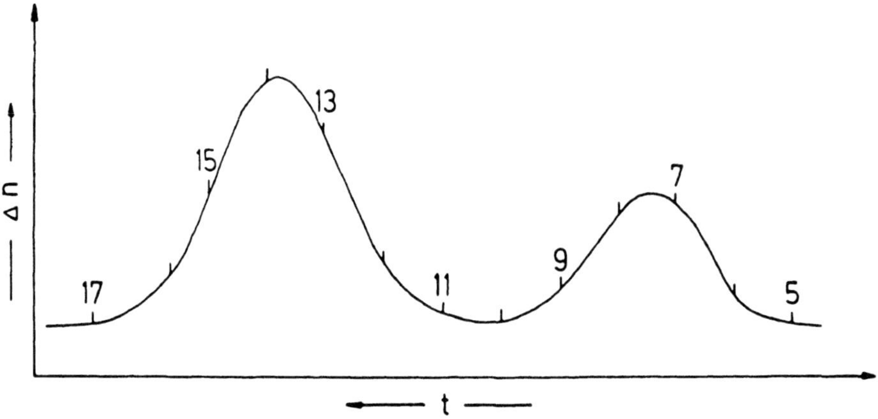

Figure 3.9. Hypothetical GPC chromatogram of a typical polymer. (Source: Elias, 1977.)

lower-molecular-weight fractions follow successively. A differential refractometer detector (and sometime an ultraviolet detector) is used to detect the molecular fractions as refractive index increases with molecular weight.

The Perkin–Elmer Gel Permeation Chromatograph is pictured in Fig. 3.6. A hypothetical bimodal GPC chromatogram of a typical polymer is given in Fig. 3.9, showing the development of peaks corresponding to change in refractive index with time of elution through the column. The numbers give the fraction numbers, which are proportional to the eluted volume (Elias, 1977). The refractive index is generally measured as a function of time. A calibration curve is necessary to correlate the events in a sample run with standard molecular weights in the same column, carrier liquid, and under the same conditions. There cannot be an accurate molecular weight determination without a reliable calibration curve.

• High-pressure liquid chromatography. HPLC is useful for identifying liquids (volatile or nonvolatile) using a calibrated column. An HPLC chromatogram of anthracene obtained with the Perkin–Elmer Integral 4000 High Performance Liquid Chromatograph is shown in Fig. 3.10.

HPLC analysis is useful for analyzing nonvolatile liquids which are suitable for gas chromatograph analysis.

• Gas chromatography. GC is useful for identifying volatile materials such as solvents using a calibrated column. A Perkin–Elmer AutoSystem XL Gas Chromatograph produced the GC chromatogram of diesel oil shown in Fig. 3.11. GC is useful for analyzing materials that will volatilize (about 15% of all organic compounds) up to about 450°C. For materials that will not volatilize, HPLC is useful.

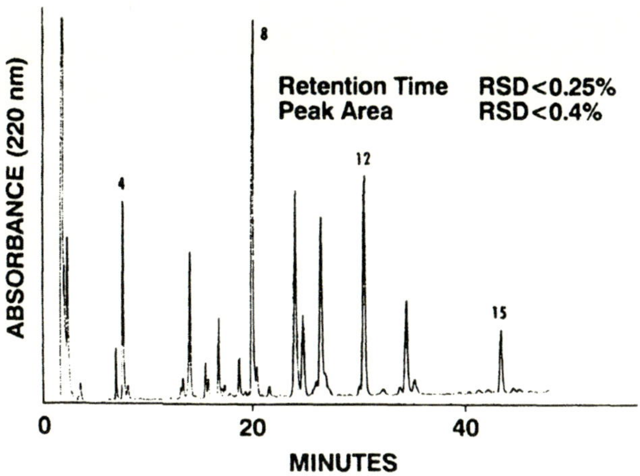

Figure 3.10. HPLC chromatogram of anthracene.

Gas–liquid chromatography accomplishes a separation by partitioning a sample between a mobile gas phase and a thin layer of nonvolatile liquid held on a solid support. Gas–solid chromatography employs a solid adsorbent as the stationary phase. The sequence of a GC separation is as follows: A sample containing the solutes is injected into a heating block where it is vaporized and swept as a plug of vapor by the carrier gas stream into the column inlet. The solutes are adsorbed at the head of the column by the stationary phase and then desorbed by fresh carrier

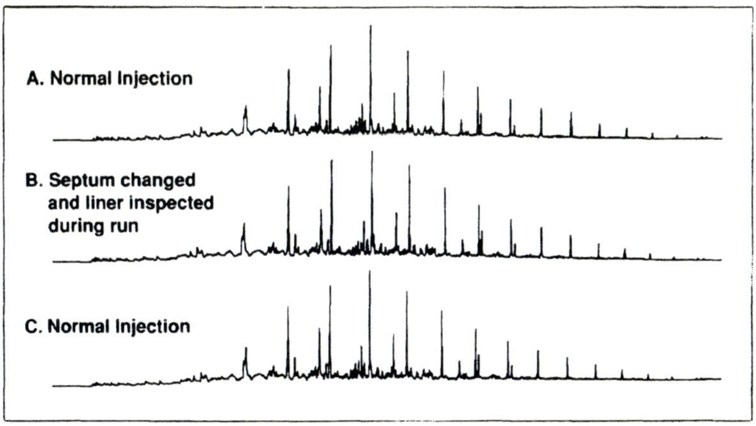

Figure 3.11. GC chromatogram of three separate injections of diesel oil.

gas. This partitioning process occurs repeatedly as the sample is moved toward the outlet by the carrier gas. Each solute will travel at its own rate through the column, and a band corresponding to each solute will form. The bands will separate to a degree that is determined by the partition ratios of the solutes and the extent of band spreading. The solutes are eluted, successively, in the increasing order of their partition ratios and enter a detector attached to the column exit. Signals are generated from an electronic detector, and the time of emergence of a peak identifies the component and the peak area reveals the concentration of the component mixture.

3.5. NUCLEAR MAGNETIC RESONANCE SPECTROSCOPY (NMR)

3.5.1. Fundamentals

The nuclei of certain atoms are considered to spin (Morrison and Boyd, 1973; Willard *et al.*, 1974). The spinning of these charged particles or circulation of charge, generates a magnetic moment along the axis of spin, so that these nuclei act like tiny magnets. The nucleus of hydrogen (^1H) is the one of greatest interest for what is referred to as ^1H-NMR, which is useful for the broad spectrum of organic molecules. However, another nucleus (^{13}C) will be discussed which forms the basis for ^{13}C-NMR, which is very useful for studying polymers and resins.

If a proton is placed in an external magnetic field, its magnetic moment, according to quantum mechanics, can be aligned in either of two ways: with or against the external field. Alignment with the field is more stable, and energy must be absorbed to "flip" the tiny proton magnetic moment over to the less stable alignment, against the field.

The amount of energy needed to flip the proton over depends on the strength of the external field: the stronger the field, the greater the tendency to remain lined up with it, and the higher the frequency ($\Delta E = h\nu$):

$$\mu = \gamma H_o / 2\pi$$

where ν is the frequency (Hz), H_o is the strength of the magnetic field (gauss), and γ is the nuclear constant, the gyromagnetic ratio, 26,750 for the proton.

In a field of 14,092 gauss, the energy required corresponds to electromagnetic radiation of frequency 60 MHz (60 megahertz or 60 million cycles per second): radiation in the radio frequency (RF) range, and much lower energy (lower frequency, longer wavelength) than even infrared light.

In principle, a substance could be placed in a magnetic field of constant strength, and then obtain a spectrum in the same way an infrared or ultraviolet spectrum is obtained: pass radiation of steadily changing frequency through the substance, and observe the frequency at which radiation is absorbed. In practice, it has been found more convenient to keep the radiation frequency constant, and to vary the magnetic field; at some value of the field strength the energy required to flip the proton matches the energy of the radiation, absorption occurs, and a signal is observed. Such a spectrum is called a *nuclear magnetic resonance spectrum*. Because the nucleus is a proton, the spectrum is sometimes called a *PMR* (proton magnetic resonance), to differentiate it from spectra involving such nuclei as ^{13}C or ^{19}F.

All of the protons in an organic molecule do not absorb at exactly the same field strength, and the spectrum would consist of a single signal that would give information about the structure of the molecule. The frequency at which a proton absorbs radiation depends on the magnetic field that that proton *feels* (i.e., has reaction to), and this effective field strength is not exactly the same as the applied field strength. The effective field strength at each proton depends on the environment of that proton including the electron density at the proton, and the presence of other nearby protons. Each proton, or each set of equivalent protons, will have a slightly different environment from every other set of protons and will require a slightly different applied field strength to produce the same effective field strength: the particular field strength at which absorption takes place.

At a given radio frequency, all protons absorb at the same effective field strength, but they absorb at different applied field strengths. It is this applied field strength that is measured, and against which the absorption is plotted.

The result is a spectrum showing many absorption peaks, whose relative positions can give an enormous amount of information about molecular structure. Aspects of the NMR spectrum are:

1. The number of signals indicate how many different kinds of protons there are in a molecule.
2. The positions of the signals indicate the electronic environment of each kind of proton.
3. The intensities of the signals indicate how many protons of each kind are present.
4. The splitting of a signal into several peaks indicates the environment of a proton with respect to other nearby protons.

• Number of NMR signals—equivalent and nonequivalent protons. In a given molecule, protons with the same environment absorb at the same (applied) field strength; protons with different environments absorb at different (applied) field strengths. A set of protons with the same environments are *equivalent*; the number of signals in the NMR spectrum indicate how many sets of equivalent protons (how many kinds of protons) a molecule contains.

Equivalent protons are chemically equivalent protons. To be chemically equivalent, protons must also be stereochemically equivalent. Observing structural formulas, ethyl chloride generates two NMR signals; isopropyl chloride, two NMR signals; and *n*-propyl chloride, three NMR signals. These conclusions are partially explained by the following terms describing different types of protons:

1. Enantiotopic protons: the environments of these two protons are mirror images of each other; in a chiral medium, these protons behave as if they were equivalent, and one NMR signal is generated for the pair.
2. Diastereotopic protons: the environments of these two protons are neither identical nor mirror images of each other; these protons are nonequivalent, and an NMR signal would be generated for each one.

• Chemical shift—position of signals. The number of signals in an NMR spectrum indicate how many kinds of protons a molecule contains, so the positions of the signals indicate what kinds of protons they are: aromatic, aliphatic, primary, secondary, tertiary, benzylic, vinylic, acetylic; adjacent to halogen to other atoms or groups.

When a molecule is placed in a magnetic field, its electrons are caused to circulate and, in circulating, they generate secondary magnetic fields, i.e., induced magnetic fields. Circulation of electrons about the proton itself generates a field aligned in such a way that, at the proton, it opposes the applied field. The field felt by the proton is thus diminished, and the proton is *shielded.* If the induced field reinforces the applied field, then the field felt by the proton is augmented, and the proton is *deshielded.*

Compared with a naked proton, a shielded proton requires a higher applied field strength, and a deshielded proton requires a lower applied field strength to absorb the particular effective field strength at which the absorption occurs. Shielding shifts the absorption upfield and deshielding shifts the absorption downfield. Shifts in the position of NMR absorptions, arising from shielding and deshielding by electrons, are called *chemical shifts.*

The unit in which a chemical shift is most conveniently expressed is *parts per million* (ppm) of the total applied magnetic field. Chemical shifts of compounds are listed in Table 3.1.

The *reference point* from which chemical shifts are measured is not the signal from a naked proton, but the signal from an actual compound, usually tetramethylsilane [$(CH_3)_4S$]. Because of the low electronegativity of silicon, the shielding of the protons in the silane is greater than in most other organic molecules; as a result, most NMR signals appear in the same direction from the tetramethylsilane signal, namely, downfield.

The most commonly used scale is the δ (delta) scale. The position of the tetramethylsilane signal is taken as 0.0 ppm. Most chemical shifts have δ values between 0 and 10 (minus 10, actually). A small δ value represents a small downfield shift and a large δ value represents a large downfield shift. An NMR signal from a particular proton appears at a different field strength than the signal from tetramethylsilane. This difference (the chemical shift) is measured not in gauss, but in the equivalent frequency units ($\nu = \gamma H_0/2\pi$), and it is divided by the frequency of the spectrometer used. For a spectrometer operating at 60 MHz (60×10^6 Hz):

$$\delta = \text{observed shift (Hz)} \times 10^6/60 \times 10^6 \text{ (Hz)}$$

The chemical shift is determined by the electronic environment of the proton. Protons with the same environments (equivalent protons) have the same chemical shift, and nonequivalent protons have different chemical shifts.

- Proton counting. The relative intensities of the peak heights are most important for counting protons. The area under an NMR signal is directly proportional to the number of protons generating the signal. This phenomenon is expected as the absorption of energy results from the flipping over of a proton in the same effective magnetic field; the more flippings, the more the energy absorbed, and the greater is the area under the absorption peak.

Areas under NMR peaks may be measured by electron integrators and are given on the spectrum chart in the form of a stepped curve; heights of steps are proportional to peak areas. NMR paper is crosshatched and step heights can be estimated by counting squares. From a calculation a set of numbers is arrived at that are in the same ratio as the numbers of different kinds of protons. This set of numbers is converted into a set of smallest whole numbers. The number of protons giving rise to each signal is equal to the whole number for that signal, or to some multiple of it.

Example. The NMR spectrum of *p-tert*-butyltoluene is shown in Fig. 3.12. The ratio of step heights $a{:}b{:}c$ is 8.8:2.9:3.8 = 3.0:1.0:1.3 = 9.0:3.0:3.9.

Alternately, as the molecular formula $C_{11}H_{16}$ is known,

$$16 \text{ H}/15.5 \text{ units} = 1.03 \text{ H per unit}$$

$$a = 1.03 \times 8.8 = 9.1$$

$$b = 1.03 \times 2.9 = 3.0$$

$$c = 1.03 \times 3.8 = 3.9$$

Either way, *a*, 9H; *b*, 3H; *c*, 4H.

The 4H of *c* (δ 7.1) are in the aromatic range, suggesting a disubstituted benzene $-C_6H_4-$. The 3H of *b* (δ 2.28) have a shift expected for benzylic protons, giving $CH_3-C_6H_4-$. There is left C_4H_9 which, in view of the 9H of *a* (δ 1.28), must

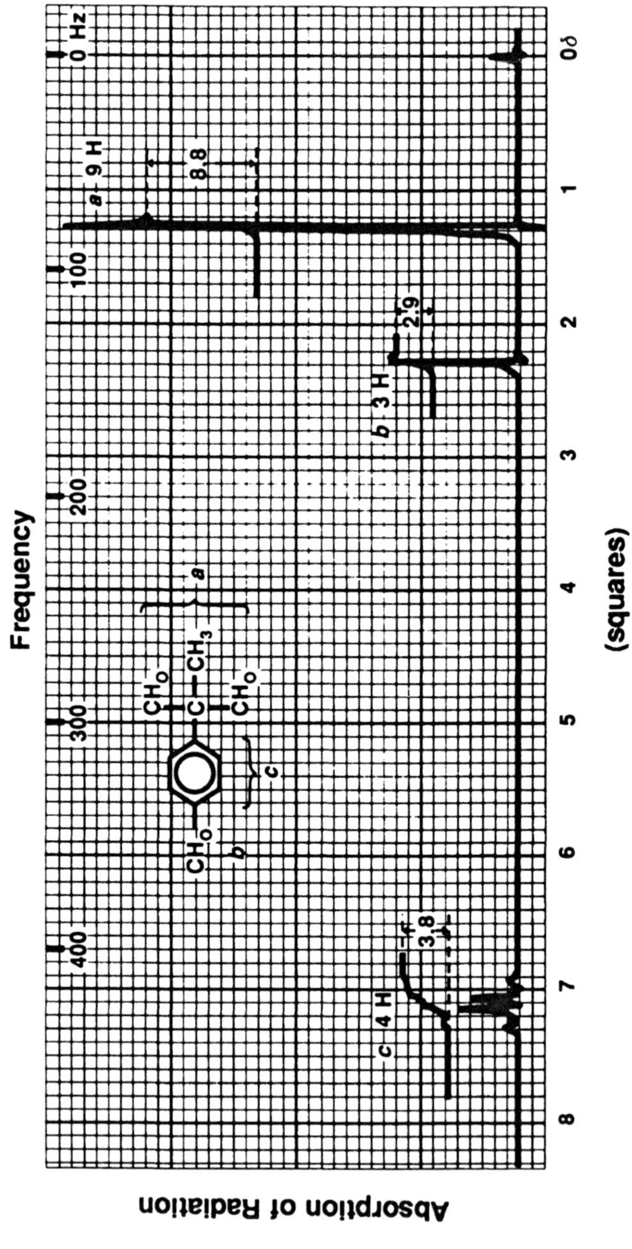

Figure 3.12. ^{1}H-NMR spectrum of *p-tert*-butyltoluene, proton counting. (Source: Morrison and Boyd, 1973.)

be $-C(CH_3)_3$; as these are once removed from the ring, their shift is nearly normal for an alkyl group. The compound is *tert*-butyltoluene (actually, as shown by the absorption pattern of the aromatic protons, the *para* isomer).

- Spin–spin coupling—splitting of signals. An NMR spectrum shows a signal for each kind of proton in a molecule. Actually, spectra are more complicated than this. Considering 1,1,2-tribromethane, 1,1-dibromethane, and ethyl bromide, each compound shows only two kinds of protons; yet, instead of two peaks, the NMR spectra show five, six, and seven peaks, respectively.

The reason for the apparent inconsistency is that *splitting* of NMR signals caused by spin–spin coupling is occurring. The signal expected from each set of equivalent protons appears not as a single peak but as a group of peaks. Splitting reflects the environment of the absorbing protons: not with respect to electrons, but with respect to other nearby protons.

- Coupling constants. The distance between peaks in a multiplet is a measure of the effectiveness of spin–spin coupling, and is called the *coupling constant, J*. Coupling, unlike chemical shift, is not a matter of induced magnetic fields. The value of the coupling constant (measured in Hz) remains the same regardless of the applied magnetic field (RF). Spin–spin coupling differs from chemical shift, and, when necessary, the two can be distinguished on this basis: the spectrum is run at a second, different RF; when measured in hertz, peak separations resulting from splitting remain constant, whereas peak separations resulting from chemical shifts change. When divided by the RF and thus converted into parts per million, the numerical value of the chemical shift would, of course, remain constant.

- Deuterium labeling and complicated spectra. Most NMR spectra that the organic chemist is likely to encounter are considerably more complicated than ones discussed above. Instrumental techniques are available to help in the analysis of complicated spectra, and to simplify the spectra actually measured. By the method of *double resonance* (or *double irradiation*), for example, the spins of two sets of protons can be *decoupled*, and a simper spectrum obtained.

The molecule is irradiated with two RF beams: the usual one, whose absorption is being measured; and a second, much stronger beam, whose frequency differs from that of the first in such a way that the following happens. When the field strength is reached at which the proton of interest absorbs and generates a signal, the splitting protons are absorbing the other, very strong radiation. These splitting protons are "stirred up" and flip over so very rapidly that the signaling proton sees them not in the various combinations of spin alignments but in a single average alignment. The spins are decoupled, and the signal appears as a single, unsplit peak. A way to simplify an NMR spectrum is by using *deuterium labeling*.

Figure 3.13. Photograph of Bruker MSL $^{1}H/^{13}C$-NMR spectrometers, tabletop configuration. Reprinted with permission of Bruker Analytical Systems.

Because a deuteron has a much smaller magnetic moment than a proton, it absorbs at a much higher field and so gives no signal in the proton NMR spectrum. As a result, the replacement of a proton by a deuteron removes from an NMR spectrum both the signal from that proton and the splitting by it of signals of other protons.

An important use of deuterium labeling is to discover which signal is produced by which proton or protons: the disappearance of a particular signal when a proton in a known location is replaced by deuterium. Another use of deuterium labeling is to simplify a complicated spectrum so that a certain set of signals can be seen more clearly.

• ^{13}C-NMR spectroscopy. This type of NMR spectroscopy utilizes the ^{13}C isotope of carbon to generate chemical shifts. The method is particularly useful for polymers and resins as the copolymers can be accurately determined with regard to carbon atoms instead of hydrogen atoms.

3.5.2. Equipment

The Bruker ^{1}H/^{13}C-NMR spectrophotometers are shown in Fig. 3.13.

3.5.3. Applications

NMR spectra complement IR spectra and the combination of NMR and IR provide a more positive identification of an organic compound. However, NMR spectra are usually generated from solutions of organic compounds, and few solid samples are used.

Where IR spectra are useful for identifying materials, NMR spectra are desired for reinforcing the qualitative analysis.

3.6. THERMAL ANALYSIS

3.6.1. Fundamentals

Thermal analysis includes the measurements of:

1. Glass transition temperature [differential scanning calorimetry (DSC)]
2. Melting temperature (DSC)
3. Heat of melting (DSC)
4. Decomposition temperature [thermogravimetric analysis (TGA)]
5. Softening temperature [thermomechanical analysis (TMA)]
6. Dynamic mechanical modulus [dynamic mechanical analysis (DMA)]

There are different and sometimes combined instruments to measure these properties (Slade *et al.*, 1970).

3.6.2. Equipment

Instruments used in thermal analysis are pictured in the following figures:

- Figure 3.14—Perkin–Elmer DSC 7 Differential Scanning Calorimeter
- Figure 3.15—Perkin–Elmer TGA 7 Thermogravimetric Analyzer

Figure 3.14. Photograph of Perkin–Elmer DSC 7 Differential Scanning Calorimeter. Reprinted with permission of Perkin–Elmer Corp.

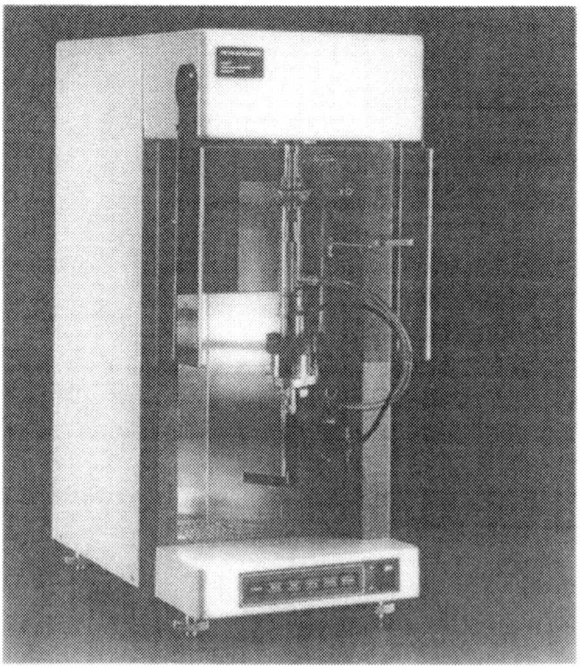

Figure 3.15. Photograph of Perkin–Elmer TGA 7 Thermogravimetric Analyzer. Reprinted with permission of Perkin–Elmer Corp.

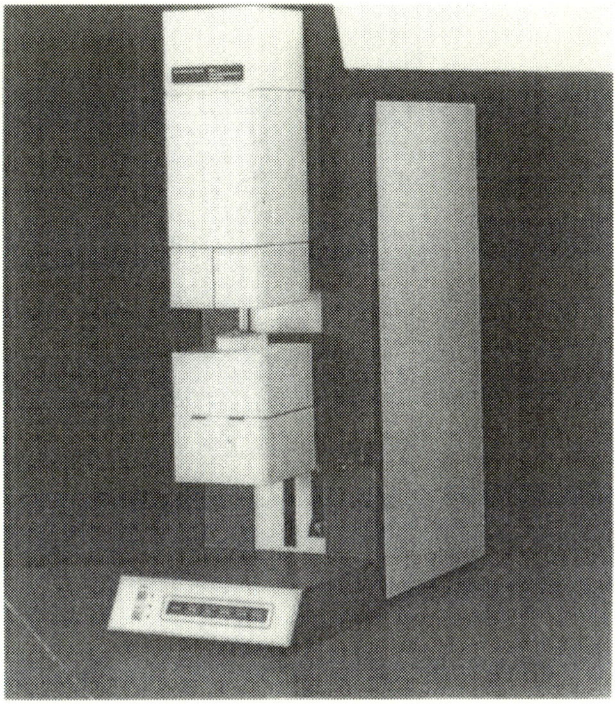

Figure 3.16. Photograph of Perkin–Elmer DMA 7 Dynamic Mechanical Analyzer. Reprinted with permission of Perkin–Elmer Corp.

- Figure 3.16—Perkin–Elmer DMA 7 Dynamic Mechanical Analyzer
- Figure 3.17—Perkin–Elmer TMA 7 Thermomechanical Analyzer
- Figure 3.18—Perkin–Elmer DTA 7 Differential Thermal Analyzer
- Figure 3.19. Perkin–Elmer computer and thermal analysis software program

3.6.3. Applications

The application of thermal analysis to paint, plastics, adhesives, and inks is for the measurement of any thermal transitions of which the important ones are discussed below.

- Glass transition temperature (T_g and T_m). This is the temperature at which an amorphous material such as polystyrene (T_g = 100°C) becomes rigid and after which, softens. Segmental motion of polymer chains is at a minimum. The instrument measures *heat versus temperature*. Epoxy paints or coatings possess a glass transition temperature which indicates the degree of curing. Amorphous

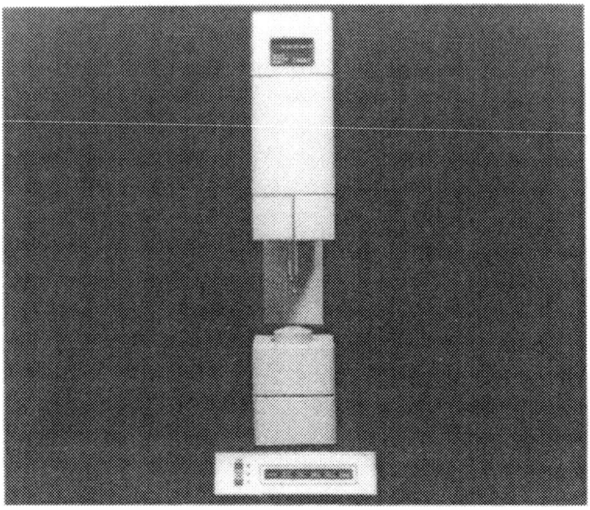

Figure 3.17. Photograph of Perkin–Elmer TMA 7 Thermomechanical Analyzer. Reprinted with permission of Perkin–Elmer Corp.

polymers have only a glass transition temperature, semicrystalline polymers have a glass transition and melting temperature, and totally crystalline materials have only a melting temperature.

- Melting temperature (T_m). Melting is the temperature (Collins *et al.*, 1973) at which crystals in a material disintegrate and liquefy, e.g., low-density polyethylene (T_m = 127°C). The instrument measures *heat versus temperature*

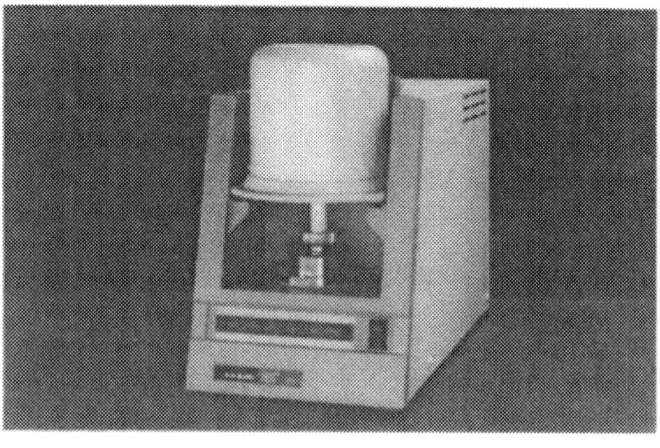

Figure 3.18. Photograph of Perkin–Elmer DTA 7 Differential Thermal Analyzer. Reprinted with permission

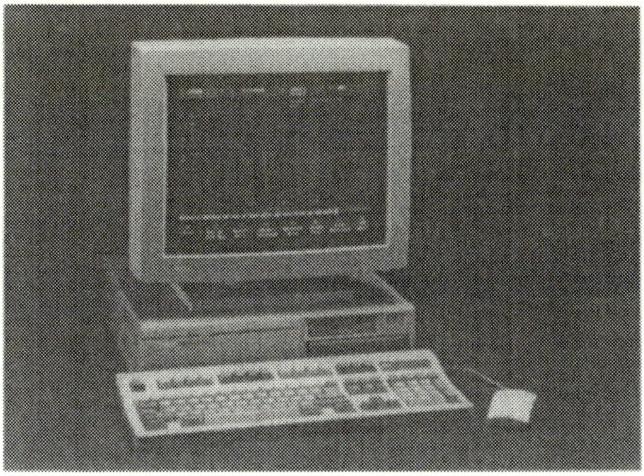

Figure 3.19. Photograph of Perkin–Elmer computer and thermal analysis software program. Reprinted with permission of Perkin–Elmer Corp. of Perkin–Elmer Corp.

(dH/dt versus ΔT) and total heat H absorbed by a sample is $c_p\Delta T$. The basic equation for DSC is

$$\Delta T = qC_p/K$$

where ΔT is the difference between sample temperature and programmed temperature, q is the heating rate, C_p is the heat capacity, and K is the thermal conductivity. Also, heat capacity (C_p) is equal to mc_p, where m is mass and c_p is specific heat.

Melting is associated with softening or melting of a resin or polymer which corresponds to a change in heat capacitance. Only a crystalline material has a true melting temperature or peak on a thermogram. This is because energy is required to disintegrate crystallites and associated structures such as in polyethylene. An amorphous material, such as polystyrene, does not exhibit a true melting temperature, but rather a glass transition temperature. The T_g is associated with a change in heat capacity when the polymer begins to flow. The heating rate is important for developing an accurate thermogram, and a rate that corresponds to 10°C/min is acceptable for most polymeric materials.

Low-density polyethylene contains about 20% amorphous and 80% crystalline regions, and a DSC thermogram will indicate both events.

A DSC thermogram of polypropylene is shown in Fig. 3.20.

- Decomposition temperature (T_d). This is the temperature at which a polymer or resin chemically decomposes into fragments and gases (i.e., smoke). The instrument measures *weight versus temperature* (dW/dt versus ΔT). The tempera-

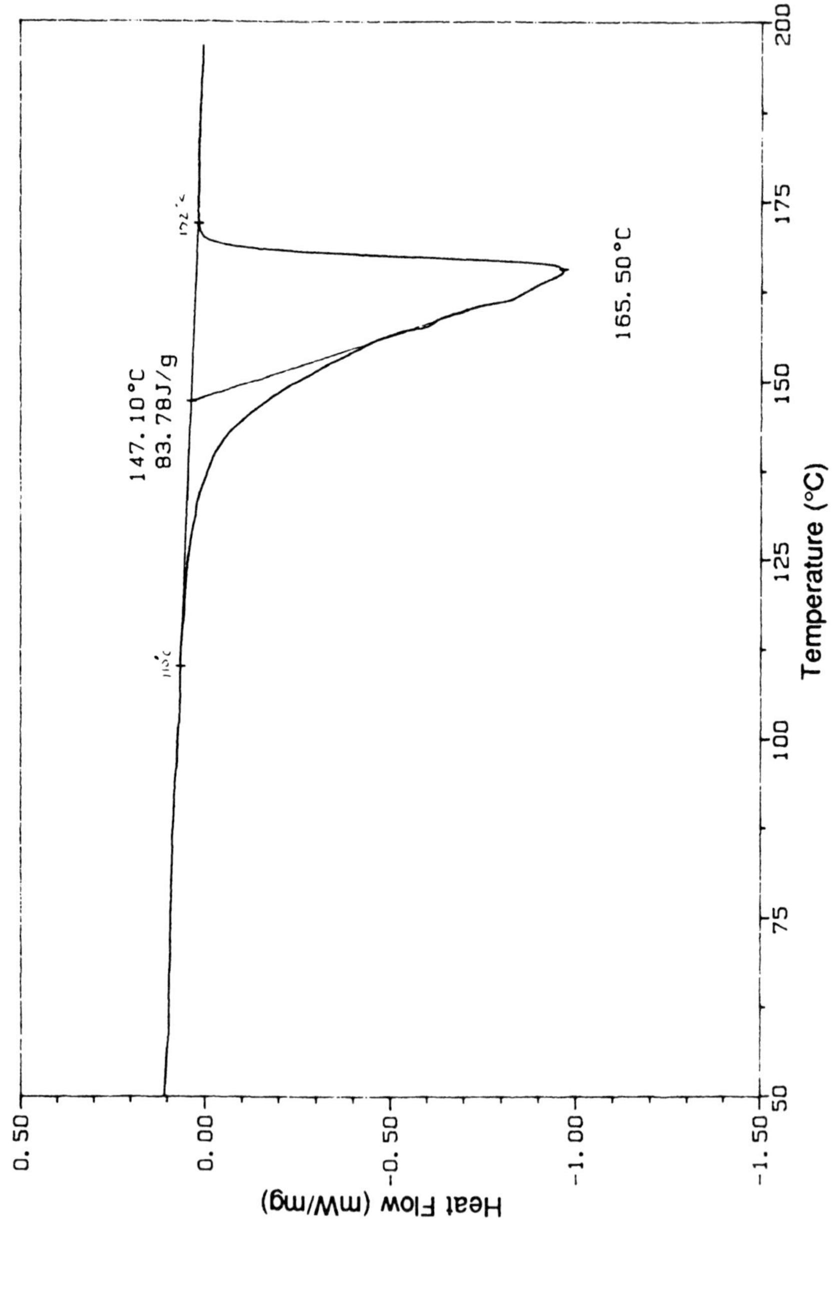

Figure 3.20. DSC thermogram of polypropylene.

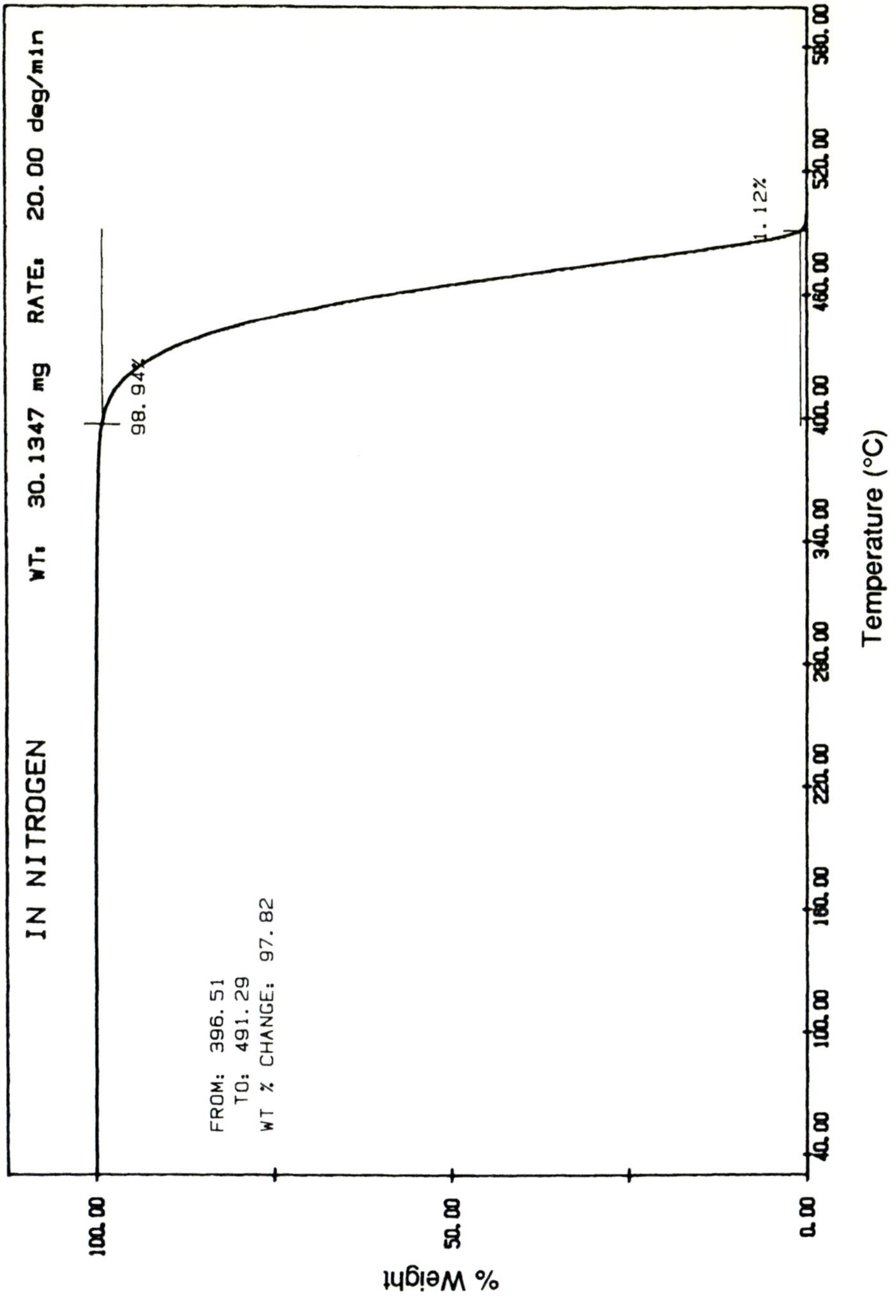

Figure 3.21. TGA thermogram of polystyrene.

ture is indicative of chemical structure as different bonds require different energies to break. Also, a mixture of materials can be detected and measured if they are chemically different. Another feature is the measurement of percent pigment or nondecomposed material. This is an effective technique for measuring percent pigment or filler. A combination of DSC and TGA data will show that a polymer will decompose after melting.

A polymer, resin, or rubber exhibits a curve that is representative of the corresponding chemical structure that is useful for identifying the unknown specimen. In the case of partially burned specimens, the "hottest" temperature that the specimen experienced can be estimated by observing the decomposition curve.

A TGA thermogram of polystyrene is shown in Fig. 3.21.

- Softening temperature (T_m). This is the glass transition and/or melting temperature of a polymer or resin. The instrument measures softening mechanically as *thickness change (cm/cm) versus temperature* which also measures the coefficient of thermal expansion.

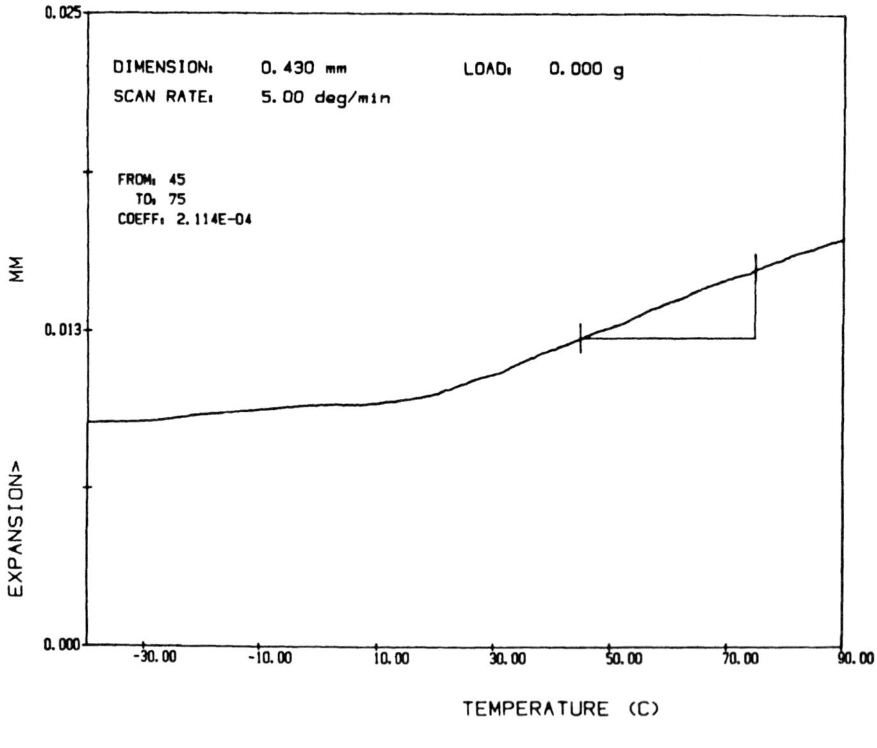

Figure 3.22. TMA thermogram of poly (styrene-co-butadiene) copolymer film (Source: Colo, 1986).

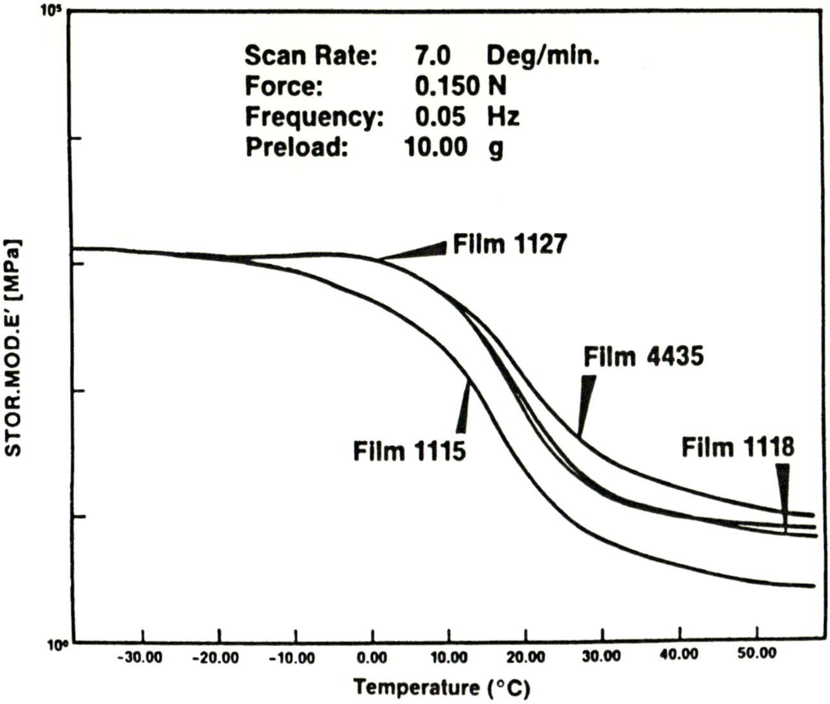

Figure 3.23. DMA thermograms of poly (styrene-co-butadiene) copolymer films of different compositions. (Reprinted with permission of Perkin–Elmer Corp.)

A TMA thermogram of polyethylene is shown in Fig. 3.22.

• Modulus (E). This is a measure of mechanical modulus (stress/strain) at a given temperature (Colo, 1986). A probe vibrates at a frequency on a specimen and measures elasticity and stored modulus with temperature. This instrument is useful for determining strength (modulus), elasticity, and an indication of hardness, nondestructively, and on a small specimen. DMA thermograms are shown in Figs. 3.23 and 3.24.

3.7. VISCOMETRIC ANALYSIS

3.7.1. Fundamentals

Viscosity refers to how thick a liquid is or how easily it flows. A viscometer measures resistance to flow of a rotating probe in a liquid. Measurement of viscosity (dyn·cm/sec^2) reveals the presence of a polymer or resin in a solvent and the concentration of which corresponds to the viscosity.

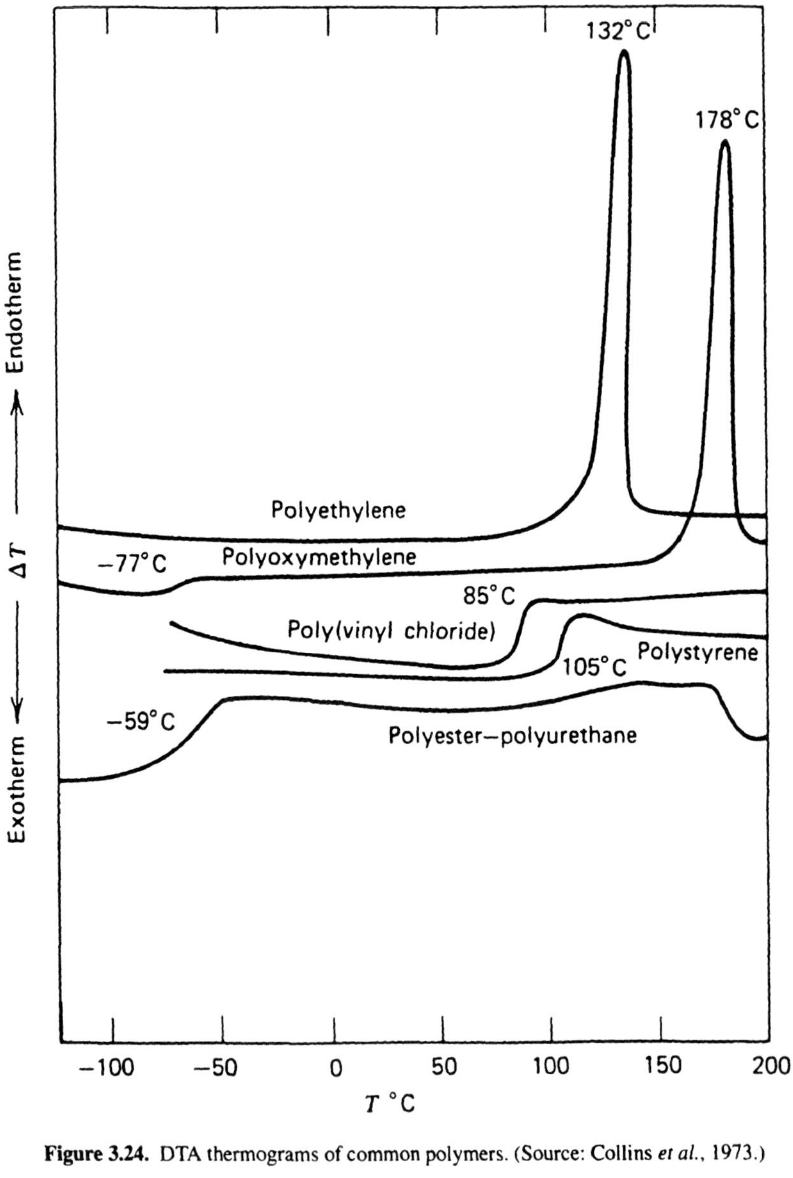

Figure 3.24. DTA thermograms of common polymers. (Source: Collins *et al.*, 1973.)

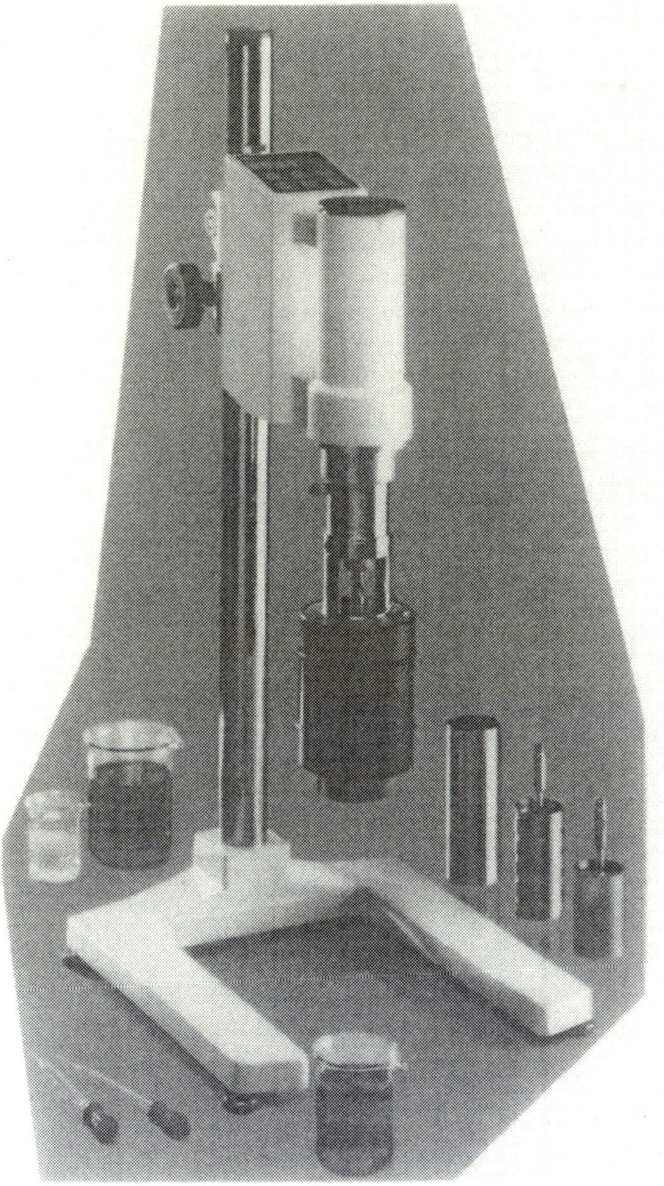

Figure 3.25. Photograph of Haake VT550 Viscometer. Reprinted with permission of Haake Corp.

3.7.2. Equipment

The Haake viscometer is shown in Fig. 3.25.

3.7.3. Applications

The concentration of a resin or polymer can be measured using viscometry. Increased concentration corresponds to increased viscosity. It is a good method for determining the difference between a solvent (low viscosity) and resin solution (high viscosity) or a mixture. Viscometry is useful for characterizing paint, adhesives, and inks as these materials are diluted with solvent or water. Viscosity of melted polymers is best measured with a melt flow index method.

Rheology curves of classic liquids and dispersions are shown in Fig. 3.26. When a liquid dispersion of paint or other is stirred, the shear rate increases with shear forces, and this is characteristic of a pseudoplastic liquid dispersion. The opposite effect is called shear-thickening or a dilatant liquid dispersion. A liquid that does exhibit a linear relationship between shear and shear rate is a Newtonian liquid such as water, silicone oil, or solvent. When a shear-thinning dispersion is sheared at a constant rate, the viscosity decreases with time, and this is a thixotropic dispersion (viscosity decreasing with shear). The opposite of a thixotropic dispersion liquid is a rheopectic dispersion, rarely encountered. These rheological effects are of great importance when formulating dispersions. For example, when a paint is sprayed or brushed, the shear-thinning and corresponding viscosity values must be suitable for the paint to flow onto a surface and provide a uniform film.

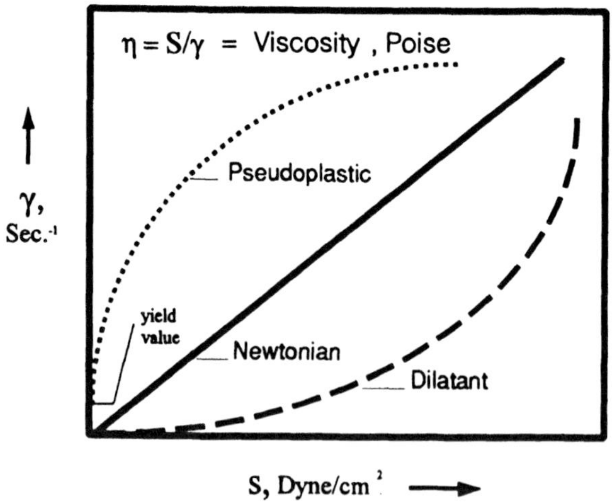

Figure 3.26. Rheology curves of liquids and dispersions.

3.8. X-RAY MICROSCOPY

3.8.1. Fundamentals

The X-ray microscope is useful for investigating a material's interior structure that is hidden from "sight." Three-dimensional images of polymeric materials can be observed for fractures, inclusions, and welds. Pigment size particles can be observed in paint, adhesives and inks. Hairline size fractures beneath the surface of a material, not visible by optical or electron microscopy, can be observed using this method. Relative to topological methods, X-ray microscopy offers analysis "beneath the surface" of a material. Generally, X-ray microscopic analysis shows differences in densities between materials (at least a difference of 5%) and the contrast between them provides an image.

According to Cunningham *et al.* (1986), X-ray microscopy denotes a form of projection radiography that employs low-energy X-ray photons emitted from a point source to generate high-resolution images. The energy of the electron beam that is focused onto the target material to generate the X-ray source is typically <10 keV and is generally lower than is conventional either in industrial contact radiography or in microfocal radiography. The reason is twofold (Cunningham *et al.*, 1986):

> First, the low-energy X-rays thus produced generate more specimen contrast, especially between light elements ($Z < 8$ (oxygen)), than do the high-energy X-rays used. Secondly, the relatively low power dissipation in the target allow a much smaller X-ray source diameter, of about 1 micron, to be achieved in comparison even with that, about 15 microns, for microfocal radiography. The small source-size allows primary image magnifications of typically ×100 without loss of image sharpness by recording the projected image at some distance from the specimen.

The sample does not have to be specially prepared by methods such as polishing and microtome, but a smooth and unobstructed surface is preferred. To observe the interior of a plastic part requires passing X rays through the bulk of the part, and removal of obstructions will produce a better image. Figure 3.27 is an X-ray micrograph of a plastic lighter showing internal parts.

Images generated by an X-ray microscope are typically recorded on VHS videotape and thermal printed paper.

In summary, X-ray microscopy offers much to the materials sciences including investigation of the microstructure of plastics, paints, adhesives, and inks. Also, fibers, pigments, ceramics, and other solid materials can be studied within composite structures. The major impact of X-ray microscopy in the application to materials is the ease and rapidity with which it conveys the underlying three-dimensional structure of the specimen examined to the eye and brain.

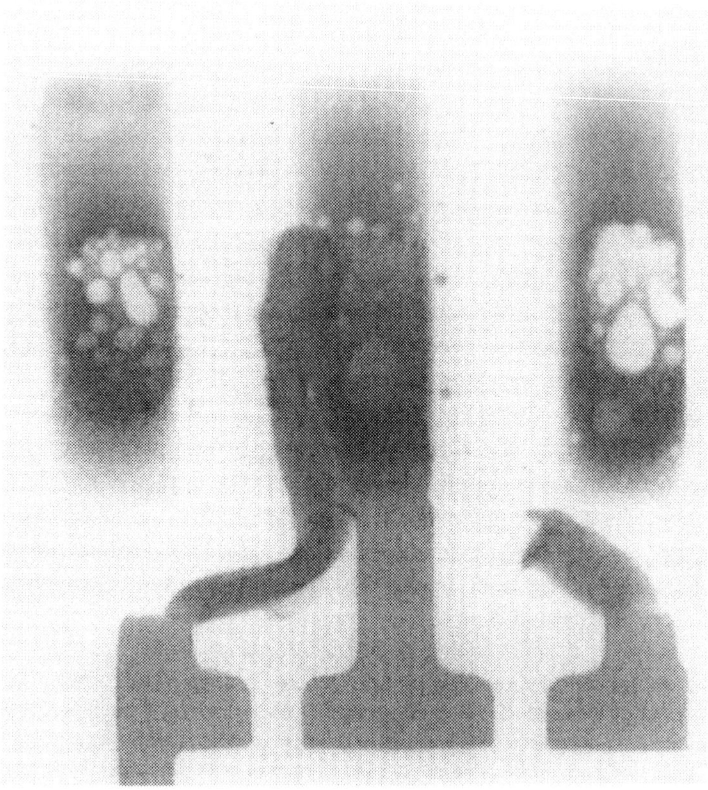

Figure 3.27. X-ray micrograph of solder joint with internal defects, voids (light areas), and broken leads. Topological view shows no voids or fractures.

3.8.2. Equipment

The author has obtained excellent results with the Series FXS-100 or -160 Microfocus X-Ray Inspection and Testing System shown in Fig. 3.28, manufactured by:

FEIN FOCUS USA Inc.
5142 N. Clareton Drive, Suite 160
Agoura Hills, CA 91301
(818) 889-1440
Fax: (818) 889-3737

Some of the operating parameters of the Model 160 X-Ray Microscope are:

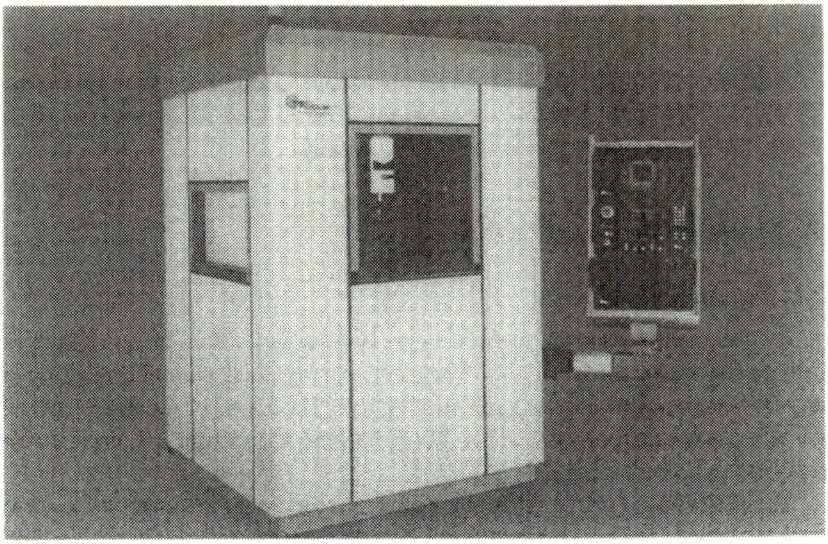

Figure 3.28. Photograph of FEIN FOCUS Microfocus FXS-160.30 X-Ray Inspection and Testing System. Reprinted with permission of FEIN FOCUS Corp.

High voltage range	10–160 kV
Tube current range	0.025–0.2 mA
Target material	Tungsten
Focus dimensions	Manual 3–200 μm
	Autofocus < 10 μm
Beam angle	100° conical
Depth of field	Extends throughout sample chamber
Minimum focus distance	1.5 mm
Geometric direct magnification	3.4–290×
Total magnification	Maximum 1000×

3.8.3. Applications

The surfaces and internal structures of plastic parts, paints, adhesives, and inks can be investigated *nondestructively* using X-ray micrography. Examples of applications are:

1. Observing fractures within plastic parts
2. Observing inclusions in paint and ink coatings and surfaces of painted substrates
3. Measuring thicknesses of coatings on surfaces

4. Estimating densities of materials and inclusions

An X-ray micrograph of a solder joint with strong external and internal defects is shown in Fig. 3.27. The internal parts are seen as dark areas because of their greater density relative to the lighter plastic images.

3.9. MASS SPECTROSCOPY

3.9.1. Fundamentals

In a mass spectrometer, molecules are bombarded with a beam of energetic electrons (Silverstein *et al.*, 1974). The molecules are ionized and broken up into many fragments, some of which are positive ions. Each kind of ion has a particular ratio of mass to charge or *m/e* value. For most ions, the charge is 1, so that *m/e* is simply the mass of the ion.

The set of ions generated from a chemical compound are analyzed in such a way that a signal is obtained for each value of *m/e* represented; the intensity of each signal reflects the relative abundance of the ion producing the signal. The largest peak is called the *base peak*; its intensity is taken as 100, and the intensities of the other peaks are expressed relative to it. A plot or list showing the relative intensities of signals at the various *m/e* values is called a *mass spectrum*, and is highly characteristic of a particular compound. The mass spectrum of toluene is shown in Fig. 3.29.

3.9.2. Equipment

A Bruker TOF-Mass Spectrometer is shown in Fig. 3.30.

3.9.3. Applications

Mass spectroscopy is useful for identifying gases, liquids, and solids (that will volatilize) of unknown composition. A mixture of materials can be individually identified. Mass spectroscopy is qualitative rather than quantitative. It is usually more expensive than chromatography or infrared spectroscopy.

3.10. ULTRAVIOLET SPECTROSCOPY

3.10.1. Fundamentals

In contrast to the infrared spectrum, the ultraviolet spectrum is not used primarily to show the presence of individual functional groups, but rather to show relationships between functional groups, chiefly conjugation: conjugation between two or more carbon–carbon double (triple) bonds; between carbon–carbon and carbon–oxygen double bonds; between double bonds and an aromatic ring; and

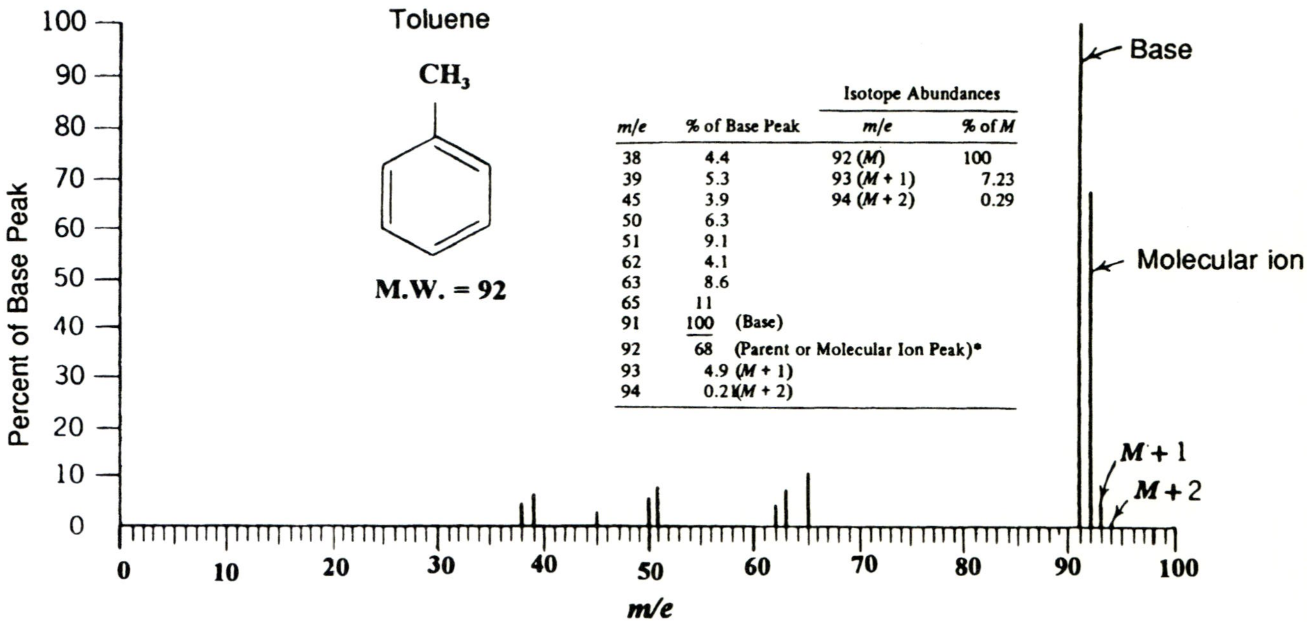

Figure 3.29. Mass spectrometer spectrum of toluene. Reprinted with permission of John Wiley & Sons.

Figure 3.30. Photograph of Bruker REFLEX MALD TOF-Mass Spectrometer. Reprinted with permission of Bruker Instruments, Inc.

even the presence of an aromatic ring itself. It can reveal the number and location of substituents attached to the carbons of the conjugated system.

Light of wavelength between about 400 and 750 nm is visible. Below the violet end (<400 nm) of the visible spectrum lies the ultraviolet region. The ultraviolet/visible (UV/VIS) spectrometers usually operate in the range of 200–750 nm. The UV spectrum is a few broad humps on a chart, and the spectrum the top of the peak or hump (λ_{max}) and the intensity of that absorption (ε_{max}, the extinction coefficient).

When a molecule is being raised to a higher electronic level, it means that an electron has been changed from one orbital to another orbital of higher energy. This electron can be of many kinds, for example, a σ electron is held tightly, and a good deal of energy is required to excite it: energy corresponding to ultraviolet light of short wavelength in a region or "far" ultraviolet outside the range of the usual spectrometer. It is chiefly excitations of the comparatively loosely held n and π electrons that appear in the near-ultraviolet spectrum, and, of these, only jumps to the lower and more stable excited states.

The transitions of most concern are:

1. $n \rightarrow \pi^*$ in which the electron of an unshared pair goes to an unstable (antibonding) π orbital
2. $\pi \rightarrow \pi^*$ in which an electron goes from a stable (bonding) π orbital to an unstable π orbital.

Figure 3.31. Photograph of Cary 1E UV-Vis-NIR Spectrophotometer. Reprinted with permission of Cary Corp.

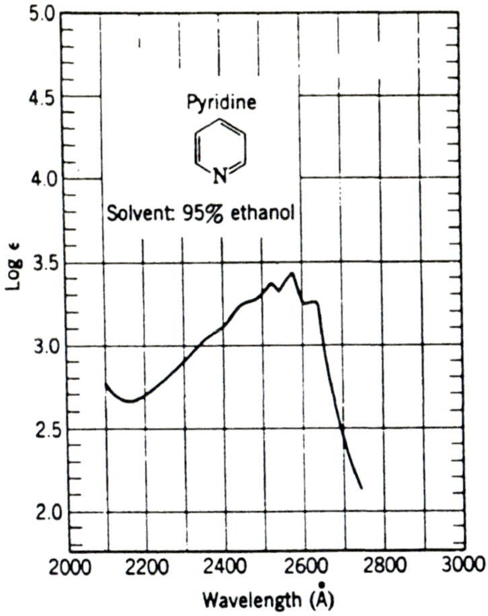

Figure 3.32. UV spectrum of pyridine. (Source: Silverstein, 1974. Reprinted with permission of John Wiley & Sons.)

3.10.2. Equipment

A Cary 1E UV-Vis-NIR Spectrophotometer is pictured in Fig. 3.31.

3.10.3. Applications

The UV spectrum of pyridine is provided in Fig. 3.32. UV spectroscopy is useful for identifying many chemical species if they are UV-absorbing, and the method is simple and inexpensive. Not all chemical species are UV-absorbing. Visible spectroscopy is useful for measuring turbidity in solutions and suspensions, as well as other uses.

Most of the methods of analysis discussed in this chapter are described in the American Standards Testing Methods publications, 1916 Race Street, Philadelphia, PA 19103-1187, telephone (215) 299-5400 and fax (215) 977-9679. Standard methods describe the procedures in greater detail than space allows here.

Not all methods of bulk analysis are represented in this chapter because economy and simplicity are stressed here. These are the tools most useful for deformulation of paint, plastics, adhesives, and inks. They will be applied to actual examples of deformulation in the following chapters.

4

Paint Formulations

4.1. GENERAL

It is necessary to be familiar with the fundamentals (Weismantel, 1981; Martens, 1974) of paint to understand and intelligently discuss paint or coatings. Like all technologies, paint technology has its own jargon. The terms *paint* and *coatings* are sometimes used interchangeably; *paint* is the older term used before the 1940s (e.g., for painting houses) after which new sophisticated synthesized materials were developed for automobiles and aircraft and called *coatings* to distinguish them from the vegetable oil-based materials.

A paint is a decorative, protective, or otherwise functional coating applied to a substrate. This substrate may be another coat of paint. Some terms (Gooch, 1993) associated with paint follow:

- Dopant (D. *doop*, adj.). Any thick liquid or pasty preparation used in preparing a surface. Any varnishlike material for water-proofing surfaces.
- Paint (M.E., *peint*, n.). A substance composed of a solid coloring matter suspended in a liquid medium.
- Coating (M.E., *cote*, n.). A layer of any substance spread over a surface; modern synthesized materials, such as polyurethane resins, that replace older paint materials.

The professional and trade organization for the paint industry is:

Federation of Societies for Coatings Technologies
Blue Bell, PA 19422
(610) 940-0777
Fax: (610) 940-0292

4.1.1. The Paint Formula

The formula lists the ingredients of the paint (Weismantel, 1981): vehicle, solvents, pigmentation, and additives. The basic paint formulation and ingredients are listed in Table 4.1. Amounts are normally stated in units of weight for accuracy.

Accurate metering equipment permits measuring the liquids in units of volume. The significant relationships among the ingredients of the dried paint film are volume relationships, not weight relationships.

The film former may be present as drying oil, as varnish, as resin solution, as dry resin, as plasticizer, or as some combination of these. Solvent may be present as free solvent or as a component of varnishes or resin solutions. The pigments and the additives are usually listed separately.

Differences between the ratios of the principal ingredients is the most important factor in the differences between types of paints. The most important of these ratios is the volume of the pigmentation in the dried film compared with the total volume of the dried film. The common types of paints, in terms of the differences in the ratios of the ingredients they contain, are: clear finishes, stains, gloss enamels, semigloss (satin) enamels, flat paints, sealers and primers, house paints (for wood siding), stucco paints, and filling and caulking compounds.

Examples of widely used paint formulations are provided in Tables 4.1–4.43.

4.1.2. Functions of Paint and Coatings

Paint is a mechanical mixture or dispersion of pigments or powders, at least some of which are normally opaque, with a liquid or medium known as the vehicle. It must be able to be applied properly, and it must adhere to the surface on which it will be applied and form the type of film desired. Paint must also perform the function for which it is being used (Weismantel, 1981): protection, decoration, or some other function.

4.1.3. Classification

Paints can be classified by many methods, and the method chosen is a function of what is to be accomplished.

The first purpose of classification is to group those paints that have the property being discussed and have it to the degree considered necessary for inclusion. In this way, they are set apart from paints not having this property or not having it to the required degree.

The second purpose of classification is to group those paints that are used in the same way or for the same purpose or for the same type of application. They are thus set apart from other paints not used in the same way or for the same type of application.

As examples of paint classification, gloss paints have a reflectance (shine) like a mirror, whereas flat paints lack this high degree of reflectance. Industrial finishes are applied to manufactured objects (e.g., automobiles, appliances, and furniture) before they are sold to the user. Trade sales paints (e.g., house paints, wall paints, and kitchen enamels) are applied to completed articles by the owner, the owner's employees, or a painter hired by the owner.

The vehicle portion of the paint normally consists of a nonvolatile portion which will remain as part of the paint film and a volatile portion which will evaporate, thus leaving the film. The dried paint film will therefore consist of pigment and nonvolatile vehicle. The volatile portion of the vehicle is normally used for proper application properties.

- Gloss. The proportion of pigment (and particle size) to nonvolatile vehicle normally determines the type of gloss that the dried film will have. If this proportion is small (e.g., less than 25% of the total nonvolatile volume), the result probably would be a glossy film, as there would be more than enough nonvolatile vehicle to cover the pigment completely. But, the pigment size must be small as well, as smaller particle size corresponds to higher gloss. Usually, as the percentage of pigment volume goes up, the gloss goes down. At a 45% pigment-volume concentration (PVC) the paint would probably be a semigloss, and at a 70% PVC the sheen is likely to be dull or flat.

- Solvent- and water-based. The general public is aware of two types of coatings: those that are solvent-based, i.e., that are reducible (soluble) by an organic solvent; and those that are water-based, i.e., that may be thinned or reduced by water. The specific properties of a coating will depend almost wholly on the specific properties of the pigments and vehicles used and on the proportions of one to the other.

There are, of course, many coatings that contain little or no pigmentation. These are the clear coatings, including clear lacquers and varnishes. They are usually used over wood when the beauty of the substrate is not to be hidden or obliterated. Also, clear acrylic coatings provide the glossy and protective covers used, for example, for attractive printed fashion magazines. Clear coatings normally dry to a high gloss, but pigmented clear coatings dry to a dull finish. Special flatting types of pigments that give no color and have no obliterating properties are normally used in these dull-finish clear coatings.

- Type of film former. Another classification of paints and coatings is by type of film former.

a. Solid Thermoplastic Film Formers. Hot-mop coatings are an old example of these vehicles. The tar is melted and resolidified on cooling. A new application of this type is the flame sprayed thermoplastic powder coating which consists of a powdered resin sprayed with a propane torch. The resin melts in the flame, adheres to the substrate, and forms a film. Another new application of this type of drying mechanism is the powder coating which can be a fluidized bed or electrostatically sprayed and baked type.

b. Lacquer-Type Film Former. In describing the curing of a lacquer, the solvent evaporates and the film dries. The most familiar type of lacquer is based on nitrocellulose. In addition to nitrocellulose, which provides fast drying and hardness, softener resins are included to provide adhesion. There are also one or more plasticizers to provide flexibility. A solvent blend is used to give a controlled evaporation rate and to ensure that all of the components stay in solution until solvent evaporation is complete.

c. Oxidizing Film Formers. These film formers are based on drying oils, which react with oxygen in air to "autoxidize" or cross-link the oil molecules and form a network polymer or gel. More specifically, the double bonds in the oil chains are attacked by diatomic oxygen via catalysis to form free radical reactions (Gooch, 1980). The oils include linseed, soybean, safflower, tung (china-wood), fish, tall (from pine tree as a by-product of kraft-paper manufacturing), and others.

d. Varnishes. These vehicles are made by heating drying oils with hard resins. The properties of the varnish are representative of the drying oil, the resin, the ratios of these to each other, and processing conditions. Among familiar resins used in varnishes are phenolic, ester-gum, maleic, and epoxy resins. Urethane varnishes are sometimes called urethane oils because of their low viscosity and great flexibility. Short-oil varnishes contain more resin and less oil, which makes them harder, more brittle, and faster drying. Medium-oil varnishes are intermediate in composition and properties. Long-oil varnishes contain more oil and less resin, which makes them softer, more flexible, and slower drying.

e. Alkyds. These vehicles consist of drying oils reacted with synthetic materials such as maleic anhydride and multifunctional alcohols to form a resin. In a varnish the resin is dispersed in oil gel. The alkyd can be used alone as a vehicle. Air-drying alkyds dry at room temperature with catalysts such as cobalt naphthalate. The amount of oil in the alkyd composition determines the drying rate and properties. Alkyds are classified as short-, medium-, and long-oil to describe the differences in drying-oil content and properties. Alkyds prepared from nondrying oils, such as coconut oil, are used in heat-cured film formers and as plasticizers.

- Room-temperature catalyzed film formers. These film formers possess chemical groups that react when catalyzed. Unlike drying oils, they do not depend on autoxidation processes. Chemical bonds are formed between reacting groups. The reaction and formation of a film is often referred to as the "curing" process. These materials could be in two parts, the curing beginning only after the two parts are mixed. After mixing, there is usually a limited amount of time for applying the material because of the onset of curing. Solvents are usually utilized to adjust the viscosity (thinning) of the parts and the mixture for ease of application. The resulting properties of these film formers are superior to drying-oil-based vehicles. Examples of room-temperature catalyzed film formers are epoxies, polyesters, and urethanes. Applications of these vehicles include hard coatings for industrial steel structures.

• Heat-cured film formers. These vehicles are similar to those in the previous section except that the catalyst is activated at higher temperatures. These vehicles are sometimes called "baked" coatings. Improved hardness and water resistance are among the properties these vehicles provide. Examples of applications are baked polyester powder coatings for appliance finishes such as refrigerators and fluidized bed coatings for pipes.

• Emulsion film formers. Emulsion systems consist of vehicles, such as acrylics, suspended in water with the assistance of a surfactant. When the water evaporates, the particles coalesce to form a film. Under magnification, the boundaries of these coalesced particles are sometimes visible, whereas the solvent systems produce very smooth films. Plasticizers are added to make the films more flexible and increase adhesion to substrates. Coalescing agents are added to the emulsion to form a smoother film. The other typical ingredients such as pigments are also present. A typical waterborne or emulsion formulation is shown in Table 4.2. These films produce lower gloss than solvent systems, but are easy to apply and are more environmentally friendly because of the lack of organic solvents.

4.2. SOLVENT SYSTEMS

Solvent systems can form a film by simple evaporation of a solvent leaving a solid vehicle/pigments such as a lacquer; or by evaporation of solvent followed by chemical reaction of components such as an epoxy with an amine.

4.3. WATERBORNE SYSTEMS

A waterborne system consists of a water-dispersible resin such as acrylic and pigment is added to provide color. The formation of a film occurs when the aqueous phase evaporates and the acrylic latex particles coalesce and form a solid layer.

4.4. POWDER SYSTEMS

A powder consists of prepolymer or resin adducts and pigments mixed with a chemical catalyst to form a fine powder. The powder is deposited on a metal substrate and oven-heated to cure the powder coating which also melts and flows out on the surface to form a smooth film.

4.5. ELECTRODEPOSITION SYSTEMS

Electrodeposition coatings (E-coatings) are deposited on a substrate by an electric current. These coatings are applied by submerging the electrically conductive substrate in a water solution of the coating and a direct current (dc) is applied which attracts the charged coating particles. The substrate serves as one electrode (anode or cathode) and an oppositely charged electrode is submerged in the solution. Pigments are usually suspended in the solution and they coat-out with the vehicle particles. The E-coat vehicle consists of a resin, such as epoxy, the pendant groups of which have been chemically modified to react to an electric current. Usually, carboxylates are added to provide a positive charge and an amine for a negative charge.

Following deposition of E-coatings on a substrate, baking the coating forces the particles of vehicle to flow together and produce a film. These films produce a medium gloss, and examples of applications are steel shelf coatings and other industrial steel coatings.

Electrodeposition is an established commercial method of painting. There are over 1500 systems worldwide. PPG supplies nearly 50% directly or 75% by license. Electrodeposition systems are used to prime or finish coat in almost every area of metal finishing, including appliance, automotive, and industrial.

The roots of the electrodeposition process were set in 1809, when the basic principle of electrophoresis was detailed. Electrophoresis is the movement of suspended particles through a fluid under the action of an electromotive force applied to electrodes in contact with the suspension.

Electrodeposition functions much like a plating process. The parts to be coated serve as one electrode and the tank or auxiliary electrodes serve as the oppositely charged electrode. The parts to be coated are immersed into a coating tank by a conveyor or program transfer system. The charged paint particles are electrolytically attracted to the parts oppositely charged and are deposited. Electrodeposition continues until sufficient coating thickness is applied so as to insulate the article being finished and then the process is complete.

The process of electrodeposition itself was first patented in 1919 and the first applications for coatings were attempted for the lacquering of food can interiors in 1935–1939. It was not until the late 1950s that this concept was genuinely investigated and applied to commercial use.

Research into using electrodeposition for automotive primer was initiated in the late 1950s. The chemistry that appeared most likely to succeed was based on knowledge of anionic soaps and the current paints of that time. The chemistry of cationic materials was theoretically desirable, but the technology was not well known.

4.5.1. Anionic Electrodeposition Coatings

1. PPG Powercron 100—general-purpose anodic coating with excellent chemical and corrosion resistance, and use as a primer.
2. PPG Powercron 150—very low cure epoxy coating. For use as a primer on temperature-sensitive substrates.

3. PPG Powercron 210—general-purpose anodic acrylic coating. Most economical system available. For use as a one-coat interior finish for products that have critical color and gloss requirements.
4. PPG Powercron 330—advanced acrylic coating.

4.5.2. Cationic Electrodeposition Coatings

1. PPG Powercron 400—high-performance cathodic epoxy coating with excellent chemical resistance. Available in corrosion-resistant whites and ultrabright colors.
2. PPG Powercron 500—cathodic epoxy for excellent corrosion resistance. Excellent primer for steel.
3. PPG Powercron 600—advanced cathodic epoxy with the lowest VOC and cure temperature. High operational flexibility levels with variable film build capabilities.
4. PPG Powercron 700—high-gloss cathodic acrylic coating with one-coat coverage, low cure economy. Bright colors with "wet look" sheen.
5. PPG Powercron 800—cathodic acrylic coatings with wide application versatility, rugged one-coat coverage. Unique combination of durability and corrosion resistance properties.
6. PPG Powercron 900—premier cathodic acrylic coatings for the broadest range of application. A very durable coating.*

The qualities of cationic electrodeposition coatings were recognized after 1960 for the appliance industry. Cationic coatings have superior corrosion protection properties for the following reasons:

1. The applied electric potential causes the positively charged polymer ions to move to the cathode. As the coating is being deposited, hydrogen gas is simultaneously evolved. There is no dissolution of metal from the substrate so the presence of metal ions in the coatings and the bath is avoided. This eliminates, undesirable by-products such as film staining or discoloration and lower chemical and salt spray resistance. In anionic systems, oxygen gas is liberated and metal from the anode is dissolved with subsequent inclusion of metal ions in the deposited coating.

2. When applied, the cationic systems are alkaline in nature and tend to be inherent corrosion inhibitors. Electrodeposited anionic coatings are acid in nature.

*Source: PPG Industries, Inc., Pittsburgh, Pennsylvania 15272.

Cationic automotive primers presently in use are waterborne, thermosetting organic coatings that are applied by cathodic electrodeposition. The cationic coating is based on an organic alkaline polymer which imparts good corrosion resistance to steel parts. Anionic coatings are based on mild, organic acid polymers that cannot provide corrosion protection.

4.6. THERMAL SPRAY POWDER COATINGS

The flame-spray powder coating technique has been developed over the last dozen years for application of thermoplastic powder coatings. Polyethylene, copolymers of ethylene and vinyl acetate, nylon and polyester powder coatings have been successfully applied by flame spraying. This technique permits powder coatings to be applied to practically any substrate, as the coated article does not undergo extensive additional heating to ensure film formation. In this way, substrates such as metal, wood, rubber, and masonry can be successfully coated with powders if the coating itself has a proper adhesion to the substrate. The technique itself is relatively simple:

1. Powder coating is fluidized by compressed air and fed into the flame gun.
2. The powder is then injected at high velocity through a flame of propane. The residence time of the powder in the flame and its vicinity is short, but just enough to allow complete melting of the powder particles.
3. The molten particles in the form of high-viscosity droplets deposit on the substrate forming high-build film on solidification.

An example of a flame spray gun was disclosed in a patent of Oxacetylene Equi (Swedish Patent 1423176, 1985). The gun has a body with air, combustion gas, and powder material supply channels. The outlet of the powder channel is axially positioned at the gun mouthpiece with the channels for the combustion gas outlet situated at equal distances on the circumference concentric to the axial powder channel. The efficiency is increased by preventing the powder from burning in the flame as the concentric circumference diameter is 2.85–4.00 times the powder outlet channel diameter. The coating quality is increased when using liquefied gas as the combustion gas outlet channel axis is at 6–9° to the powder channel axis, forming a diverging flame. The amounts of air and combustion gas are regulated by valves. The air passes through rough ejectors creating a refraction in the channel. The air and liquefied gas mix in chambers forming a combustible mixture which flows to the mouthpiece nozzles. The powder particles entering the flame are heated and in a molten form are supplied onto the surface being coated.

Because the flame spray process does not involve oven heating, it is very suitable for field application on workpieces that are large or permanently fixed and thus not able to fit inside an oven. It has been reported that objects such as bridges, pipelines, storage tanks, and rail cars are suitable surfaces to be coated by this technique. The nominal coating thicknesses reported are 3–5 mils and 6+ mils for most applications.

The flame spray equipment vendors are:

Canadian Flamecoat Co.
Plastic Flamecoat Systems, Inc.
UTP Welding Technology Co.

The technology from Applied Polymer Systems, Inc. is an electrically generated arc-type plasma rather than a combustible gas flame like the others. The most active of the above vendors appear to be Canadian Flamecoat Co. and Plastic Flamecoat Systems, Inc.

Suppliers

Suppliers of the TPC (ethylene-acrylic acid copolymers) coatings are Dow Chemical Co. and DuPont Polymers Co. These two are the leading suppliers of TPC materials, and vendors purchase these materials and customize them for their own specific uses. Non-ethylene-acrylic acid copolymer TPC coatings are supplied by Hoechst-Celanese Co., Atochem, and others.

4.7. PLASMA SPRAY COATINGS

4.7.1. Principles of Operation

Thermoplastic polymers can be sprayed onto substrates without the use of solvents, postbaking cure, or being dispersed in water. The principle consists of passing a mixture of inert gas and fine thermoplastic polymer powder through an arc which melts the powder without oxidation. The method is different from flame spray methods as no flame is employed, much better control of film thickness is possible, and a wide range of polymeric materials is available.

Plasma is often considered the fourth state of matter after solid, liquid, and gas. This extremely hot substance consists of free electrons and positive ions. Although the plasma conducts electricity, it is electrically neutral. The plasma spray system utilizes argon gas passing through an electric arc between an anode and cathode. The carrier gas loses one of its electrons and becomes a highly energetic, extremely hot plasma. As the plasma leaves the internally water-cooled plasma generator in the gun, powdered thermoplastic formulations and inert gas are introduced into the stream in a precisely controlled manner. As the temperature of the polymer increases in the plasma stream, it becomes a liquid and is projected against the surface being coated which causes the liquid polymer particles to flow, coalesce, and form a coherent film.

4.7.2. Plasma Sprayable Thermoplastic Polymers

1. Linear polyethylene
2. Ultrahigh-molecular-weight polyethylene
3. Polypropylene
4. Polyetheramide copolymer
5. Flexible nylon
6. 6,12 copolyamide nylon
7. Polyester
8. Polyvinylidene fluoride
9. Polyvinylidene fluoride/hexafluoropropylene copolymer
10. Polytetrafluoroethylene and copolymers

4.7.3. Advantages of Plasma Sprayed Coatings

1. Elimination of preheating
2. High deposition spraying rates
3. Multilayered coatings, unlimited film thickness
4. Inert atmosphere
5. Minimal surface preparation
6. Low- and No-VOC
7. Materials not sprayable by other methods

4.8. FLUIDIZED BED COATINGS

These coatings are deposited on preheated metal parts in an air-agitated suspension of fine particles. The particles adhere to the metal and form a thick film (10–30 mils). These coatings are usually applied on industrial pipe, and other heavy-duty industrial parts.

4.9. VAPOR DEPOSITION COATINGS

This type of coating has specialized applications. Thin films of metal (e.g., aluminum, gold, titanium) or other materials vaporized in a vacuum chamber can be deposited on solid surfaces in thicknesses from a few angstroms to a few micrometers. This type of coating is useful for making surfaces electrically conductive and aluminized reflective plastic film.

4.10. PLASMA POLYMERIZED COATINGS

Ethylene gas in a strong electromagnetic film will polymerize and precipitate on a surface to form a film. A chamber must be under medium vacuum and the parts to be coated are small because of the size of the chamber. The usefulness of this type of coating is limited to special effects from polymers with low surface energy or dielectric properties. Polyethylene and polytetrafluoroethylene have been successfully plasma polymerized.

Popular industrial and trade-sale formulations for paints and coatings are given in Tables 4.3–4.43.

5

Paint Materials

5.1. OILS

Oils (Martens, 1974) are used in coatings either by themselves, as a portion of the nonvolatile vehicle, or as an integral part of a varnish, when combined with resin, or of a synthetic liquid, when combined with the resinous portion of the synthetic.

1. Oil improves the flexibility of the paint film: eliminating oil from certain formulations would cause the film to crack.
2. In exterior finishes, oil gives durability.
3. As part of the nonvolatile vehicle, oil improves gloss.
4. Some oils give moderate resistance to water, soap, chemicals, and other corrosive products.
5. Some oils give specialty properties such as wrinkling (for wrinkle finishes).
6. With special treatments, oils can be used to improve leveling and the flow, nonpenetration, and wetting properties of the vehicle. They also have other desirable characteristics.

5.1.1. Composition

Most of the oils are triglycerides of fatty acids. Glycerin, $C_3H_5(OH)_3$, has three OH groups, each of which can react with the carboxyl group of a fatty acid. Such a reaction will result in water being split off and a triglyceride being formed. This is the oil as it is found in nature.

5.1.2. Properties

The properties of the specific oil depend largely on the type of fatty acids in the oil molecule. Thus, highly unsaturated fatty acids will give improved drying properties but have a greater tendency toward yellowing. Drying is especially improved if the double bonds are in a conjugate system in which two double bonds are separated by a single bond. Such oils also have a faster bodying rate when heated and somewhat better water and chemical resistance.

5.1.3. Oil Treatments

Many of the oils cannot be used in the raw state, as they are produced by the crushing of seeds, nuts, fish, etc., and must be treated to make them usable. Others can be used in the raw state, but are often treated to give them special properties (Gooch, 1980). Among these treatments are the following:

1. Alkali refining. The oil is treated with alkali, which lowers its acidity and makes it less reactive and also improves its color.
2. Kettle bodying. The oil, usually refined, is heated to a high temperature for several hours to polymerize it. This increases its viscosity and improves its dry, color retention, flow, gloss, wetting properties, and nonpenetration. However, the process impairs brushability.
3. Blowing. Air or oxygen is passed through the oil at elevated temperatures. The resultant oil has improved wetting, flow, gloss, drying, and setting properties, but brushability and, often, color and color retention are impaired. In addition, paints containing blown oils have a greater tendency toward pigment settling.

Among the more important paint oils are the following.

5.1.4. Linseed Oil

This is the largest-volume oil used by the coatings industry. It is very durable, yellows in interior finishes, but bleaches in exterior paints, and has good nonsagging properties, easy brushing, good drying, fair water resistance, medium gloss, a medium bodying rate, and poor resistance to acids and alkalies. It is used largely in house paints, trim paints, and color-in-oil pastes. Alkali-refined and kettle-bodied linseed oil is used in varnishes and interior paints. Linseed oil is an important modifying oil in synthetic alkyds.

5.1.5. Soybean Oil

This is a semidrying oil that can be used only with modifying oils and resins to improve its drying properties The refined oil has excellent color and color retention. Soybean oil is one of the most important modifying oils in alkyds and is used in nonyellowing types of paint.

5.1.6. Tung Oil (China-Wood Oil)

This oil contains conjugated double bonds and cannot be used in its raw state as it would dry to a soft, cheesy type of film. In its kettle-bodied state, it gives the best-drying and most resistant film of any of the common paint oils. It has a good gloss and good durability and is used in finishes for which dry and resistance are important: spar varnishes, quick-drying enamels, floor, porch, and deck paints, concrete paints, and others.

5.1.7. Oiticica Oil

This oil is similar to tung oil in its properties, but its drying, flexibility, and resistance characteristics are not quite as good. It also has somewhat poorer color and color retention. However, it has better gloss and better leveling qualities than tung oil. Oiticica oil is normally used as a substitute for tung oil when there is a large price difference between them.

5.1.8. Fish Oil

This is a poor-drying oil that cannot be used in it raw state because of its odor. In its kettle-bodied state, it has relatively easy-brushing and good nonsagging properties. It also has fairly good heat resistance. Fish oil is used in low-cost paints as it is usually lower-priced than the other oils.

5.1.9. Dehydrated Castor Oil

Raw castor oil is a nondrying oil that is used in lacquers as a plasticizing agent to make them more flexible. When it is treated chemically to remove water from the molecule, additional double bonds are formed; this makes it a drying oil. The dehydrated oil dries better than linseed oil, although paints made with it sometimes have a residual tack that is difficult to remove. Dehydrated castor oil has very good water and alkali resistance—almost as good as that of tung oil. It also has excellent color and color retention, on a par with that of soybean oil. The oil is used in finishes for which color and dry are important: alkyds, varnishes, and quick-drying paints.

5.1.10. Safflower Oil

This oil, a relative newcomer to the coatings industry, has some of the good properties of both soybean oil and linseed oil. It has the excellent nonyellowing features of soybean oil and dries almost as well as linseed oil. Safflower oil can therefore be used as a substitute for linseed oil in many white formulations for which color retention is important, especially kitchen and bathroom enamels.

5.1.11. Tall Oils

This is not really an oil, but it is often used as an oil or as a combination of an oil and a resin. Tall oil is a combination of fatty acids and rosin. Normally it is separated into its separate ingredients, which are used as such. The rosin is used for the rosin properties, and the tall-oil fatty acids are used for the fatty-acid properties. As a component in alkyds, the fatty acids give vehicles similar to those made with soybean fatty acids. When limed, tall oil gives a liquid that is low in cost and high in gloss, has poor flexibility, and tends to yellow very badly on aging.

5.2. RESINS

5.2.1. General

If coatings were made with oil (Weismantel, 1981) as the only nonvolatile component with the exception of driers, the result would be a relatively soft, slow-drying film. Such a film would be satisfactory for house paints, ceiling paints, or other surfaces for which hardness and fast dry are not important, but totally unsatisfactory for many trade sales and maintenance coatings and for most industrial or chemical coatings. In addition to improving hardness and speeding drying time, specific resins give other important properties. Thus, they often improve gloss and gloss retention, and they also usually improve adhesion to the substrate. Resistance to all types of agents such as chemicals, water alkalies, and acids would not be obtained without the use of different types of resins. Low-cost resins are used to reduce the raw-material cost of a coating. Following are properties of the more popular resins.

5.2.2. Rosin

This low-cost natural resin, derived from the sap of trees, is essentially abietic acid,$C_{20}H_{30}O_2$. It must be largely neutralized before it can be used. This is normally done by reacting the rosin with lime, in which case it is known as limed rosin, with glycerin, which gives ester gum, or with pentaerythritol, which yields pentaresin. Liming rosin gives a resin with a high gloss, excellent gloss retention, and fine adhesion. However, the resin is relatively poor in drying time and in resistance to water and chemicals. Because it tolerates large quantities of water, it is popular for low-cost finishes. A solution of limed rosin in mineral spirits, called gloss oil, is popular in low-cost floor paints, barn paints, and general utility varnishes.

5.2.3. Ester Gum

This resin, made by reacting rosin with glycerol, $C_3H_5(OH)_3$, which neutralizes or esterifies the abietic acid, might be considered the first synthetic resin. Ester gum dries somewhat more slowly than limed rosin but has much-improved color-retention and resistance characteristics. It gives a very high gloss and has excellent adhesion. The higher-acid-number ester gums are compatible with nitrocellulose and therefore are used in lower-cost gloss lacquers.

5.2.4. Pentaresin

When pentaerythritol, $C(CH_2OH)_4$ is the alcohol used to react with rosin, the result is a resin with a higher melting point that has good heat stability, color, and color retention and gives a high gloss. When the resin is cooked into varnishes with different oils, good drying properties and a moderate degree of water and alkali resistance are obtained. Similar to the other resin esters, pentaresin has good adhesion to all types of surfaces.

5.2.5. Coumarone-Indene (Cumar) Resins

These resins, derived from coal tar, are essentially high polymers of the complex cyclic and ring compounds of coumarone and indene. They are completely neutral and thus are ideal for leafing types of aluminum paints. In addition, they have good alcohol and electrical breakdown properties. They also are resistant to corrosive agents such as brine, dilute acids, and water. On the negative side, they have poor color retention and only fair drying properties and gloss. Their cost is normally quite low.

5.2.6. Pure Phenolic Resins

These are pure synthetic resins (Fry *et al.*, 1985) made by reacting phenol with formaldehyde. There are two essential types: a type that is cooked into oil and is used largely in trade sales and marine paints and a type that is sold dissolved in a solvent and is applied in that form and baked. The first type has excellent water resistance and durability, making it ideal for exterior, floor, porch, deck, and marine paints or varnishes. As it also has fine chemical, alkali, and alcohol resistance, it can be used for furniture, bars, patios, and similar applications. In some instances, adhesion is rather poor. The solvent type is heat-reactive and becomes extremely hard and resistant to chemicals when properly cured. It is used for can linings, linings for the interior of tanks, and similar applications. All phenolics tend to yellow.

5.2.7. Modified Phenolic Resins

Combinations of ester gum and pure phenolics, these resins have properties between those of their components. They have very good water, alkali, and chemical resistance, and the ester-gum portion gives them good adhesion. They offer a good dry and a high gloss. These resins are fine for floors, porches, and decks, in sealers, for spar varnishes, and for any other uses for which a combination of good resistance, a hard film, and fast drying is desirable and for which yellowing can be tolerated.

5.2.8. Maleic Resins

These resins are made by reacting maleic acid or anhydride with a polyhydric alcohol such as glycerin in the presence of rosin or ester gum. They have very fast solvent release, good compatibility with nitrocellulose, and good sanding properties. This combination makes them ideal resins for sanding lacquers. Maleic resins also have a fast dry and good color retention so that they can be used in quick-drying white coatings. They should be used only in shorter oil lengths, for in longer oil lengths they have some tendency to lose dry as they age.

5.2.9. Alkyd Resins

These resins (Martens, 1974), which are made by reacting a polybasic acid such as phthalic acid or anhydride with a polyhydric alcohol such as glycerin and pentaerythritol and which are further modified with drying or nondrying oils, are probably the most important resins used in solvent-based trade sales paints and in many industrial coatings. Those that are modified with large percentages of drying oils are normally used in trade sales paints; they are known as long- or medium-oil alkyds. Those that are modified with smaller percentages of oil or with nondrying oils are used in industrials, baking finishes, and lacquers; they are known as short-oil or nondrying alkyds. Normally, the larger the percentage of glyceryl phthalate, or 15 resinous portion, the faster is the dry, the more brittle the finish, and the better the baking properties. Other properties depend on the type of modifying oil and the type of polybasic acid used.

Generally, alkyds have excellent drying properties combined with good flexibility and resultant excellent durability. Color retention, when modified with nondrying oil or with oil having good retention such as soybean or safflower oils, is very good. Gloss and gloss retention in alkyd paints are unusually good. In baking finishes, alkyds are normally combined with other resins such as urea and melamine to obtain top-grade films. The resistance characteristics of alkyds, though good, do not compare with those of pure phenolics and are not equal to those of modified phenolics. If high-resistance characteristics are not required, however, alkyds are second to none in good overall properties. Thus, they are ideal for all types of interior, exterior, and marine paints and for a large percentage of industrial coatings.

5.2.10. Urea Resins

The short-oil, high-phthalic alkyds previously mentioned are combined with ureas and melamines in baking finishes. Urea resins can be used only in baking types of coatings because they convert from a liquid to a solid form under the influence of heat, in a type of polymerization often called curing. The ureas, a product obtained from the reaction of urea and formaldehyde, give a film that is hard, fairly brittle, and colorless. This brittleness and rather poor adhesion can be corrected by combining them with alkyd resins or plasticizers. The ureas have excellent color retention and fine resistance to alcohol, grease, oils, and many corrosive agents. They make excellent finishes for many metallic surfaces such as those of refrigerators, metal furniture, automobiles, and toys.

5.2.11. Melamine Resins

These resins, synthesized from melamine, a ring compound, and formaldehyde, act much as urea resins do (Williams *et al.*, 1985). However, they cure more quickly or at lower temperatures and give a somewhat harder, more durable film with higher gloss and better heat stability. Although they are more expensive, they are to be preferred for high-quality white finishes because their shorter baking cycle produces a film that is whiter and has the best color retention.

5.2.12. Vinyl Resins

Solvent-based vinyl resins (Park, 1985) are normally copolymers of polyvinyl chloride and polyvinyl acetate, though they are available as polymers of either one. They are usually sold as white powders to be dissolved in strong solvents such as esters or ketones, but may be sold already dissolved in such solvents. They are plasticized to make an acceptable film. The chloride is very difficult to dissolve but has extreme resistance to chemicals, acids, alkalies, and solvents. The acetate is not as resistant, but is much more soluble. The more practical copolymer still exhibits exceptional resistance to corrosive agents, chemicals, water, alcohol, acids, and alkalies. Vinyl resins do an exceptionally fine job in coatings for cables, swimming pools, cans, masonry, or any surface requiring very high resistance.

5.2.13. Petroleum Resins

These completely neutral, rather low-cost resins are obtained by removing the monomers during the cracking of gasoline and polymerizing them. They have good resistance to water, alkalies, alcohol, and heat. Some have good initial color, but they all tend to yellow on aging. Petroleum resins are very good for aluminum paints, and they make good finishes for bars, concrete, and floors when cooked into tung or oiticica oil.

5.2.14. Epoxy Resins

These resins, more correctly called epichlorohydrin bisphenol resins, are chain-structure compounds composed of aromatic groups and glycerol, joined by ether linkages. Various modifying agents are used to give epoxies of different properties, but all such resins generally have excellent durability, hardness, and chemical resistance. They can be employed for high-quality air-drying and baking coatings, and some can even be used with nitrocellulose in lacquers.

5.2.15. Polyester Resins

In addition to the alkyd resins, which are polyesters modified with oil, there are other types of polyesters, such as polyester polymers, that have a light color and good color retention, excellent hardness combined with good flexibility, and very good adhesion to metals. They are useful in many industrial-type coatings for which such properties are important.

5.2.16. Polystyrene Resins

Resins of this group made by the polymerization of styrene, are available with a variety of melting points that depend on the degree of polymerization. They are thermoplastic. The higher-melting-point resins are incompatible with drying oils, but the lower polymers are compatible to some degree. Polystyrene resins have high electrical resistance, good film strength, high resistance to moisture, and good flexibility when combined with oils or plasticizers. They are useful in insulating varnishes, waterproofing paper, and similar applications.

5.2.17. Acrylic Resins

These thermoplastic resins, obtained by the polymerization or copolymerization of acrylic and methacrylic esters, may be combined with melamine, epoxy, alkyd, acrylamide, etc., to give systems that bake to a film with excellent resistance to water, acids, alkalies, chemicals, and other corrosives. They find use in such applications as coatings for all types of appliances, cans, and automotive parts and for all types of metals.

5.2.18. Silicone Resins

These polymerized resins of organic polysiloxanes combine excellent chemical-resistance properties with high heat resistance (Cahn, 1974). They are expensive and therefore are not usually used for their chemical-resistance properties, which can be obtained from lower-priced resins, but for their very important heat- and electrical-resistance properties, which are superior to those of other resins. At a lower cost, they can be copolymerized with alkyds and still retain some of their important properties.

5.2.19. Rubber-Based Resins

These resins, based on synthetic rubber, give a film, when properly plasticized, that has high resistance to water, chemicals, and alkalies. They are excellent for use in swimming pool paints, concrete floor finishes, exterior stucco and asbestos shingle paints, and other coatings requiring a high degree of flexibility and resistance to corrosion.

5.2.20. Chlorinated Resins

Paraffin can be chlorinated at any level from 42% which gives a liquid resin, to 70% which gives a solid resin. Chlorinated resins are popularly used in fire-retardant paints. The 70% resin is also used in house paints and in synthetic nonyellowing enamels for improved color and gloss retention. Chlorinated biphenyls with high resistance characteristics can also be made; they are often combined with rubber-based resins for coatings requiring a high degree of alkali resistance. Rubber also is chlorinated and is sold as a white granular powder containing about 67% chlorine. It is quite compatible with alkyds, oils, and other resins such as phenolics or cumars. It has high resistance to acids, alkalies, and chemicals and is useful for alkaline surfaces such as concrete, stucco, plaster, and swimming pools.

5.2.21. Urethanes

Three general classes of urethane resins or vehicles (Frisch and Kordomenos, 1985) are available today: amine-catalyzed, two-container systems, moisture-cured urethane, and urethane oils and alkyds. The first and second types contain unreacted isocyanate groups which are available to achieve final cure in the coating. In the first case, an amine is used to catalyze a cross-linking reaction that results in a hard, insoluble film; in the second, the moisture in the air acts as a cross-linking agent.

Urethane oils and alkyds, on the other hand, are cured by oxidation in the same way as alkyds and oils, and require driers or drying catalysts. However, cure occurs more quickly and the resultant film is very hard and abrasion-resistant and has greatly improved resistance to water and alkalies. However, color retention is somewhat poorer. Because of their advantages, urethane oils and alkyds are widely used in premium floor finishes and for exterior clear finishes on wood. The hardness of the film tends to impair intercoat adhesion, and care must be exercised to sand the surface lightly between coats to provide tooth.

5.3. LACQUERS

Lacquers dry essentially by evaporation of the solvent, and they are dry as soon as the solvent is gone. Raw materials consist of substances that form a dry film, or that can become part of a dry film, without the necessity of going through oxidation or polymerization steps, and of the solvents in which these film formers are dissolved.

The basic film formers of lacquers are the cellulosics. In addition, most lacquers also contain resin for improved adhesion, build, and gloss and plasticizers for improved flexibility. Each of these three types of lacquer film formers is briefly examined.

By far the most important cellulosic is nitrocellulose; second is ethyl cellulose. Cellulose acetate is also of some importance.

Nitrocellulose, made by nitrating cotton linters, comes in two grades: RS (regular soluble types) and SS (spirit- or alcohol-soluble types). Both are available in a variety of viscosities and form a film that is hard, tough, clear, and almost colorless.

Ethyl cellulose, made by reacting alkali cellulose with ethyl chloride, also comes in different viscosities. It has greater compatibility with waxes, better flexibility, better chemical resistance, less flammability, and a higher dielectric constant. It is also somewhat softer, tends to become brittle when exposed to sunlight and heat, and is more expensive. These disadvantages can be partially overcome by the use of proper modifying agents and solvents.

Cellulose acetate lacquers are tough and stable to light and heat. They also have good resistance to oils and greases and are durable. However, they have poor solubility and compatibility, and this defect partially limits their usefulness.

In most instances, the lacquer film will contain a larger percentage of resin than the cellulosic. The reason is that resins add many important properties to lacquer films and usually are lower in cost. The most valuable property they add is adhesion; this is of particular importance, as nitrocellulose by itself has rather poor adhesion. In addition, resins give higher solids and therefore a thicker film, improve gloss, reduce shrinkage, and improve heat-seal properties.

In choosing a resin, make certain that it is compatible with the cellulosic being used. It must also be soluble in a mixture of esters, alcohols, and hydrocarbons so as to give a clear, transparent film.

Among the resins in common use are rosin esters such as ester gum, used for its low cost; maleic resin, used in wood finishes for its good sanding properties; and alkyds, employed for their good resistance and durability. Alkyds modified with coconut oil are often used; they may be further modified with other resins such as terpenes for good heat-seal properties and phenolics for good water resistance.

5.4. PLASTICIZERS

Without plasticizers, most lacquers would be much too brittle, would tend to crack, and therefore would not be durable. In addition to giving flexibility, plasticizers increase the solids content so as to produce films of practical thickness, and they also tend to improve gloss, especially of pigmented lacquers. Another plus feature, especially of chemical plasticizers, is that they act as a solvent for the cellulosic and thus enable more of this cellulosic to be used. In addition, they help slow the settling time for the lacquer, enabling it to level out satisfactorily.

Plasticizers must be completely nonvolatile so that they remain in the film permanently. There are some exceptions to this requirement, in lacquers such as nail polish which do not remain on the surface permanently.

As most plasticizers are lower in cost on a solids basis than cellulosics, there might be a tendency to use excessive amounts. This would be dangerous, for the result would be a tacky, soft film with poor chemical and water resistance and poor abrasion resistance.

Two types of plasticizers, the oil type and the chemical type, are generally used in lacquers. A good example of the nonsolvent oil type is raw and blown castor oil, which gives perpetual flexibility, is low in cost, has good color and color retention, and is sensitive to temperature change. Excessive amounts tend, however, to spew from the film. Solvent-type chemical plasticizers such as dibutyl phthalate, triphenyl phosphate, and dioctyl phthalate have excellent compatibility and good heat-seal properties. The chlorinated polyphenyls have good resistance characteristics. All tend to produce a good, tight film.

5.5. WATER-BASED POLYMERS AND EMULSIONS

Manufacture of these types of coatings (Stevens, 1980) is the fastest-growing part of the coatings industry. Most of the trade sales and architectural paints are not water-based. Even in the industrial field, more water-based or water-thinnable paints are being manufactured. The major advantages of these coatings are that they can be thinned with water and, in the case of trade sales paints, that there is little odor, a fast dry, better nonpenetration and holdout, very good alkali resistance, excellent stain resistance, and easy cleanup with water. In all cases, they practically eliminate the release of solvent fumes into the atmosphere—a big plus in view of environmental restrictions.

5.5.1. Styrene-Butadiene

This is the oldest and initially was the only polymer available for latex paints. It is a copolymer of polystyrene, a hard, colorless resin, and butadiene, a soft, tacky, rubberlike polymer. Paints based on the polymers of styrene-addition have some disadvantages in their tendency toward poor freeze–thaw stability and low critical PVC. There is also a greater tendency toward efflorescence, the appearance of a white crystalline deposit on a painted surface. The use of styrene-butadiene polymer is now very limited.

5.5.2. Polyvinyl Acetate

This is one of the most popular polymers used in the manufacture of latex paints. The polymer itself is a thermoplastic, hard, resinous, colorless product having good water resistance. Normally it is bought as a water emulsion containing surface-active agents, protective colloids, and a catalyst. It is much more stable and easier to use than styrene-butadiene and therefore has largely replaced it in latex paints. The film is clear, colorless, and odorless and has very good water and alkali resistance. The polymer gives a breathing type of film which prevents blisters if applied over somewhat moist surfaces. Because by itself the film would be too brittle, it must be plasticized, either internally or in the paint formulation. Polyvinyl acetate (PVA) types have advantages over styrene-butadiene types of durability, stability to light aging, and nonblistering properties. The emulsion tends to be acidic, and formulating with it requires some caution.

5.5.3. Acrylics

Acrylic polymers are probably the best in quality of the emulsions popularly used in the manufacture of latex paints. They are made essentially by polymerization or copolymerization of acrylic acid, methacrylic acid, acrylonitrile, and the esterification of them. The properties of acrylic polymers depend to a large degree on the type of alcohol from which the esters are prepared. Normally, alcohols of lower molecular weight produce harder polymers. The acrylates are generally softer than the methacrylates.

The acrylics differ from the PVAs in being basic (i.e., nonacid). The danger of their causing containers to rust is thus reduced. Moreover, because the acrylics are almost completely polymerized prior to application as a paint film, there is practically no embrittlement or yellowing on aging. This factor improves the durability of acrylic paints; in fact, durability is a special feature of acrylics. They are the most stable of the polymers and require a minimum of such stabilizers as protective colloids, dispersing agents, and thickeners. They will also withstand extremes of temperature to a high degree.

The acrylics have excellent resistance to both scrubbing and wet abrasion. Moreover, the extreme insolubility of the dried paint film gives it excellent resistance to oil and grease. As a result, oil stains and other dirt marks can easily be removed without injuring the film.

The major disadvantage of acrylics is cost, which is higher than that of other latices. In partial compensation, acrylics will take higher pigmentation, and more low-cost extenders may therefore be used.

5.5.4. Other Polymers and Emulsions

Though most trade sales paint is water-based, this is not true of industrials. Because of the special requirements of industrial coatings, satisfactory water-based polymers with the required properties have not yet been developed. Nevertheless, much progress has been made, and satisfactory water-reducible coatings have been made for many industrial applications.

- Water-reducible resins. The most popular general type of aqueous industrial vehicles is the so-called water-soluble resin. The basic approach is to prepare the resin at a relatively high acid number and then to neutralize it with an amine such as ammonia or dimethylaminoethanol. A wide variety of resins, including alkyds, maleinized oils, epoxy esters, oil-free polyesters, and acrylics, is produced in this manner. These resins may be either air-dried or baked vehicles. Driers such as cobalt, manganese, calcium, or zirconium may be added as cross-linkers to the baking vehicles. Coatings made with these vehicles are competitive with solvent-based industrials in terms of gloss, film properties, and overall resistance. There is a problem with air-drying efficiency on aging because of the complexing of the driers with the amines used.

- Emulsion vehicles. Emulsion vehicles, particularly acrylic and styrene-acrylic types, are also being promoted for baking industrial finishes. These cure by cross-linking mechanisms, generally through the use of melamine or urea resins. It is more difficult to obtain high gloss with emulsions as compared with water-soluble resins, but because of their higher molecular weight, emulsions may offer advantages in film strength and resistance properties.

- Copolymers. Some types of polymers can be copolymerized. The types of acrylics are the acrylates, methacrylates, and acrylonitriles. To obtain special properties, polymers are frequently blended or copolymerized.

5.6. DRIERS

The basic difference between lacquer and solvent-based paint is that lacquer dries by evaporation of the solvent and paint by a combination of oxidation and polymerization. To speed the drying action of a paint, driers are required. Without them paint would dry in days instead of in hours, and, in many cases, the film would be softer and have poorer resistance properties.

Most driers are organometallic compounds (e.g., resinates, linoleates, and naphthenates) that act as polymerization or oxidation agents, or both. The soaps must be in such form that they are soluble in the vehicle. Everything being equal, the more soluble the soaps are, the more effective they are as driers. Tall-oil driers, based on tall-oil fatty acids, are somewhat less soluble than naphthenates based on naphthenic acid. Synthetic acid driers based on octoic, neodecanoic, and similar acids are now the most popular. In addition, the metal portion of the more active driers is normally oxidizable. One theory is that these driers, especially the oxidation catalysts, act in their reduced form by taking oxygen from the air, become oxidized, pass the oxygen on to the oil or other oxidizable molecule, become reduced again, and are therefore in a position to take on additional oxygen to pass on to the oxidizable vehicle. This process is repeated until the film is completely oxidized.

5.6.1. Cobalt

The cobalt drier, sold containing 6 to 12% cobalt as metal, is the most powerful drier used by the coatings industry. It acts as an oxidation catalyst and is known as a top drier, drying the top of the film. Excessive amounts of cobalt drier will set up stresses and strains in the paint film that can result in wrinkling. Though purple in color, cobalt has low tinting strength and will not discolor a paint.

5.6.2. Lead

This drier is normally sold in strengths containing 24 or 36% lead as metal. It is very light in color and thus will not discolor a paint. Lead is a polymerization catalyst and therefore makes an ideal combination with cobalt, as it tends to harden or dry the bottom of the film. Because of lead laws, this type of drier is gradually being replaced by calcium, zirconium, or both. Some lead driers are lead resinates, lead linoleates, and lead naphthenate.

5.6.3. Manganese

This drier, sold normally in strengths of 6, 9, or 12% metal, is what is known as a through drier, acting on both the top and the bottom of the film. Mainly, however, it is an oxidation rather than a polymerization catalyst and can therefore cause wrinkling if employed in excessive amounts. It is often used in combination with cobalt and lead to cut the cobalt content and reduce skinning. At other times, it is used with lead as a manganese–lead drier combination. It is brownish in color and tends to discolor paints if used in large amounts.

5.6.4. Calcium

This very light-colored drier, which has no tendency to discolor paints, acts as a polymerization agent similar to lead. It also tends to improve the solubility of lead if used in combination with it and thus makes lead more effective as a drier. It is sold in metal contents of 4, 5, and 6%. Calcium is becoming increasingly popular for use as a substitute for lead in lead-free paints.

5.6.5. Zirconium

Like calcium, zirconium is light in color and acts usually as a polymerization catalyst. In lead-free paints it is often used with cobalt or in combination with cobalt and calcium. Zirconium is light in color and sold in concentrations of 6, 12, and 18%.

5.6.6. Other Metals

Other metals are sometimes used as driers. Among the most popular are iron, useful in colored baking finishes, and zinc, useful as wetting and hardening agent. Zinc is also used to reduce skinning tendencies in a paint. Sometimes cerium is used as a drier.

- Nonmetallic driers. The elimination of lead has focused attention on nonmetallic driers. The most popular of these is orthophenanthroline, which often gives excellent drying properties, sometimes superior to those of standard combinations, when used with manganese and sometimes with cobalt.

5.7. PAINT ADDITIVES

5.7.1. General

This group of raw materials is used in relatively small amounts to give coatings certain necessary properties. (Driers actually belong in this category.) Because additive compositions are not normally revealed by manufacturers, the following discussion refers to trade names. On occasion, additives are used on the job site if problems arise. In such cases, there should be close coordination and supervision by the paint manufacturer to avoid even bigger problems.

5.7.2. Antisettling Agents

This group of agents is used to prevent the separation or settling of the pigment from the vehicle. Most commonly this is done by using additives that set up a gel structure with the vehicle, trapping the pigment within the gel and preventing it from settling to the bottom.

5.7.3. Antiskinning Agents

These are essentially volatile antioxidants that prevent oxidation, drying, or skinning of the paint while it is in the can but volatilize and leave the paint film, allowing it to dry properly once it has been applied. The most common antiskinning agents are methyl ethyl ketoximine, very effective in alkyds, and butyaldoxine, effective in oleoresinous liquids. Phenolics are sometimes used, but they can slow the drying time of the coating.

5.7.4. Bodying and Puffing Agents

These products increase the viscosity of a paint. Without them, paint is often too thin to be used. In solvent-based paints, gelling or thixotropic agents may be used. There are also liquid bodying agents that are based largely on overpolymerized oils. In water-based paints, the most common bodying agents are methyl cellulose, hydroxyethyl cellulose, the acrylates, and the bentonites. These agents also tend to improve the stability of the emulsion.

5.7.5. Antifloating Agents

Most colors used in the paint industry are a blend of colors. Thus, to form a gray some black is added to a white paint. It is important that one color not separate from the other, and antifloating agents are used for this purpose. Silicones are sometimes used, but they pose serious bubbling and recoatability problems. Special antifloating agents are sold under various trade names.

5.7.6. Loss of Dry Inhibitors

Certain colors such as blacks, organic reds, and even titanium dioxide tend to inactivate the drier, and the paint loses drying on aging. Agents are therefore introduced to react slowly with the vehicle and feed additional drier to replace what was lost. In the past, most of the agents have been lead compounds such as litharge, but these are now being replaced by agents based on cobalt.

5.7.7. Leveling Agents

Sometimes a paint does not flow properly and shows brush or roller marks. These can often be corrected by special wetting agents that cause the vehicle to set the pigment better.

5.7.8. Foaming

This is much more of a problem in water-based than in solvent-based paints. The presence of bubbles not only makes for an unsightly paint when applied, but results in a partially filled paint can when the bubbles leave the paint while it is in the can.

5.7.9. Grinding of Pigments

Unless pigment is properly ground, the result is a coarse film of poorer opacity and, in a gloss-finish type of paint, usually in a poorer gloss. Certain types of wetting agents tend to improve the ability of the disperser or mill to separate these pigment particles more easily and thus to obtain better grind.

5.7.10. Preservatives

Almost every formulation based on water must have a preservative for can stability. Until recently, most preservatives have been mercurials, but these are being partially replaced by complex organics.

5.7.11. Mildewcides

Most exterior paints will suffer a blackish-greenish discoloration caused by the growth of fungi or mildew on the surface. Until now this condition has been prevented by the inclusion of a mercurial in the paint, often in combination with zinc oxide. Today nonmercurials also are available.

5.7.12. Antisagging Agents

When applied, a paint sometimes flows excessively so that it causes what are known as curtains, runs, or sags. Most bodying or antisettling agents prevent this tendency. Some of them prevent sag without increasing paint body.

5.7.13. Glossing Agents

Sometimes the gloss in a solvent-thinned gloss-type formulation is low. Though it can usually be increased by changing vehicles or pigmentation or by increasing the ratio of nonvolatile vehicle to pigment, the use of an additive may be a simpler step.

5.7.14. Flatting Agents

Just as gloss is desirable in gloss finishes, flatness is needed in flat finishes. Flatness is easy to obtain in regular flat paints, but in clear coatings such as flat varnishes or lacquers this goal is much more of a problem. It can be accomplished by the use of special flatting agents such as amorphous silica.

5.7.15. Penetration

In some systems, the paint is supposed to penetrate the surface. Penetration is important in stains and in paint that will be applied to a poor surface. Most paints, however, require good nonpenetration for improved sealing properties and good color and sheen uniformity. This goal is accomplished mainly by agents that set up a gel structure in the paint.

5.7.16. Wetting Agents for Water-Based Paint

Many different types of wetting agents are necessary in water-based paints. Some are used for improved pigment dispersions, whereas others are employed to improve adhesion to a poor surface such as a slick surface.

5.7.17. Freeze–Thaw Stabilizers

These are necessary in water-based paints to prevent coagulating or flocculating when the paints are subjected to freezing temperatures. The stabilizers, such as ethylene or propylene glycol, lower the temperature at which the paint will freeze. Another way of accomplishing this goal is to use an additive that improves the stability of the emulsion.

5.7.18. Coalescing Agents

The purpose of these agents in water-based paints is to soften and solvate partially the latex particles in order to help them flow together and form a more nearly continuous film, particularly at low temperatures. This can be done with ether alcohols such as butyl Cellosolve and butyl carbitol.

5.8. SOLVENTS

There are essentially three types of volatile solvents (Tess, 1985): a true solvent, which tends to dissolve the basic film former; a latent solvent, which acts as though it were a true solvent when used with a true solvent; and a diluent, a nonsolvent that is tolerated by the coating. Thus, in a lacquer, ethyl acetate is the true solvent, ethyl alcohol is the latent solvent, and petroleum hydrocarbon is the diluent. In a latex paint water might be considered a true solvent, but in an alkyd enamel it would be a diluent.

To apply the paint, some materials (Weismantel, 1981) must be used which do not become part of the paint film. With the exception of the newer 100% solids coatings such as powder coatings, paint simply could not be applied without a solvent, for in most instances the result would be a semisolid mass. It can therefore be said that the most important property of a solvent is to reduce viscosity sufficiently so that the coating can be applied, whether by brush, roller, dipping, or spraying. Besides this most important property, the solvent has other significant features. It controls the setting time of the paint film, which, in turn, controls the ability of one panel of paint to blend with another panel applied later. In addition, it controls important properties such as leveling or flow, gloss, drying time, durability, sagging tendencies, and other good or bad features in the wet paint or paint film. The use of solubility parameters (Brandrup *et al.*, 1975) is useful for selecting a proper solvent.

5.8.1. Petroleum Solvents

These constitute by far the most popular group of solvents used in the coatings industry. They consist of a blend of hydrocarbons obtained by the distillation and refining of crude petroleum oil. The faster-evaporating types, which come off first, are used as diluents in lacquers or as solvents in special industrials. Solvents of the intermediate group are used in trade sales paints. Members of the slowest group, beginning with kerosine and going into fuel oils, are used for heating, lubrication, and other applications.

The most important group used in trade sales paints and varnishes consists of mineral spirits and heavy mineral spirits. Mineral spirits are petroleum solvents with a distillation range of 300 to 400°F (149 to 204°C). They are sometimes considered a turpentine substitute because the distillation ranges are approximately the same. Because of their low price, proper solvency, and correct evaporation rate, mineral spirits are probably the most popular solvents used by the coatings industry. Normally they are the sole solvents in all interior and exterior paints with the exception of flat finishes. Special grades that pass antipollution regulations are now being sold. Heavy mineral spirits are a slower-evaporating petroleum hydrocarbon and an ideal solvent for flat-type finishes. During cold winter weather, the formulator might use a combination of regular and heavy mineral spirits.

The U.S. Environmental Protection Agency has set new guidelines, based on regulations already adopted in California, that severely limit the amount of solvent in architectural coatings. The recommended limit is 250 g of volatile organic material per liter of paint. This limit also affects water-based paints containing organic freeze–thaw agents and additives. Architects switching to new high-solids coatings should work closely with the manufacturer to assure proper performance and be certain that application personnel are properly trained to handle the more complex systems.

A faster-evaporating petroleum solvent with a distillation range of 200 to 300°F (93 to 149°C), known as VM&P naphtha, is sometimes used by painters as an all-purpose thinner. Its fast evaporation rate might cause the paint to set too quickly. It is also used by some manufacturers in traffic paints, for which a fast setting time and dry are desirable.

In some industrials and lacquers, a still faster-evaporating type, having a distillation range of 200 to 270°F (93 to 132°C), is desired. In many coatings it gives satisfactory spraying and dipping properties. An even faster-evaporating type, with a distillation range of 130 to 200°F (54 to 93°C), is sometimes used when very fast evaporation and drying are desired, but it might cause blushing or flatting of the paint or lacquer film.

Because of regulations regarding air pollution, the straight types of hydrocarbon solvents that hitherto have been the backbone of the coatings industry are being phased out and replaced by mixtures that will pass the stringent regulations of various states including California, Illinois, and New York.

5.8.2. Aromatic Solvents

This group of cyclic hydrocarbons is obtained normally from coal-tar distillation or from the distillation of special petroleum fractions. These hydrocarbons are almost pure chemical compounds and are much stronger solvents than petroleum hydrocarbons. With the exception of high-flash naphtha, they are rarely used in trade sales coatings but are employed in industrial and chemical coatings for which vehicles having weak solvent requirements are not normally used. Because aromatic solvents are pure chemicals, they have regular boiling points rather than distillation ranges. Naturally, those with the lowest boiling points will evaporate more quickly and thus give a faster dry. The most popular of these are as follows:

1. Benzene C_6H_6; boiling point, 175°F (79°C). Quite toxic, it is used in paint and varnish removers. It can cause blushing or whitening of a clear film.
2. Toluene, $C_6H_5(CH_3)$; boiling point, 230 °F (110°C). It is very popular in fast-drying industrials and in lacquers.
3. Xylene, $C_6H_4(CH_3)_2$; boiling point, 280°F (138°C). It is popular in industrials and lacquers for which slower evaporation is acceptable.
4. High-flash naphtha, a blend of slower-evaporating aromatics. The distillation range is 300 to 350°F (149 to 177°C) for brushing-type industrials and lacquers.

These products also are slowly being replaced by others that can pass stringent air pollution requirements.

5.8.3. Alcohols, Esters, and Ketones

A great many of these types of solvents are used in industrials and, especially, in lacquers. Among the more popular solvents of this type are the following:

1. Acetone, CH_3COCH_3. Very strong and very fast evaporating; it can cause blushing. It is used in paint and varnish removers.
2. Ethyl acetate, $CH_3COOC_2H_5$. This is a standard fast-evaporating solvent for lacquers. It is relatively low in cost.
3. Butylacetate, $CH_3COOC_4H_9$. This is a very good medium-boiling solvent for lacquers. It has good blush resistance.
4. Ethyl alcohol, C_2H_5OH. Used only in a denatured form, it is a good latent solvent for lacquers and also is used to dissolve shellac. It is relatively low in cost.
5. Butyl alcohol, C_4H_9OH. This is a medium-boiling popular latent solvent for lacquers.

Other popular ketones used in lacquers are methyl ethyl ketone and the slower-evaporating methyl isobutyl ketone. They are very strong and relatively low in cost.

Solvents that evaporate slowly are sometimes used in lacquers to prevent blushing or for brushing application. Among popular products are the lactates, Cellosolve, and carbitol.

5.9. PIGMENTS

5.9.1. General

All of the raw materials discussed thus far form portions of the vehicle. In nonpigmented clear coatings these raw materials are all that would be used. In pigmented coatings, or paints (Lerner and Salzman, 1985), it would be necessary to add a pigment or pigments to obtain the essential important properties of the paint that differentiate it from the clear coating. Paints may contain both a hiding, or obliterating, type of pigment and a nonhiding or, as it is sometimes known, an extender type of pigment.

One of the most important properties of pigments is to obliterate the surface being painted. This property is often known as hiding power, coverage, or opacity. The hiding power improves with increasing refractive index. We frequently hear such terms as "one-coat hiding power." This simply means that one coat of paint, normally applied, will completely cover the substrate or surface that is being painted. Sometimes, however, especially if a radical change in color is made, two or even three coats of paint may be required to do so, especially if the paint lacks good hiding power.

A type of classification for pigments is the Color Index established under the joint partnership of the American Association of Textile Chemist and Colorist (AATC) in the United States and the Society of Dyes and Colorist in the United Kingdom. For example:

Titanium Dioxide, Rutile. TiO_2. Pigment. White 6 (77891)

C.I. Pigment	White	6	77891
(general category)	(hue)	(consecutive number)	(chemical class)

The *color matching functions* refer to relative amounts of three additive primaries required to match each wavelength of light. The term is generally used to refer to the CIE Standard Observer color matching functions designated x + y + z. A colorimeter which can measure tristimulus values is used to measure color and differences between color.

Another important reason for using pigments is their decorative effect. This means giving the desired color to the surface being painted. Usually when paint is applied, great care is taken about the color scheme so as to make the surface as attractive as possible. Pigments are also used because they protect the surface being painted. Everyone will recognize red lead as a pigment used to protect steel from rusting. Not so well known are zinc chromate, zinc dust, and lead suboxide.

Still other pigments are used to give a paint special properties. For example, cuprous oxide and tributyl tin oxide are used in ship-bottom paints to kill barnacles, and antimony oxide is used to give fire retardance to paint. Pigments may also give the desired degree of gloss in a paint. Everything else being equal, the higher the pigmentation, the lower is the gloss. In addition, pigments are used to give other desirable properties. Thus, they can be employed to give a coating the desired viscosity, to control the degree of flow or leveling, to improve brushability by enabling the use of additional easy-brushing solvent, and to give very specific properties such as fire retardance, fluorescence and phosphorescence, and electrical conductance or insulation.

5.9.2. White Hiding Pigments

White is important not only as a color in its own right, but also because it forms the basis for a great many shades and tints in which it constitutes a large or small percentage of the color. The number of important white pigments being used by the paint industry has been dwindling. Thus, pigments such as lithopone, basic lead, sulfate, titanium-barium pigment, titanium-calcium pigment, zinc sulfide, and many leaded zinc oxides have practically disappeared. Of the white pigments now being used, the most important by far is titanium dioxide (Martens, 1974).

It is important to understand that lead carbonate and other lead pigments not only are useful pigments because of their colors and whitening/hiding properties, but also are effective mildewcides. Incorporating lead pigments into a paint formulation usually ensures against the troublesome growth of microorganisms.

- Titanium dioxide, TiO_2. This pigment comes in two crystalline forms (Weismantel, 1981). The older anatase form has about 75% of the opacity, or hiding power, of the present rutile form. Both forms are excellent for interior and exterior use. Titanium dioxide is used in both trade sales and chemical coatings. Very little anatase is not being used except in some specialty coatings. The rutile comes in types designed for use in enamels and flats, for solvent- and water-based coatings. Normally 2 to 3 lb/gal (240 to 359 kg/m^3) of rutile titanium dioxide will give adequate coverage in most formulations. Anatase is less chalk-resistant.

• Zinc oxide, ZnO. Despite its rather poor hiding power (only about 15% of that of TiO_2), zinc oxide still maintains its importance in the coatings industry. This is the result of unusually good properties which more than offset the relatively high cost of the pigment per unit of hiding power. Zinc oxide's most important use is in exterior finishes; it tends to reduce chalking and the growth of mildew in house paints. In enamels it tends to improve the color retention of the film on aging. Zinc oxide also is sometimes used to improve the hardness of a film.

• Extender pigments. These pigments, though they have practically no hiding power, are used in large quantities with both white and colored hiding-power pigments. An important property of some extender pigments is to lower the raw-material cost (RMC) of the paint. Most of these pigments are so-called nonhiding pigments such as whiting, talc, and clay. If prime, or hiding, pigments had to be used to lower the gloss so as to obtain a flat finish, the RMC would be extremely high in most instances. Instead, extender pigments are used to accomplish this task at a small fraction of the cost.

• Whiting (calcium carbonate). This is probably the most important extender pigment in use. It comes in a variety of particle sizes and surface treatments, and it can be dry-ground, water-ground, or chemically precipitated. Normally quite low in cost, it can be used to control such properties as sheen, nonpenetration, degree of flow, degree of flatting, tint retention, and RMC.

• Talc (magnesium silicate). Though used widely as an extender in interior finishes, this pigment finds its greatest use in exterior solvent-based coatings, especially house paints. This is related largely to a combination of durability and low cost. Most grades of talc tend to have good nonsettling properties and give a rather low sheen.

• China clay (aluminum silicate). This extender, though used to some degree in solvent-based coatings, finds its greatest use in water-based paints. It disperses readily with high-speed dispersers, in the normal method of manufacturing latex finishes, and does not impair the flow characteristics of the paint. Some grades will improve the dry hiding power of water-thinnable or solvent-based paint.

• Other extenders. Among extenders that are sometimes used are diatomaceous silica, used to reduce sheen and gloss; regular silica, which gives a rough surface; barites, used to minimize the effect of the extender; and mica, which because of its platelike structure is used to prevent the bleeding of colors.

5.9.3. Black Pigments

Next to whites, blacks are probably the most important colors used in the coatings industry. The reason for their wide use is twofold. First, black is a very popular color and is often used in industrial finishes, trim paints, toy enamels, and quick-drying enamels, among others. Second, it is also very popular as a tinting color, particularly for all shades of gray, which are made by adding black to white.

The two most popular blacks in use consist of finely divided forms of carbon; they are known as carbon black and lampblack. Carbon black, the most widely used of the blacks, is sometimes called furnace black. It is made by the incomplete combustion of oil injected into the combustion zone of a furnace. Lampblack, or channel black, is made by the impingement of gas on the channel irons of burner houses. Both types of black come in a variety of pigment sizes and jetness. Practically all black colors are made with carbon black. They have tremendous opacity; only 2 to 4 oz/gal (15 to 30 kg/m^3) of paint is necessary in most instances for proper coverage. They also have excellent durability, resistance to all types of chemicals, and lightfastness. Even the most expensive, darkest jet blacks are inexpensive to use because only a small amount is needed.

Whereas carbon black is used principally as a straight color, lampblack, a course furnace black made from oil, is used mainly as a tinting color for grays, olive shades, and so forth. Largely because of its coarseness, lampblack has little tendency to separate from the TiO_2 or other pigments with which it is used and to float up to the surface, as do the carbon blacks with their much finer particle size. Floating, a partial color float to the surface of the film, and flooding, a more nearly complete and uniform color float, are, of course, undesirable, and for this reason carbon black is rarely used as a tinting color. Lampblack has very poor jetness but gives a nice bluish shade of gray. It also has excellent heat and chemical resistance.

Other blacks that are sometimes used are black iron oxide, used as a tinting black having brown tones and in primers, and mineral and thermal blacks, used as low-cost black extenders.

5.9.4. Red Pigments

In discussing white or black colors, everyone knows what colors are meant and what they look like. Other colors, however, come in different shades. Thus, there are a great variety of reds, some of which are briefly mentioned below.

• Red tone oxides. These are good representatives of a series of metallic oxides that have very important properties. Though relatively low in cost, they have such fine opacity that 2 lb/gal (240 kg/m^3) is normally adequate, and they also possess high tinting strength. In addition, they have good chemical resistance and colorfastness, and they disperse easily in both water and oil so that high-speed dispersers can be used in manufacturing paints based on iron oxide pigments. Red iron oxides give a series of rather dull colors having excellent heat resistance. These colors are used popularly in floor paints, marine paints, barn paints, and metal primers and as popular tinting colors.

- Toluidine reds. These popular, very bright azo pigments come in colors ranging from a light to a deep red. They have excellent opacity, so that 3/4 to 1 lb/gal (90 to 120 kg/m^3) of paint normally gives adequate hiding power. As they also have fine durability and lightfastness, they are used in such finishes as storefront enamels, pump enamels, automotive enamels, bulletin paints, and similar types of finishes. The toluidines tend to be somewhat soluble in aromatics, which should therefore be kept to a minimum. They are also not the best pigments for baking finishes as they sometimes bronze, or for tinting colors, as they are somewhat fugitive in very low concentrations. They also bleed.

- Para red. This azo pigment is deeper in color than toluidine and not quite as bright. It has very good coverage, about 1 lb/gal (120 kg/m^3) giving adequate coverage. Para red is not as lightfast as toluidine and tends to bleed in oil to a greater degree. Moreover, it has poor heat resistance and cannot be used in baked coatings. Its lower cost makes it attractive for bright interior finishes and some exterior finishes.

- Rubine reds. These bright reds, sometimes known as BON (β-oxynaphthoic acid) reds, are available in both resinated and nonresinated forms. They have good bleed resistance but only fair alkali resistance.

- Lithol red. This complex organic red has very good coverage, 1 lb/gal (120 kg/m^3) giving adequate coverage in most instances. It is bright and has a bluish cast. Lithol red is relatively nonbleeding in oil but tends to bleed in water, and its durability and lightfastness are only fair. Because it is relatively low in cost, it is used in such applications as toy and novelty enamels.

- Naphthol reds. These arylide pigments have excellent alkali resistance and are relatively low in cost. They bleed in organic solvents and are more useful in emulsion than in oil-based paints.

- Quinacridone reds. These pigments come in a variety of shades, ranging from light reds to deep maroons and even violets. They have good durability and lightfastness and high resistance in alkalies. They also tend to be nonbleeding and show good resistance to heat.

• Other Reds. Among other reds sometimes used are alizarin (madder lake) red for deep, transparent finishes, pyrazolone reds for high heat and alkali resistance, and a larger series of vat colors.

5.9.5. Violet Pigments

The demand for violets is small because they are expensive and often have poor opacity. However, several violets may be mentioned.

• Quinacridone violets. These pigments are durable and have good resistance to alkalies and to heat.

• Carbazole violets. These pigments have very good heat resistance and lightfastness. They also are nonbleeding, and their high tinting strength makes them useful for violet shades.

• Other violets. Violets in use include tungstate and molybdate violets for brilliant colors and violanthrone violet for high resistance and good lightfastness.

5.9.6. Blue Pigments

Blues not only are important as straight and tinting colors but also are popular for use in combination with other colors to produce different shades and colors.

• Iron blue. This popular blue, a complex iron compound also known as Prussian blue, Milori blue, and Chinese blue, is one of the most widely used blue pigments in the coatings industry. It combines low cost, good opacity, high tinting strength, good durability, and good heat resistance. However, it has very poor resistance to alkalies and cannot be used in water paints or in any paints that require alkali resistance.

• Ultramarine blue. This color, sometimes known as cobalt blue, is popularly used as a tinting color. It gives an attractive reddish cast when added to whites. Ultramarine blue has poor opacity, high heat resistance, and good alkali resistance. Although it can be used in latex paints, special grades low in water-soluble salts must be obtained. It is often used for whites to give extra opacity and make them look whiter by lending them a bluish cast.

• Phthalocyanine blue. This blue is becoming increasingly popular because of its excellent properties. It gives a bright blue color and has excellent opacity, durability, and lightfastness. In addition, it is relatively nonbleeding and gives a greenish blue shade when used as a tinting color. Its high chemical and alkali resistance makes it satisfactory for water-based coatings as well as for all types of interior and exterior finishes. The price, though high, is not so high as to prohibit the use of this blue in most finishes.

• Other blues. Sometimes used are indanthrone blue, which has a reddish cast and high resistance; and molybdate blue, which is used when a very brilliant blue is desired.

5.9.7. Yellow Pigments

• Yellow iron oxide. Although yellow iron oxide pigments give a series of rather dull colors, they have excellent properties. They are relatively easy to disperse, are nonbleeding, and have good opacity despite low cost. They also have fine heat resistance. Their chemical and alkali resistance is excellent, and thus they may be used in both water- and solvent-based paints. Excellent durability makes them useful for all types of exterior coatings. They are also popular shading colors, for when added to white they give such popular shades as ivory, cream, and buff.

• Chrome yellow. This once-popular bright yellow comes in a variety of shades, from a very light greenish yellow to dark reddish yellow. Chrome yellow paints have good opacity and are easy to disperse, but they tend to darken under sunlight. Because they are lead pigments, they are gradually being phased out of use.

• Cadmium yellow. Largely a combination of cadmium and zinc sulfides plus barites, cadmium yellow pigments are sold in a variety of shades. They have good hiding and lightfastness if used as straight colors. They also are bright and nonbleeding, bake well, and have good resistance except to acids. They are toxic, however, and are being phased out of use.

• Hansa yellow. With the elimination of chrome yellow and cadmium yellow, Hansa yellow pigments are becoming increasingly important as bright yellows. They come in several shades, from a light to a reddish yellow. Hansa yellow pigments have excellent lightfastness when used straight but are somewhat deficient in tints. Although their hiding power is only fair, they have excellent tinting strength, which makes them good tinting pigments, especially in water-based coatings, for which they have excellent alkali resistance. However, they bleed in solvents and do not bake well.

• Benzidine yellow. Along with Hansa yellow pigments, benzidine yellow pigments are finding increasing application as the use of lead-containing yellows becomes illegal. They are stronger than Hansa yellows and have good alkali and heat resistance. Their resistance to bleeding is also better. Their lightfastness is poorer, however, and thus they are unsatisfactory for exterior coatings.

• Other yellows. Among other yellows in use are nickel yellows, which have good resistance and make greenish yellow colors; monarch gold and yellow lakes, which are used for transparent metallic gold colors; and vat yellow, which has extremely good lightfastness and good resistance to heat and to bleeding.

5.9.8. Orange Pigments

- Molybdate orange. This very popular bright orange, with its reasonable cost, hiding power, brightness, and colorfastness, is being phased out because of its lead content.

- Chrome orange. This lead pigment is also being phased out. In money value it is inferior to molybdate orange.

- Benzidine orange. Benzidine orange pigments are bright and have good alkali resistance and high hiding power. They also have good heat resistance and resistance to bleeding and can be used in both water- and solvent-based paints. Because their lightfastness is only fair, they are not the best pigments for outside use.

- Dinitroaniline orange. This bright orange has very good lightfastness and good alkali resistance, making it a good exterior pigment for aqueous systems. It tends to bleed in paint solvents.

- Other oranges. Among oranges sometimes used are orthonitroaniline orange, which is lower in cost but inferior in most properties to dinitroaniline orange; transparent orange lakes, which are used for brilliant transparents and metallics; and vat orange, which is high in overall properties but also high in price.

5.9.9. Green Pigments

- Chrome green. Until recently the most popular of all greens for its brightness, durability, hiding power, and low cost, chrome green is gradually being replaced by other greens because of its lead content. It comes in a combination of shades from a yellowish light green to a bluish dark green. Chrome green has poor alkali resistance and cannot be used in latex paints.

- Phthalocyanine green. This is fast becoming the most important green pigment of the coatings industry. A complex copper compound of bluish green cast, it has excellent opacity, chemical resistance, and lightfastness. It also is nonbleeding and can be used in both solvent- and water-based coatings, both as a straight color and for tints. It is rather expensive.

- Chromium oxide green. This rather dull green pigment has excellent durability and resistance characteristics and can be used for both water and oil, in both interior and exterior paints. It has moderate hiding power and is easy to emulsify. Its high infrared reflection makes it an important green in camouflage paints.

- Pigment green B. This pigment is used mainly in water-based paints because of its excellent alkali resistance, but it can also be used in solvent-based paints. Its lightfastness is only fair, so that it is not satisfactory for exterior paint use. It does not give clean shade of green but is satisfactory in most instances.

5.9.10. Brown Pigments

- Brown iron oxide. Most of the browns used by the coatings industry are iron oxide colors. Essentially combinations of red and black iron oxides, they have very good coverage, excellent durability, good light resistance, and good resistance to alkalies. They are suitable for both water- and solvent-based paints and for both interior and exterior finishes.

- Van Dyke brown. This essentially organic brown gives a purplish brown color. Lightfast and nonbleeding, it is used largely in glazes and stains.

5.9.11. Metallic Pigments

- Aluminum. By far the most important of the metallic pigments, aluminum is platelike in structure and silvery in color and comes in a variety of meshes and in leafing and nonleafing grades. The coarser grades are more durable and brighter, and the finer grades are more chromelike in appearance. Aluminum powder has high opacity, excellent durability, and high heat resistance. The nonleafing grade is used when a metallic luster is wanted by itself or with other pigments. The leafing grade is used when a silvery color is desired. This grade is highly reflective, making it ideal for storage tanks, as it tends to keep the contents cooler. It is also very popular for structural steel, automobiles, radiators, and other products with metallic surfaces. The nonleafing grade is used for so-called hammertone finishes.

- Bronze. Gold-colored bronze powders consist mainly of mixtures of copper, zinc, antimony, and tin. They come in a variety of colors, from a bright yellowing gold to a dark brown antique type of gold. Bronze powders are used mainly for decorative purposes. Their opacity is poorer and their price higher than those of aluminum.

- Zinc. Zinc dust is assuming increasing importance as a protective pigment for metal, especially as lead is gradually being eliminated. It is used in primers for the prevention of corrosion on steel when employed as the sole pigment in so-called zinc-rich paints, and it is used in combination with zinc oxide in zinc dust–zinc oxide primers. Zinc dust–zinc oxide paints are satisfactory for both regular and galvanized iron surfaces. Zinc-rich paints are used with both inorganic vehicles such as sodium silicate and organic vehicles such as epoxies and chlorinated rubber. Both types have excellent rust inhibition and show good resistance to weather.

- Lead. Lead flake has found useful application in exterior primers, in which it exhibits excellent durability and rust inhibition.

5.9.12. Special-Purpose Pigments

Some pigments are used not for their color or opacity but for the special properties (Weismantel, 1980) that they give a coating. Two of these have been mentioned in the metallic-pigment category: zinc dust and lead flake, which are used primarily for rust inhibition. Others are mentioned below.

- Red lead. This bright orange pigment is used almost exclusively for corrosion-inhibiting metal primers, especially on large structures such as bridges, steel tanks, and structural steel. Because it has poor opacity, it is sometimes combined with red iron oxide for improved opacity and low cost. With restrictions on the use of lead, its employment is being phased out.

- Basic lead silicochromate. This also is a bright orange pigment that is used primarily as a rust-inhibiting pigment for steel structures. Because of its low opacity, it can be combined with other pigments to give topcoats of different colors that still have rust-inhibiting properties.

- Lead silicate. This pigment is used mainly in water-based primers for wood, in which it reacts with tannates and prevents them from coming through and discoloring succeeding coats of paint. It may be eliminated from home use because of restrictions on the use of lead.

- Zinc yellow. This hydrated double salt of zinc and potassium chromate is used principally in corrosion-inhibiting metal primers. It is becoming one of the few permissible pigments to use on steel connected with houses or apartments. It is greenish yellow in color and has poor opacity.

- Basic zinc chromate. This pigment has properties somewhat similar to those of zinc yellow. It is used in metal pretreatments, especially in the well-known "wash primer" government specification for conditioning metals, in which capacity it promotes adhesion and corrosion resistance for steel and aluminum.

- Cuprous oxide. This red pigment is used almost exclusively in autifouling ship-bottom paints to kill barnacles that would normally attach themselves to a ship below the waterline.

• Antimony oxide. This white pigment is used almost entirely in fire-retardant paints, in which it has been very effective, especially in combination with whiting and chlorinated paraffin.

A list of materials and suppliers is provided in Table 5.1 in the Appendix.

6

Deformulation of Paint

6.1. INTRODUCTION

The analytical approach to deformulation of paint and coatings depends largely on the form in which the specimen occurs. Paint and coatings are found in the solid dry films and liquid forms. Components in a liquid paint specimen are separated prior to examination using centrifugation as shown in Fig. 1.2; and components comprising a solid paint film are not so easily separated. So, a different analytical approach is taken for solid specimens including surface analysis, and methods to separate the pigments/fillers from the vehicle followed by analysis of each. Regardless of the form in which a paint specimen is found, a method can be found to deformulate it. An extensive review of analytical methods and equipment is presented in Chapters 1–3, and the reader should refer to these chapters for detailed information when an analytical method or instrument is mentioned.

6.2. DEFORMULATION OF SOLID PAINT SPECIMENS

Sources of solid specimens of paint are shown in Fig. 6.1. These include paint chips from automobiles and houses. Although a liquid paint specimen is far preferable, a solid paint specimen can be analyzed using the basic scheme for analysis in Fig. 6.2. Paint and coatings are pigmented/filled up to about 35% by volume of the dry film. A liquid sample is always preferable because individual components can be separated, whereas solid specimens require significant sample preparation before individual components can be separated.

Example 1. A paint chip was taken from the exterior surface of an 80-year-old residential structure; a SEM micrograph of the cross-sectional view is presented in Fig. 6.3. There are six different layers of paint, the first layer being adjacent to the wood substrate. The specimen in Fig. 6.3 was broken in liquid nitrogen and placed in acrylic resin followed by polishing to prepare a mounted specimen. The mounted specimen was coated with carbon, and then palladium before being placed

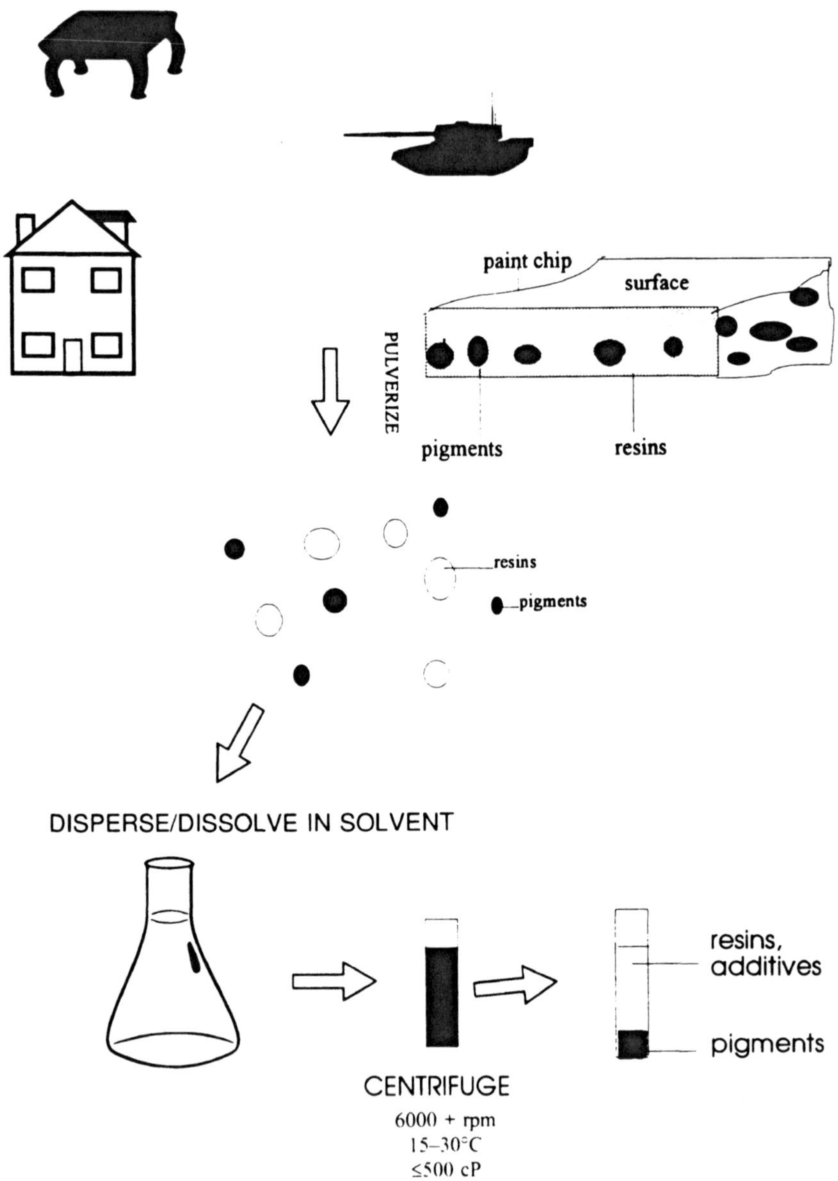

Figure 6.1. Sources of paint and preparation of solid paint specimens for deformulation.

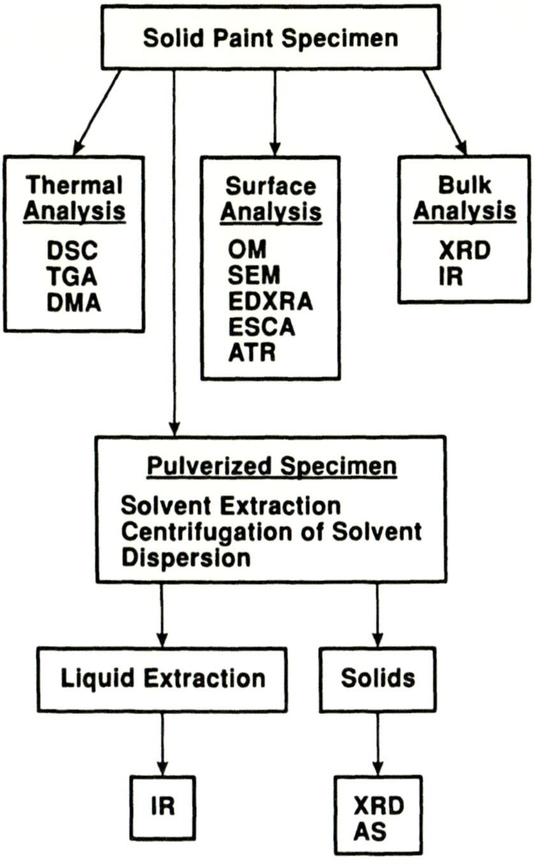

Figure 6.2. Scheme for deformulation of a solid paint specimen.

in the SEM microscope. Each layer was investigated by EDXRA. A separate uncoated mounted specimen was analyzed by microscopic IR spectroscopy.

The results of the investigation are:

- Layer 1—Basic lead carbonate, calcium carbonate, and zinc oxide in a vegetable oil matrix.
- Layer 2—Basic lead carbonate and zinc oxide in an alkyd resin.
- Layer 3—Lead oxide in an alkyd resin.
- Layer 4—Titanium dioxide and zinc oxide in an alkyd resin. This layer shows severe internal cracking which must have contributed to its early failure.

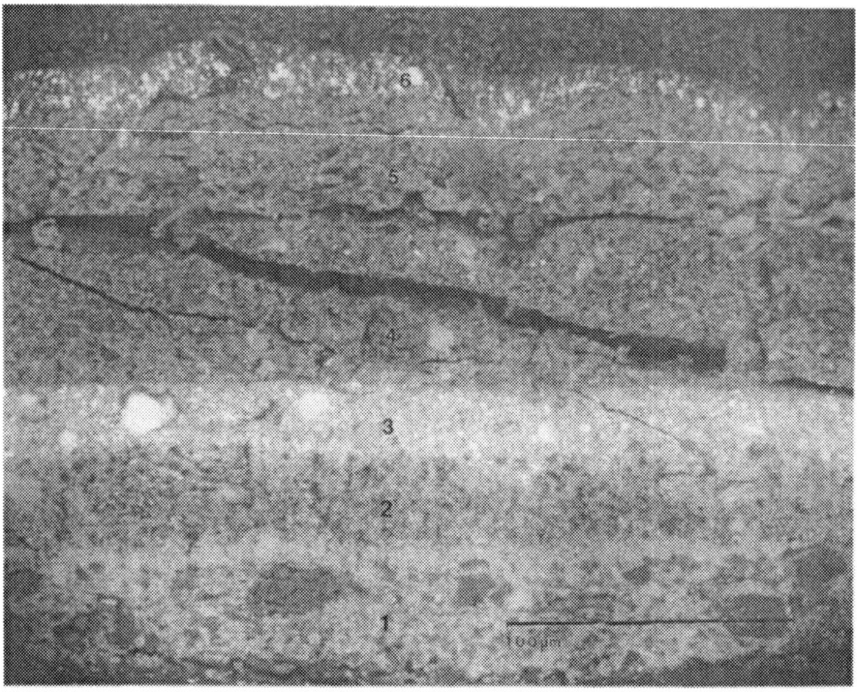

Figure 6.3. SEM micrograph (cross section) of a paint chip.

- Layer 5—Titanium dioxide and calcium carbonate in an acrylic resin.
- Layer 6—Titanium dioxide, magnesium oxide, and zinc oxide in an acrylic resin.

Further analysis by ESCA confirmed the presence of the basic lead carbonate pigments. The dimensions of pigment particles are obvious using the bar scale. TGA will determine the total amount of pigment in the chip, but a microtomed separation of each layer will determine the percent pigment/filler weight in each layer.

Using the above method one paint can be compared with another. For example, a specific layer of paint on a house can be matched with a manufactured source of paint; or different paints can be assigned to different automobiles involved in an accident.

Pigments and fillers are separated from the vehicle by refluxing in solvents over a period of hours. It is best to first pulverize the paint chip in a device or with a simple mortar/pestle. While in hot refluxing solvent (see Fig. 6.5), the vehicle will swell (not dissolve), disintegrate into gel particles, and release pigments. The dispersion is centrifuged and treated like a liquid specimen. Also, ultrasonic probes will disintegrate a paint chip. The gelled particles of vehicle are analyzed by IR and pigments by XRD.

In a nondestructive analysis, a paint chip specimen can be placed in an SEM equipped with an EDXRA. Immediately, the number of paint layers and the shape of pigments can be observed by SEM, and the elemental analysis of pigments can be accomplished by EDXRA. If the resin matrix in the coating contains elements other than carbon and hydrogen, then some information about the resin can be

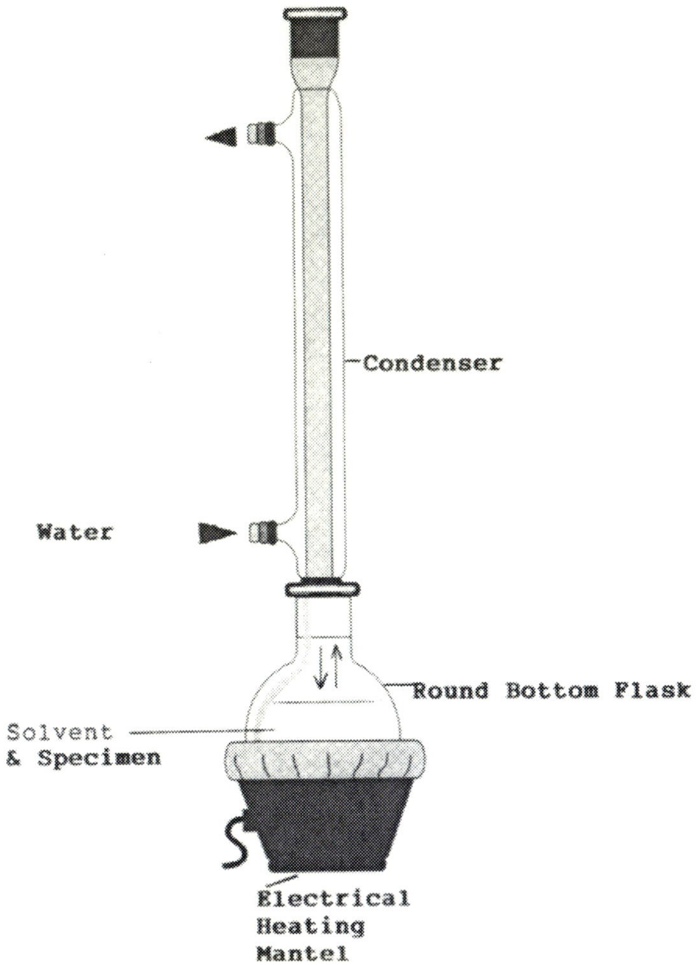

Figure 6.4. Solvent refluxing apparatus for separating vehicle from pigments in paint chips.

generated. Resins do not have shapes as pigments or very distinctive spectra in the EDXRA. A microscopic IR will provide a spectrum of the resin matrix without interference from pigment particles. Another instrument for nondestructive examination (and under magnification) of a paint chip is ESCA, which will generate the composition of the resin matrix and pigments.

Most dried or cured paint is thermoset or cross-linked, which means that it does not melt with heat and is insoluble in solvents. Some thermoplastic acrylic paints and coatings can be dissolved in solvents. A method for disintegrating the paint chip is refluxing the specimen in hot solvent as shown in Fig. 6.4. The vehicle swells and parts from the pigment during refluxing, and the suspension is separated by centrifugation. Because resins decompose according to composition in a TGA instrument, thermal analysis of dried paint specimens is valuable. The glass transition temperature is distinctive for epoxy coatings and can be determined by DCS analysis. Other thermal analysis can be used if enough of the specimen is available.

6.3. DEFORMULATION OF LIQUID PAINT SPECIMENS

Figure 6.5 shows a scheme for preparation of a liquid specimen. The components in a liquid specimen are ready to be separated by centrifugation with an adjustment in viscosity. A scheme for deformulation of a liquid specimen is shown in Fig. 6.6. Every material in a liquid paint formulation can be isolated and identified using this method. Because laboratories do not possess the same equipment, substitution of equipment and modification of the methods are permissible as far as comparable results are obtained.

6.3.1. Measurements and Preparation of Liquid Paint Specimen

A liquid paint is viscous and components do not separate without centrifugation or filtration. Referring to Fig. 6.5, separation of a liquid specimen is accomplished using centrifugation. The viscosity of the specimen can be measured using a viscometer (Chapter 3) which corresponds to the percent solids or concentration of components in the formulation. As solvent is added to the formulation, the viscosity decreases. Referring to Fig. 6.5, the liquid sample is centrifuged (above 6000 rpm at 15–30°C) to separate the heaviest components such as pigments from the paint. A polypropylene tube is preferred as both materials are insoluble and unbreakable. Solvents, resins, and soluble additives will reside in the upper portion of the centrifuge tube. Typically, the solids portion of a paint is about 10–25% of the total liquid volume. The solids will include colored pigments as well as fillers such as silica. The lightest or upper portion of the centrifuge tube will be colorless unless a soluble organic dye is part of the formulation. The components should be separated individually as follows:

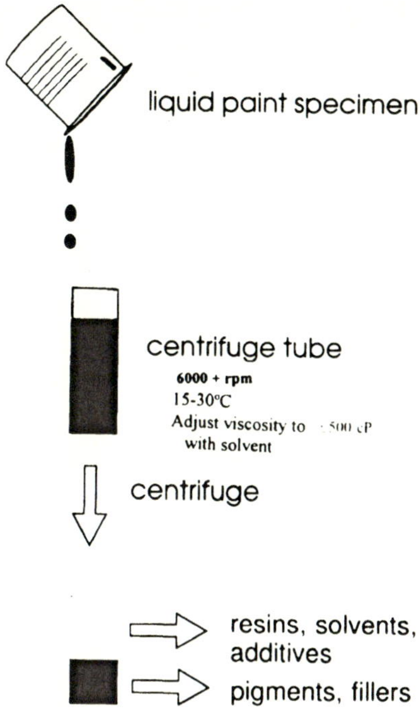

Figure 6.5. Scheme for preparation of liquid paint specimen for deformulation.

1. Remove the individual liquid layers from the upper part of the tube using a syringe.
2. Remove individual solid layers using a small spatula.
3. Weigh each component using an analytical balance (± 0.01 g).

6.3.2. Separated Liquid Fraction of Specimen

The liquid fraction of the specimen will contain polymers, resins, solvents, water, and additives. The distillation method shown in Fig. 6.7 is recommended for separation of solvents and other volatile materials from resins and polymers. The volatile liquids including water will distill according to vapor pressure (boiling temperature) and each component can be collected and weighed. After separation, each individual component can be analyzed by IR and NMR. This is an accurate and economical method of qualitatively and quantitatively characterizing the solvents and other liquid materials in the formulation.

High-vapor-pressure materials such as solvents can be analyzed with a calibrated GC instrument, but HPLC can separate and quantify all but the high-molecular-weight materials.

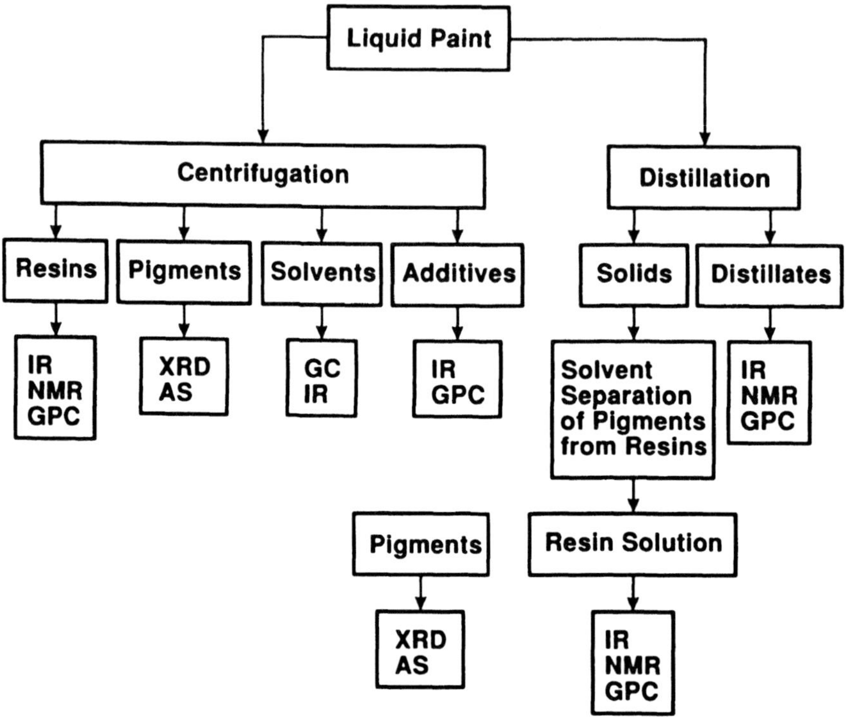

Figure 6.6. Scheme for deformulation of liquid paint specimen.

Low-vapor-pressure materials such as resins are best analyzed with GPC to determine the range of molecular weights which indicates the number of species. Usually, only one or two resins will be present, but GPC is an excellent method to scan an unknown sample. Another part of the resin fraction can be analyzed by IR and NMR.

A calibrated HPLC will separate most organic liquid components, but water is run on a column designed for aqueous systems.

6.3.3. Separated Solid Fraction of Specimen

The solid fraction of the specimen will contain solid pigments and fillers and these will separate according to density, heaviest on the bottom of the centrifuge tube. A well-centrifuged specimen will form consecutive individual layers of pigments and fillers in the tube. Of course, these solid materials will contain some amounts of liquid (pasty texture) which must be removed for accurate analysis. Each layer of solid can be removed, weighed, and washed with solvent followed by oven drying to remove the solvent and render a pure material.

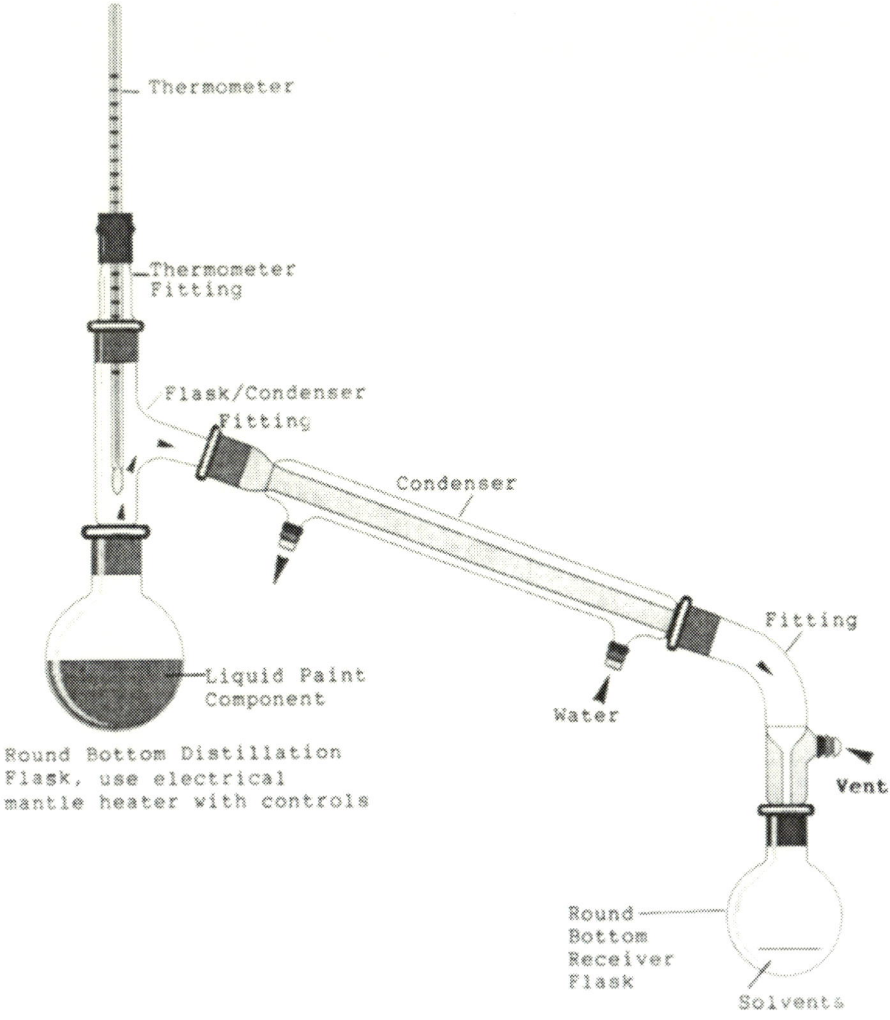

Figure 6.7. Distillation apparatus for separation of solvents from liquid paint specimens.

Another method of washing the solids is to remove the liquid fraction, then add new solvent and recentrifuge. The solids on the bottom of the tube will be reseparated, free from resin and other contaminants. After decanting the solvent, the solids are ready to be removed for oven drying. The process can be repeated to further purify the solids.

Each separated and dry solid material can be examined in this scheme, but it is suggested that a preliminary EDXRA scan be performed to quickly determine the major elements present. Some pigments and filler possess an IR spectrum, but more definitive methods include XRD and AS.

Example 2. A liquid sample of a light brown water-based paint was centrifuged at 10,000 rpm at 15°C for 4 hours in six tared 60-cm^3 polypropylene tubes. The tubes were gently removed from the centrifuge, and each layer was measured with a milliliter scale and marked on the tube. Each layer was removed and the tubes were reweighed to provide a gram weight for each layer. Water was on the surface, followed by resin, then pigments.

The pigments were titanium dioxide (white) and iron oxide (red) as confirmed by XRD spectroscopy. The resin was polymethyl methacrylate with an ester plasticizer as confirmed by IR spectroscopy.

Also, a nonionic surfactant was present in the aqueous phase. A fresh 500 cm^3 of paint was distilled in the apparatus shown in Fig. 6.7. The water was distilled, collected, and a surfactant (emulsifier) was left in the distillation flask after the water was distilled. The surfactant was weighed, and analyzed by IR and was identified as nonylphenolethylene oxide.

The formulation is by percent weight:

Titanium dioxide	13.3%
Iron oxide	6.5%
Polymethyl methacrylate	25.5%
Nonionic surfactant	2.1%
Water	52.6%

Most of the analytical methods discussed above are described in the American Standard Testing and Methods (ASTM) publications.

Examples of paint and coating formulations are shown in Tables 4.1–4.43. These are selected popular formulations, as there are literally thousands of such formulations. However, with the proper tools, an investigator can deformulate any composition.

6.4. REFORMULATION

After all components have been analyzed, create a table and list each material with percent by weight. An additional column of percent by volume is sometimes useful, which requires that the densities of all materials be known. When finished, this table is the formulation of the original mixture. To confirm the results, acquire materials from the included materials and suppliers information (see Table 5.1) to reformulate the original recipe from the generated table and compare the properties of both formulations.

7

Plastics Formulations

7.1. GENERAL

Formulation of plastics materials consists of the polymeric or resin material and additives for affecting specific functions such as foaming. Solvents can be added for cast molding (Rubin, 1974) of parts, but cannot be used for injection molding because heating the solvent would cause explosive pressures. Also, nonsolvent or thermally melted resins can be poured into a mold and are said to be cast molded.

Plastics formulations are usually simple compared to paint, adhesives, and inks. The resins for molding, for example, are usually preformulated and sold to the molder. Injected, extruded, or blow molded parts contain small amounts of pigment or other fillers, no solvents, and small amounts of additives. The additives are introduced for specific purposes as plasticizer for PVC, gas releasing/foaming agents for low-density parts, and others. The basic material in a molded part is the resin. The resin is usually thermoplastic and to a lesser degree, thermoset. Detailed material information is presented in Chapter 8.

Formulations for molded plastic parts are simpler than paint, adhesive, or ink dispersions. The resin/polymer is usually over 95% of the formulation. A typical plastics formulation consists primarily of some or all of the following components:

1. Resin/polymer
2. Pigment/dye
3. Flow agent
4. Mold release agent
5. Plasticizer
6. Antioxidant
7. UV stabilizer

7.2. THERMOPLASTICS

7.2.1. Homopolymers

Polymers and resins that flow when heated and do not chemically react or cross-link are called thermoplastics materials. Examples of thermoplastics are polyethylene (PE) and nylon. After injection molding parts from these materials, they can be reheated above their melting temperatures, and they will melt. An example of a homopolymer is PE.

7.2.2. Copolymers

A polymer polymerized from two or more monomers is called a copolymer. An example of a copolymer is poly(styrene-co-acrylonitrile) (SAN).

7.2.3. Alloys

Alloys of thermoplastic materials (Uihlein, 1992) are employed for developing useful properties of two or more polymers. Examples of alloys are:

1. PPO/PS, Noryl by GE Plastics
2. Nylon/ABS, Triax 1000 by Monsanto
3. PPO/nylon, Noryl GTX by GE Plastics
4. PET/PBT, Valox by GE Plastics
5. PEEK/PES, Victrex by ICI
6. Nylon/PE, Selar RB by DuPont

7.3. THERMOSETS

Polymers and resins that chemically react or cured form parts that will not remelt. The molecular chains are attached to each other and will not reflow. An example of a thermoset material is an amine cured epoxy.

7.4. FIBERS

Synthetic polymeric fibers are usually spin-formed from molten materials and often undergo posttreatment to achieve optimum results (Joseph, 1986). The fibers are usually thermoplastic such as polyester, but can be thermoplastic. Fibers are drawn or oriented in one direction, along the axis of the fiber, and synthetic fibers are usually semicrystalline. Fibers vary in diameter from a few micrometers to millimeters.

- Cellulosic fibers (e.g., rayon, acetate, and triacetate)
- Polyamide fibers (e.g., nylon and aramid)
- Polyester fibers

- Acrylic fibers
- Olefin fibers
- Elastomeric fibers (e.g., spandex and rubber)
- Noncellulosic fibers (e.g., saran, vinal, novoloid, azlon, nytril)
- Miscellaneous fibers (e.g., Teflon, polybenzimidazole, polycarbonate, polyurea, polyphenylene sulfide).

7.5. FILMS

Films, or sheets, are usually heat-extruded thermoplastic polymers, e.g., polypropylene and polyethylene terephthalate or cast molded (e.g., acrylic). They may contain additives depending on the end use. For example, polyethylene plastic bags may contain an antistatic agent to prevent buildup of static electricity. Film is usually oriented in two directions (biaxially), which means that is pulled or stretched in one direction. Film can also be blow molded.

7.6. FOAMS

A foaming agent can be added to thermoplastic resin and injection molded to form a part with gas microbubbles. The entrained gas bubbles create a less dense part and use less resin. An example of a foamed plastic product is polyurethane foam.

7.7. GELS

A large amount of plasticizer mixed with a polymer or resin will yield a soft or semi-solid. An example of a gel is a plastisol, dioctyl phthalate in polyvinyl chloride. Plastisols are used for beverage bottle cap seals.

Two-part urethanes can be plasticized with dioctyl phthalate to provide a very soft and gelatinous filling material for electrical cables to eliminate water.

7.8. ELASTOMERS, RUBBERS, AND SEALANTS

Rubber is compounded by incorporating a selection of additives into a rubber material followed by vulcanization. Styrene-butadiene rubber (SBR) and other synthetic rubbers produced by emulsion polymerization are in the form of a latex. The rubber particles are coagulated from the latex and dried. They may be oil-extended by diluting with compatible oils which plasticize and soften the rubber. Vulcanization consists of heating the mixture with sulfur, which cures or cross-links the rubber chains to develop an extensible material with physical return.

A thermoplastic elastomer is a material that combines the processibility of a thermoplastic with the functional performance of a conventional thermoset rubber.

The major advantage of thermoplastic elastomers is the wide range of properties and ease of fabrication. Elastomeric alloys exhibit a broad range of performance.

There are many types of materials that exhibit properties useful for elastomers, rubbers, and sealants. Because there is overlap between adhesives and elastomers, formulations for elastomers, rubbers, and sealants are discussed in Chapter 10.

Tables 7.1–7.8 contain formulations for plastics and other materials.

8

Plastics Materials

8.1. GENERAL

Plastics consist of polymers and sometimes resins. The polymers are usually thermoplastic and the resins can be thermoplastic or thermoset. Major categories of polymers and resins are discussed below.

8.1.1. Carbon Polymers

Carbon occurs in several allotropic forms or isomers with different bonds between the carbon atoms. In diamond all atoms are equidistant from each other and bonded together in the form of a tetrahedron (Elias, 1977).

Coal is a fossilized vegetable product containing mostly C, H, O, and N.

Carbon black is formed from the burning of gaseous or liquid hydrocarbons under conditions of restricted air access. Carbon black has a microporosity.

Bitumen is a naturally occurring black material that is also obtained in mineral-oil refining. It consists of high-molecular-weight hydrocarbons dispersed in oillike material.

Asphalt is a brown or pitch-black, naturally occurring or artificially produced mixture of bitumen with minerals.

Graphite is moderately stable to oxidation and this property yields high-temperature stable fibers. Graphite fibers are crystalline and carbon fibers are not, although they are similar in appearance. Both fibers are usually made from polyacrylonitrile precursors which undergo an internal rearrangement at high temperatures.

Paraffin is a low-molecular-weight polyethylene and usually a by-product of petroleum refining. It is petroleum jelly or better known by the trade name Vasoline.

8.1.2. Amino Resins

Amino resins (aminoplasts) are condensation products from compounds containing –NH groups, which are joined by a Mannich reaction to a nucleophilic component via the carbonyl atom of an aldehyde or ketone. An example of an amino resin is melamine, and urea-formaldehyde (Martens, 1974; Elias, 1977) for cross-linking baking-type resins. They may be used with alkyds, epoxies, thermosetting acrylics, phenolics, and heat-reactive resins.

8.1.3. Polyacetals

Polyacetals (e.g., polyoxymethylene $-CH_2-O-$) are highly crystalline, rigid, and cold-flow resistant, solvent resistant, fatigue resistant, mechanically tough and strong, and self-lubricating (Elias, 1977). They tend to absorb less water and are not plasticized by water to the same degree as the polyamides (Fox and Peters, 1985).

- Polyoxymethylene. Hoechst-Cellanese Celcon is polymerized from trioxane and DuPont Delrin is polymerized from formaldehyde.

Major examples of polyacetals are:

- Polyacetaldehyde
- Polyhalogenoacetals
- Polyspiroacetal
- Polythioacetal
- Polyvinyl acetal
- Polyformaldehyde
- Polyparaformaldehyde
- Polyformal

8.1.4. Polyacrylics

Acrylic monomers are derived from acrylic acid $CH_2-CH-CO-OH$ where the $-OH$ group can be replaced by $-OCH_3$ and others. Acrylics have many uses including the manufacture of polymethyl methacrylate, commonly referred to as DuPont Lucite, or Rohm & Haas Plexiglas used for clear plastic sheeting and plastic parts. Also, acrylic latex paint is made from a mixture of emulsion polymerized acrylic monomers. Thermosetting baked acrylic resins are used for appliance coatings and they are cross-linked with amino or epoxy resins. Examples are:

- Polyacrylic acid
- Polyacrylic esters
- Polyacrolein
- Polyacrylamide
- Polyacrylonitrile
- Poly(α-cyanoacrylate)
- Polymethyl methacrylate
- Polymethylacrylimide

8.1.5. Polyallyls

Allyl compounds $CH_2{=}CH{-}CH_2Y$ with Y = OH, $OCHOCH_3$ can be polymerized free radically only to low degrees of polymerization. Examples are di- and triallyl ester monomers produced by the reaction of allyl alcohol with acids, acid anhydrides, or acid chlorides (Elias, 1977). Examples of this are the reaction of phthalic anhydride with allyl alcohol to diallyl phthalate, and the conversion of trichloro-s-triazine (by trimerization of ClCN) to triallyl cyanurate.

The monomers are polymerized free radically up to yields of about 25% via the vinyl group give products of 10,000–25,000 g/m. Then, the prepolymers are cross-linked or cured. Polydiethylene glycol bisallyl carbonate is used for sunglass lenses, and as molding resin for related optical articles (the transparency is similar to that of polymethyl methacrylate, but the abrasion resistance is 30–40 times greater). The cured resins have an electrical resistivity between polytetrafluoroethylene and porcelain, which makes them useful for electrical insulation.

8.1.6. Polyamides

Polyamides contain the amide group –NH–CO– and can be classified in two homologous series. In the Perlon series, monomeric and repeat units are identical, for these polyamides occur either by the polymerization of lactams (cyclic amides) or by the polycondensation of ω-amino carboxylic acids. In contrast, the polyamides in the nylon series are formed by the polycondensation of diamines and dicarboxylic acids and two monomeric units form one repeat unit (Miller *et al.*, 1985). An example of a polyamide is poly (hexamethylene adipamide) commonly known as DuPont nylon 6,6. Examples of aliphatic polyamides are:

Nylon 6 (polycaprolactam)
Nylon 6,6 (polyhexamethylene adipamide)
Nylon 6,9 (polyhexamethylene nonanediamide)
Nylon 6,10 (polyhexamethylene sebacamide)
Nylon 6,12 (polyhexamethylene dodecanediamide)
Nylon 6,T (polyhexamethylene terephthalamide)
Nylon 11 (polyundecanamide)
Nylon 12 (polyuryllactam)

• Poly(*p*-benzamide). The simplest aromatic polyamide is poly(*p*-benzamide). Polycondensation of terephthalic acid with hexamethylene diamine leads to a high-melting polyamide that can only be fiber spun from concentrated sulfuric acid because of its high melting point of 370°C.

• Polycycloamides. Alicyclic polyamides or polycycloamides result from 1,4-bis(aminomethyl)cyclohexane and aliphatic dicarboxylic acids (Elias, 1977) such as suberic acid.

• Versimides. Versimides are obtained by polycondensation of the ester group of "polymerized" vegetable oils with diamines and triamines.

• Polyamide(imide-co-amide). Poly(imide-co-amides) are easier to produce and to process than aromatic polyamides or polyimides. They are used in electrical insulation. One method of producing these polymers is to react dianhydrides with excess diamines, the prepolymer subsequently being converted with dicarbacyl chlorides (Elias, 1977).

8.1.7. Polydienes

Polydienes are produced by the polymerization of dienes such as butadiene, isoprene, and chloroprene (Elias, 1977).

• Polybutadiene. The polymer is synthesized from the monomer butadiene, $CH_2{=}CH{-}CH{=}CH_2$. Vulcanization (an ionic reaction) cross-links the polymer (BUNA S and BUNA N) through a reaction with sulfur to form a rubber.

• Polyisoprenes. Polyisoprene occurs naturally as *cis*-1,4-polyisoprene, which is commonly referred to as natural rubber, and as *trans*-1,4-polyisoprene, referred to as gutta percha and balata. Both isomers can be prepared synthetically.

• Polydimethyl butadiene. During World War I, polydimethyl butadiene (methyl rubber) was manufactured as a substitute for the natural rubber that the Allies lacked.

• Polychloroprene. The first generation of synthetic elastomers included polychloroprene, which was marketed in 1931 and developed at DuPont. Chloroprene is produced from monovinyl acetylene, butadiene, butene or butane, and HCl and CuCl.

• Polycyanoprene. Cyanoprene can be polymerized in the same way as chloroprene.

• Polypentenamer. A class of polyenes is obtained from the ring-expansion or ring-extension polymerization of cyclo-olefins.

8.1.8. Miscellaneous Polyhydrocarbons

• Polyphenylenes. Black, insoluble polyphenylenes with the monomeric unit $-C_6H_4-$ can be produced from benzene with $AlCl_3/CuCl$ as catalyst. The polymer is branched, but not cross-linked to a network.

• Poly(*p*-xylenes). These polymers are obtained from xylene; two examples are poly(*p*-xylene) and poly(*p*-monochloroxylene).

• Polyalkylidenes. These polymers are produced by a polyalkylation of alkyldienes. The catalytically effective $AlCl_3$ must be complexed and the complex must be suitably stabilized (Elias, 1977).

• Polyarylmethylenes. Prepolymers are produced by the condensation of aryl alkyl ethers or aryl alkyl halides or other aromatic, heterocyclic, or metallo-organic compounds in the presence of Friedel–Crafts catalysts. The prepolymers can be cross-linked with diepoxides or polyepoxides, or hexamethylene tetramine.

• Diels–Alder polymers. In the Diels–Alder synthesis, idineophile adds on to a diene in a reversible reaction. Commercial production starts from cyclopentadiene.

• Coumarone-indene resins. The tar fraction of petroleum (bp 150–200°C) contains 20–30% coumarone (benzofuran), significant amounts of indene and naphtha which is a cyclic-paraffin-rich fraction. The polymerization proceeds via the double bond of the five-membered ring. The naphtha is evaporated after the polymerization.

α-Pinene and β-pinene are present in turpentine oil. They can be polymerized to oligomeric resins.

8.1.9. Polyesters

• Dicarboxylic acids with diols. Polyesters are polymerized from the condensation of dicarboxylic acids with diols (Miller and Zimmerman, 1985). Poly(ethylene terephthalate) is a polyester condensed from ethylene glycol and terephthalic acid. Other examples are:

Poly(1,4-butylene terephthalate)
Poly(diallyl phthalate)
Poly(1,4-cyclohexanedimethylene terephthalate)
Poly(diallyl isophthalate)

Major examples of polyesters are:

- Acid anhydrides with diols
- Alcoholysis or transesterification (ester exchange)
- Condensation of acyl chlorides with hydroxyl groups (Schotten–Baumann reaction)
- Copolymerization of anhydrides with simple cyclic ethers
- Polymerization of lactones (ring opening), e.g., poly(ε-caprolactone)

• Polycarbonates. The simplest polyesters are the polycarbonates, being carbonic esters. Reaction of bisphenol A with diacids forms the polycarbonates of the greatest commercial interest. A major trade name is Lexan manufactured by the General Electric Co. (Fox and Peters, 1985).

• Aliphatic saturated polyesters. Examples of these polymers are poly(ethylene oxalate) [poly(ethylene glycol oxalate)], polyesters based on ethylene glycol and sebacic or adipic acid, poly(ethylene adipate), polyglycolide, and poly(ε-caprolactone).

• Unsaturated polyesters. These polymers are made by condensing maleic anhydride or phthalic anhydride with ethylene glycol or propylene glycol.

• Aromatic polyesters. Aromatic polyesters can contain either terephthalic acid or *p*-hydroxybenzoic acid as the acid component and ethylene glycol to condense poly(ethylene terephthalate) or poly(*p*-hydroxybenzoic acid) (Elias, 1977).

Poly(butylene terephthalate) is condensed from 1,4-butane diol and terephthalic acid.

• Alkyd resins (see Chapter 5). Alkyd or glyptal resins (glycerine + phthalic acid) occur through the conversion of alcohols with a functionality of three or more (glycerine, trimethylol propane, pentaerythritol, sorbitol) with bivalent acids (phthalic acid, succinic acid, maleic acid, fumaric acid, adipic acid), fatty acids (from linseed oil, soybean oil, castor oil), or anhydrides (phthalic anhydride) at temperatures between 200 and 250°C. Cross-linking occurs during autoxidation of the olefinic groups after application.

• Polyanhydrides. Polyanhydrides are produced by the self-condensation of certain aromatic dicarboxylic acids.

8.1.10. Polyethers

Polyethers have the functional unit –C–O–C– and are very useful materials (Elias, 1977). Examples include:

- Polyethylene oxide
- Polypropylene oxide
- Epoxide resins
- Polyepichlorohydrin
- Phenoxy resins
- Perfluorinated epoxides
- Poly[3,3-bis(chloromethyl)oxacyclobutane]
- Polytetrahydrofuran

- Polyphenylene oxide
- Copolyketones

8.1.11. Polyhydrazines

Polyhydrazines are produced from terephthaloyl dichloride and *p*-amino benzhydrazine.

8.1.12. Polyhalogenohydrocarbons and Fluoroplastics

This class of polymers has the functional unit $-CH_2CH_2-$, but with the H atoms replaced by halogens such as F and Cl. The class of halogenated polymers referred to as "fluoroplastics" (Fifoot, 1992) include polytetrafluoroethylene (PTFE). The resulting polymers are halogenated, which generally lowers surface energy and moisture permeation, increases chemical resistance (Lupinski, 1985), and lowers dielectric constant. Examples include:

- Polytetrafluoroethylene
- Fluorinated polyethylene and -propylene
- Chlorinated polyethylene and polyvinyl chloride
- Polyvinylidene fluoride
- Polyvinylidene chloride

8.1.13. Polyimides

Polyamides contain the group –CO–NR–CO–. The basic member of this series arises from the spontaneous polymerization of isocyanic acid H–N=C=O in benzene at 15°C.

Polyimides retain good mechanical properties up to 350°C in air and can be used for a limited time up to 425°C. Above 425°C sublimation evaporation takes over and is complete after 5 hours at 485°C. Polyimides do not deform at higher application temperatures.

- Aromatic polyimides. A high-temperature stable polyimide occurs from the reaction of pyromellitic anhydride with aromatic diamines such as *p,p'*-diaminodiphenyl ether.

- Poly(imide-co-amides). Poly(imide-co-amides) are easier to produce and to process than aromatic polyamides or polyimides. They are used particularly in electrical insulation.

- Poly(imide-co-esters). These copolymers are synthesized in the same way as poly(imide-co-amides), except in this case the precursor is a dianhydride with aromatic ester bonds, which is obtained by conversion of trimellitic acid anhydride with phenol esters.

• Poly(imide-co-amines). Thermosetting polymers are produced by the addition of aromatic amines to the double bonds of bismaleimides.

8.1.14. Polyimines

The polymerization products of ethylene imine are known as polyimines. The ring-opening polymerization of ethylene imine can be initiated by acids HA or alkylating agents RX; e.g., unbranched poly(ethylene imines) can be produced by the isomerization polymerization of unsubstituted 2-oxazolines.

8.1.15. Polyolefins

Polyolefins include polymers synthesized from monomers containing the olefin group –CH=CH–, and are very diversified. Some examples are:

- Polyethylene
- Polypropylene
- Poly(butene-1)
- Poly(4-methyl pentene-1)
- Polyisobutylene
- Polystyrene
- Polyvinyl pyridine
- Ionomers

An ionomer is a polyethylene molecule with ionic groups, cations and anions, positioned on the chain. The cations serve to provide interchain bonding. The primary commercial product from ionomer is DuPont Surlyn.

Ionomers are tough, durable, transparent thermoplastics widely used as films, molded products, foams, etc., for a wide range of consumer products.

8.1.16. Polysulfides

The simplest chain structure $-(CH_2-S-)_n$ occurs through the polymerization (Elias, 1977) of thioformaldehyde, CH_2S, or its cyclic trimer (trithiane). Aliphatic polysulfides with two or more carbon atoms per monomeric unit are available through the polymerization of cyclic sulfides. Commercial grades of polysulfides are synthesized by using dichloroethylene (Thiokol Chemical Corp., sulfur grade 4), bis(2-chloroethyl)-formaolin (Thiokol FA, sulfur grade 2), or a mixture of these two compounds (Thiokol ST, sulfur grade 2.2) as the dihalogen compounds.

Aromatic polysulfides include poly(phenylene sulfide) and poly(thio-1,4-phenylene).

8.1.17. Polysulfones

Polysulfone is a thermoplastic copolymer of the sodium salt of bisphenol A and *p,p'*-dichlorodiphenyl sulfone. Polysulfones can be produced by Friedel–Crafts-type reactions.

Polythiocarbonyl fluoride, thiocarbonyl fluoride, or difluorothioformaldehyde, CF_2S, can be polymerized by initiators such as amines, phosphines, tetraalkyl titanates, or dimethyl foramide.

8.1.18. Polyureas

Polyureas have the repeat unit $(-R-NH-CO-NH-)_n$. Conversion of various diamines with urea yields predominately amorphous copolymers, which can be processed by injection molding, extrusion, blowing, or fluidized-bed sintering.

8.1.19. Polyazoles

Polyazoles are polymers with five-membered rings in the main chain, the rings containing at least one tertiary nitrogen atom. Examples of polyazoles include:

- Polybenzimidazoles
- Polyterephthaloyl oxamidrazone
- Polytriazoles and polyoxadiazoles
- Polyhydantoins
- Polyparabanic acids

8.1.20. Polyurethanes

Polyurethanes possess the characteristic group (–NHCOO–) within the repeat unit of the polymer (Elias, 1977). They are manufactured by the conversion of diisocyanates (triisocyanates) with diol compounds.

The C=N double bond of the isocyanate group can either polymerize, or oligomerize at higher temperature, or add functional groups containing an active hydrogen atom (water, alcohols, phenols, thiols, amines, amides, and carboxylic acids). A typical polyurethane can be synthesized from toluene diisocyanate and 1,4-butanediol.

Polyurethanes are used for fibers, films, paints, lacquers, adhesives, foams, and elastomers.

Allophanates are formed by addition of an excess of isocyanate groups to alcohols. Trimerization of isocyanate produces an isocyanurate.

Biurets are prepared by addition of an excess of isocyanate groups to amines.

Polyureas are prepared from reactions of diamines and diisocyanates.

Polythiocarbamates are prepared by the addition of dimercaptans to diisocyanates.

Polyureylenes are prepared from desiccant addition to dihydrazides.

Polyimine-oxides result from diisocyanate addition to dioximes.

8.1.21. Polyvinyls

Polyvinyl compounds are produced either by the polymerization of vinyl compounds $CH_2{=}CHX$ (where X is substitution group) or by polymer analogue reactions on polyvinyl compounds. Examples of commercially useful polymers are

Polyvinyl alcohols, $[-CH_2-CH(OH)-]_n$
Polyvinyl halides, $(CH_2-CHX-)_2$
Polyvinyl amines, $[-CH_2CH(NR_1R_2-)]_n$

and polyvinyl sulfides, $[-CH_2-CH(SR)]_n$, are a developing material.

• Polyvinyl acetate. Polyvinyl acetate (Elias, 1977) is used for adhesives and for wood size (40% solution), as a raw material in lacquers and varnishes (dispersions), and as a concrete additive in the form of a line, dispersible powder obtained from spray drying. It swells in water, but does not readily dissolve in water.

• Polyvinyl acetate copolymers. Polyvinyl acetate grades (Elias, 1977) that are resistant to hydrolysis are obtained by copolymerization with vinyl stearate and vinyl pivalate (vinyl ester of trimethyl acetic acid), since the saponification rate is reduced by the bulkier side groups.

• Polyvinyl alcohol. Polyvinyl alcohol (Elias, 1977) is produced by the deesterification or transesterification of polyvinyl acetate with methanol or butanol. Methyl acetate and the valuable butyl acetate are useful solvents. It has many applications as sizing for nylon/rayon fibers and protective colloids, as a component in printing inks, toothpastes and chemotic preparation. Polyvinyl alcohol is soluble in water.

• Polyvinyl acetals. Conversion of polyvinyl alcohol with butyraldehyde in a suitable solvent that dissolves polyvinyl butyral well produces polyvinyl butyral (Elias, 1977). It is used for sandwiching between two layers of glass to make safety glasses and other applications. Polyvinyl acetals are used in mechanical engineering as rubber for moldings, since the gapped impact strength and the flexural modulus. Polyvinyl formals are compatible with phenolic resins and produce elastic high tension electrical cables.

• Polyvinyl ethers. Polyvinyl ethers form soft resins which are very resistant to saponification and have good light stability. They are used as adhesives, plasticizers, and additives for the textile industry.

• Poly (N-vinyl compounds)

• Poly (N-vinyl carbazole). These polymers retain their shape up to 160°C, and they are brittle. The brittleness can be reduced by copolymerization with isoprene. It is used for insulation layers in high frequency electrical cables.

• Poly (N-vinyl pyrrolidone). These polymers are soluble in water or in polar, organic solvents such as chloroform. They serve as protective colloids, emulsifiers, hair spray components, and a blood plasma substitute.

• Polyhalogenohydrocarbons

• Polyvinyl fluoride. This polymer is partially crystalline and is more similar in its properties to polyethylene than to polyvinyl chloride. Since the melting temperature is about 200°C, it is processed at temperatures of about 210°C (Elias, 1977). Films of polyvinyl fluoride are more stable to weathering than those of either polyethylene or polyvinylchloride. Polyvinyl fluoride is usually used for coating wood and metals.

• Polyvinylidene fluoride. Polyvinylidene fluoride is polymorphous. The glass transition temperature is –40°C and the melting temperature lies between 158 and 197°C. The polymer is thermoplastic and more similar to polyethylene than to polyvinylidene chloride. It can be extruded and injection-molded. Because of its good weathering and chemical stability, it is used for packaging, cable covering, and protective coatings in chemical apparatus in building materials. It crosslinks under with exposure to ionizing radiation unlike other fluorinated polymers.

• Polytrifluorochloroethylene . Polytrifluorochloroethylene is more susceptible to chemical attack than polytetrafluoroethylene due to C–Cl bonding. The larger size of the chlorine atom lead to a less tightly packed crystal structure and a lower melting temperature (220°C) and better solubility compared to polyetrafluoroethylene (Elias, 1977). The glass transition temperature is 50°C. It can be processed under restraint, e.g., by sawing or drilling and with the usual plastic fabrication equipment. Films and coatings can be obtained from dispersions with carriers at sintering temperatures 220°C. Nonporous final coatings are formed after 8–10 intermediate coatings and a final sintering of 300°C.

• Polytetrafluoroethylene. Polytetrafluoroethylene is chemically stable, resistant to oxidation, and of low flammability. These properties result from the high bond energy C–F bond. The polymer has few polar groups, it has a low dielectric loss factor and is therefore a good electrical insulator. It has a melting temperature of 327°C and a glass transition temperature of 120°C. A transition temperature below 30°C is responsible for the cold flow of the material. The regularity in structure and the helical conformation are evident in the high crystallinity (93–98%) of the polymer. Film and parts formed from the polymer require unique processing since the polymer does melt and flow (melt viscosity of 10^{10} at 380°C). This polymer is used for low surface energy, high chemical resistance and low coefficient of friction applications.

• Polyvinyl chloride. Polyvinyl chloride is the plastic that is produced in largest quantity in Europe and Japan, and second to polyolefins in the United States. Since the melting temperature of a completely syndioatactic polymer is 273°C and that of a commercial product is 173°C, pure polyvinyl chloride is brittle and difficult to process. Plasticized polyvinyl chloride is predominantly used commercial for film and pipe products. Polyvinyl chloride discolors thermally at the processing temperature and by light induced oxidation. In thermal degradation, HCl is eliminated with the formation of conjugated bond system.

• Polyvinylidene chloride. Polyvinylidene chloride has a melting temperature of 220°C and a glass transition temperature of 23°C (Elias, 1977). It is chemically unstable at the high processing temperatures that are required. The tendency to crystallize is decreased by copolymerization of 85–90% vinylidene chloride with 10–15% vinyl chloride. The lower melting temperature of 120°C of the copolymer (glass temperature of –5°C) enables the product to be processed into food wrapping films which are only slightly permeable to water and air. Pipes and filter cloths made from polyvinylidene chloride are resistant to solvents. The high abrasion strength of this polymer is useful for long wearing seat covers.

8.1.22. Phenolic Resins

Phenol-formaldehyde resins are a condensation of phenol with formaldehyde. Examples are Novalacs, Resoles, and Bakelites A, B, and C (Fry *et al.*, 1985).

8.1.23. Cellulose and Cellulosics

Cellulose in its natural form is usually derived from cotton fibers, and some of its most useful derivatives are used for fiber production (Joseph, 1986) such as cellulose acetate fibers (rayon). Some examples are:

- Cellulose acetate
- Cellulose acetate butyrate
- Cellulose propionate
- Cellulose triacetate
- Ethyl cellulose
- Cellulose sulfate, sodium salt
- Hydroxybutyl methyl cellulose
- Hydroxy propyl cellulose

8.1.24. Hetero Chain Polymers

• Polysiloxanes. Organopolysiloxanes (trivial name: silicones) are composed of organosilicone compounds containing the group –Si–C– in the polymer chain structure $-Si(R_2)-O-$.

• Polyphosphates. Polyphosphate refers to the oligomeric, cyclic metaphosphate as well as the high-molecular-weight, branched, unbranched, and cross-linked network polymers. The phosphates are formed by controlled dehydration of alkali metal dihydrogen phosphates, e.g., NaH_2PO_4.

• Polyphosphazenes. The phosphonitrile chloride (polydichlorophosphazene) series is obtained by heating phosphorous pentachloride and ammonium chloride in solvents such as chlorobenzene and tetrachloroethane.

• Polycarborane siloxanes. Polycarborane siloxanes contain *m*-carborane groups as well as siloxane groups in the main chain.

• Polyorganometallic. Polymers with metals in the side groups can be produced by polymerization of the corresponding monomers or by polymer analogue conversion. An example is poly(*p*-chloromethyl styrene).

8.1.25. Natural Polymers

Examples of natural polymers are fibers such as cellulose (cotton), flax, linen, and hemp. Low-molecular-weight products are produced from unsaturated natural oils by cross-linking reactions. Special attention is given to these materials in Chapter 5.

8.2. MONOMERS AND RELATED MATERIALS

Polymers and resins are synthesized from monomers and other reactants usually with an initiator and/or catalyst. A list of such materials follows.

- Acrylates and methacrylates
- Alcohols
- Aldehydes
- Amides
- Amines
- Anhydrides
- Aromatic hydrocarbons
- Carboxylic acid chlorides
- Carboxylic acids
- Compounds containing halogen
- Compounds containing nitrogen
- Compounds containing phosphorus
- Isocyanates
- Ketones
- Organometallics

- Oxides and peroxides
- Oximes
- Phthalates
- Quinones
- Ultraviolet light absorbers

8.3. ADDITIVES FOR PLASTICS

8.3.1. Polymerization Materials

• Catalysts. Catalysts include materials that affect the synthesis of urethanes (Wasilczyk, 1992), reactions between diisocyanates and multifunctional alcohols such as tertiary amines and others.

• Coupling agents. Coupling agents (Monte, 1992) are usually silane types and function by tieing together dissimilar surfaces such as glass fiber and epoxy resin. Generally silane coupling agents are represented by the formula YRSiX, where X is a hydrolyzable group (alkoxy) and Y is a functional organic group (e.g., amino, methacryloxy, epoxy). R is a small aliphatic linkage such as $(–CH_2–)_n$ that serves to attach the functional organic group to silicon (Si).

Titanates or titanium-derived coupling agents (titanium alkoxides) react with free protons at the inorganic interface, resulting in the formation of an organic monomolecular layer on the inorganic surface. Another application of titanates of unfilled polymers is 0.3% neoalkoxy dodecylbenzene sulfonyl functional titanate for reducing moisture in poly(ethyl cellulose).

• Cross-linking agents. Cross-linking agents are materials that cause two polymer chains to "tie" or link together. Examples of this wide range of chemical compounds include multifunctional monomers (e.g., alcohols with diisocyanates) which react with two similar or dissimilar polymer chains; and peroxides which create free radicals and cause polymers (e.g., polyethylene) to react with each other.

• Curing agents. Curing agents include catalysts and polymerization initiators. A catalyst causes a reaction to occur, but does not participate in the reaction. A polymerization initiator causes a reaction to occur and becomes part of the polymer chain.

Organic peroxides (Kamath, 1992) are initiators and sources of free radicals. Examples are benzoyl peroxide, methyl ethyl ketone peroxide, peroxyesters, peroxycarbonates, peroxyketals, and dialkyl peroxides. Peroxides and hydroperoxides are selected primarily on their specific half-life at a temperature for a specific polymerization. Examples of hydroperoxides are *t*-amyl hydroperoxide, *t*-butyl hydroperoxide, and cumene hydroperoxide.

• Dispersants/surface-active agents. The processing of polymers often requires the use of surface-active agents and dispersants (Friedman, 1992). Particles are better dispersed in a polymer if first coated with a dispersing aid. Uses for dispersing agents include the compounding of thermoplastics and elastomers, latices for textile sizing, paper coating, pigment in paint, and adhesives.

The surface-active agent consists of a medium-molecular-weight polymer with chemically different ends. These ends possess chemical groups that are chosen to provide compatibility between dissimilar materials like inorganic pigment and polymer resin.

• Free radical initiators. Peroxides and hydroperoxides provide the bulk of products that generate free radicals for initiation and cross-linking of resins and polymers. (See curing agents and cross-linking agents.)

• Fragrances. Fragrance concentrates (Rutherford, 1992) are compounded mixtures of aromatic chemicals dispersed in a thermoplastic resin. A fragrance additive improves the odor of a product such as a polyethylene garbage bag. An example of a fragrance additive is a citrus oil, e.g., lemon oil.

8.3.2. Protective Materials

• Antioxidants. Antioxidants (Fisch, 1992) inhibit atmospheric oxidation and its degradative effects on a polymer system, and degradation during processing and storage. Polymers deteriorate through a complex sequence of chemical reactions including chain scission or cross-linking.

Chemical bonds are broken in polymers to form free radicals by heat, ionizing radiation, mechanical stress, and chemical reactions. They are two main classes of antioxidants. First are those that inhibit oxidation through reaction with chain-propagating alkyl or hydroperoxy free radicals. These materials are free radical scavengers or primary antioxidants. Second are those that decompose peroxide molecules into non-radical, stable products. Compounds in this class are secondary antioxidants, synergists, or peroxide decomposers.

Examples of antioxidants are polypropylene-low volatility hindered phenol and phosphite; polyethylene-hindered phenols, polyphenols (LDPE), polystyrene-hindered phenols; polyvinyl chloride-organometallic compounds and salts derived from lead, barium, cadmium, zinc, and tin, as well as epoxide and phosphites are the most common stabilizers.

• Antistatic agents. Static electricity on plastic products can be generated in many ways (Van Drumpt, 1992). Usually, friction is involved during extrusion, injection molding, or when leading plastic film at high speed along rollers. In the absence of movement, static electricity may even build by friction with ambient air.

Examples of antistatic agents are cationic antistats (long-chain alkyl quaternary ammonium, phosphonium, or sulfonium salts with counterions such as chloride); anionic antistats (alkali salts of alkyl sulfonic, phosphonic, dithiocarbamic, or carboxylic acids); nonionic antistats (ethoxylated fatty amines, fatty acid esters or ethanolamides, polyethylene glycol esters or ethers, and mono- and triglycerides).

- Preservatives. Preservatives (Lenhart, 1992) are often called antimicrobials, mildewcides, fungicides, or bacteriocides (biocides). Preservatives serve to protect polymeric materials from attack by microorganisms. Microorganisms affect the appearance, and cause mildew odors, embrittlement, and premature product failure.

There are several different preservative additives for polymeric materials. The most commonly used are 2-*n*-octyl-4-isothiazolin-3-one and 10,10′-oxybisphenoxarsine. Preservatives for polymers are considered pesticides and are registered with the Environmental Protection Agency under the Federal Insecticide, Fungicide, and Rodenticide Act.

- Heat stabilizers. Heat stabilizers (Ringwood, 1992) are used for polyvinyl chloride and other compounds because of their poor thermal stability (e.g., heat, radiation). Examples of heat stabilizers are dibutyltin (isooctyl mercaptan) acetate, dibutyltin bis(alkyl maleate), mercaptides, mercapto acid esters, mercapto alcohol esters, dibasic lead stearate, and dibasic lead phthalate.

- Ultraviolet light stabilizers. Ultraviolet (UV) light stabilizers (Son, 1992) are used in plastic parts and related polymer products to reduce the rate of photooxidation reaction on the polymer chain. Scavenging of free radicals is the mechanism of reducing photodegradation, as UV radiation generates reactive free radicals. Examples of UV inhibitor/scavenger agents are hindered amine light stabilizers are alkoxy hindered amine light stabilizers.

- Degradability additives. Because of ecological factors, the degradability of a plastic product is important for the environment. The primary mechanisms of degradation are thermal, photooxidative, hydrolytic, chemical, mechanical, and biological. The most important of these are photooxidative and biological. Surface degradation of plastics by biological methods can be enhanced by the addition of a corn starch additive (6–15%). Bulk degradation occurs at 15% concentrations.

Natural photooxidation of plastic products can be accelerated by additives (Ennis, 1992) containing single-component vinyl ketone polymers. These additives are added to polystyrene or polyethylene at letdowns of 5 p.h.r. or greater. Another additive is an organometallic such as caprylate or benzophenone compounds supplied by Dow Chemical, Du Pont, Union Carbide, Ampacet, Princeton Polymer, Atlantic International Group, and Rhone-Poulenc.

• Flame retardants. Flame retardant chemicals (Braksmayer, 1992) are used to make plastic products ignition- or flame resistant. The active species in fire retarding are the halogens chlorine and bromine, phosphorus, and water. Flame retardants perform in different ways. Some help to develop a protective char (phosphorus based) which separates the flame from the polymer (fuel). Others change the flame chemistry by inhibiting free radical formation in the vapor phase (halogen based). Alumina trihydrate releases water during a fire, cooling the fire.

The polybrominated diphenyloxides are the most widely used halogenated additive for ABS, HIPS, other styrenes, polyesters, polyamides, and polyolefins. Brominated phthalate esters are nonblooming and thermally stable flame retardants.

Reactive flame retardants include chlorenic anhydride, tetrabromophthalic anhydride, and diol derivatives. Reactive retardants used in urethane foams include polyols containing halogens, phosphorus, and/or nitrogen.

8.3.3. Processing Materials

• Chemical blowing agents (foamers). Addition of a blowing, foaming, or gassing agent (Geelan, 1992) to a plastic product reduces the density and material consumption of the product. The hardness can also be adjusted with these additives. There are two classes of foamers: physical (liquid to gas) and chemical (chemical reaction to produce gas). The gases are carbon dioxide or nitrogen.

Examples of a physical foaming agent are the chlorofluorocarbons including products called CFCs, e.g., CFC-11, -12, -22. Chemical foaming agents range from low to high temperature. Also, they are endothermic or exothermic. Most of the modern foaming agents are based on polycarbonic acids. An example of a low-temperature foaming agent is toluene sulfonyl hydrazide; a high-temperature foaming agent, toluene sulfonyl semicarbazide. Azodicarbonamide (azobisformamide, azo or az) is the most widely used foaming agent for plastic parts which produces nitrogen and lesser amounts of carbon dioxide.

• Fillers/extenders. Fillers or extenders (Washabaugh, 1992) include materials such as silica to replace the resins, usually for reasons of cost. A filler is chosen according to cost and compatibility with the host resin.

• Plasticizers. Additives that soften and flexibilize inherently rigid, and brittle polymers are plasticizers (Dieckmann, 1992). For example, polyvinyl chloride (PVC) is a rigid host polymer and is semicrystalline. A preferred plasticizer for PVC is an organic ester. Ophthalates (benzenedicarbooxylates) led by di(2-ethylhexyl)phthalate (DOP or DEHP) are the preeminent family of monomeric plasticizers.

• Lubricants. Lubricant additives (Mesch, 1992) aid the processing of polymers. They perform primarily by reducing the friction from within the polymer (internal lubricants) and from polymer to the equipment (external lubricants). Examples of lubricants are metal stearates, paraffins, fatty acids, amides, and combinations of lubricants.

• Colorants. Color can be a critical part of the appearance of molded parts (Gordon, 1992). The most widely used colorants are dyes and pigments. A pigment is a colorant that is insoluble and dispersed as particles throughout a resin to induce a specific color. A dye is a colorant that is soluble in a resin and is usually an organic compound. Organic colorants tend to be stronger and brighter than duller and more opaque inorganic colorants. A wide range of colors are produced from color concentrates (Hattori, 1992).

Carbon black is the most common black pigment. Titanium dioxide and zinc sulfide are white pigments; iron oxides are black, brown, red, and yellow; lead chromates and lead chromate molybdates include bright yellows and oranges; cadmium pigments are red, yellow, orange, and maroon; chromium oxides are green; ultramarines are blue, pink, and violet. Mixed metal oxides include yellow nickel titanates and blue and green cobalt aluminates.

Red organic pigments include quinacridone, diazo, azo condensation, monoazo, naphthol, and perylene types. Yellow pigments include disazo, benzimideazalone, isoindolinone, diarylide, and quinophthalone. A blue pigment is phthalocyanine, and violet pigments include quinacridone and dioxazine. Quinacridones are also available in magenta.

Dyes are often used when good transparency is necessary in a molded plastic part. Dye classes include azo, perinone, quinoline, and anthraquinone types.

Special colorants include pearlescent pigments (titanium dioxide-coated mica and ferric oxide-coated mica); metallic flake (aluminum and brass); fluorescent pigments; and phosphorescent pigments (zinc sulfide with partial substitution of the zinc with cadmium, calcium, or strontium).

• Mold release agents. A mold release agent is an interfacial coating applied between two surfaces that would otherwise stick together (Axel, 1992). The release agent enhances the separation of the plastic part from the mold. Examples of release agents are fluorotelomers, polydimethylsiloxanes, silicones, and vegetable derivatives.

• Smoke suppressants. Smoke evolution from burning polymers and compounds has become an important issue in various applications (Levesque, 1992). A common test for smoke density is ASTM E 662, using the NIST Smoke Chamber and ASTM E 84 using the Steiner Tunnel.

The addition of zinc borates, tin oxides, and molybdenum compounds to polymer formulations has been examined. The most effective of these additives is nickel molybdate, molybdenum trioxide, and ammonium octyl molybdate.

8.4. STANDARDS FOR PROPERTIES OF PLASTIC MATERIALS

The following organizations provide standards for testing and specifying properties of plastic materials:

- American Society for Testing Materials (ASTM)
- U.S. Government
  U.S. Department of Commerce
  General Services Administration
  Military Specifications
- American Standards Association and the International Organization for Standardization (ISO)
- Society of the Plastics Industry (SPI)
- Underwriters Laboratory (UL)
- Society of Plastics Engineers (SPE)

Melting and glass temperatures of some plastic materials are provided in Table 8.1. Plastic materials and suppliers are listed in Table 8.2 in the Appendix.

9

Deformulation of Plastics

9.1. SOLID SPECIMENS

Solid specimens of plastic or polymeric materials usually consist of less than 5% by weight of pigments and fillers, the remainder being polymers. Small amounts of additives may be present. A scheme for the preliminary preparation of solid specimens is shown in Fig. 9.1. Most plastic products and related materials are not heavily pigmented or filled, with some exceptions. A thin section cut from these

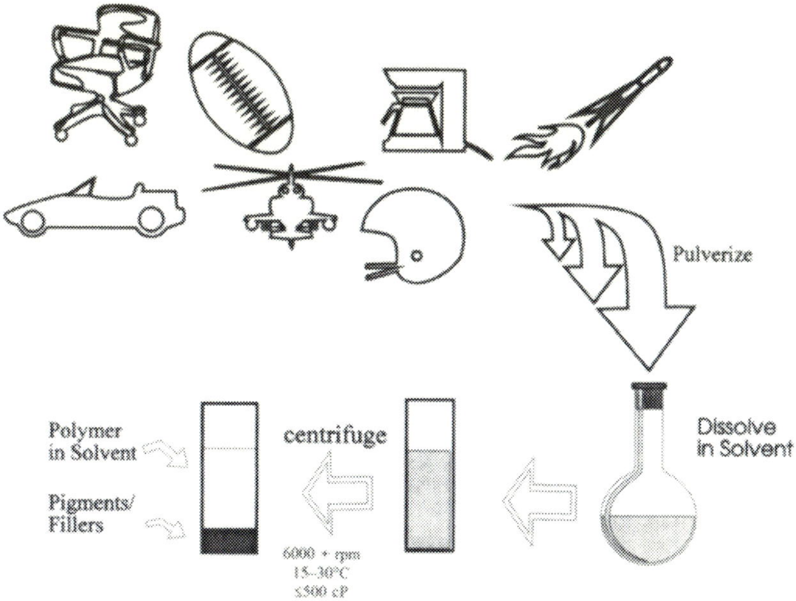

Figure 9.1. Scheme for preparation of solid plastic specimen.

materials will usually suffice for IR analysis to provide an identification of the plastic material.

Krause *et al.* (1979) discusses identification of plastics by combustibility and solubility properties. An effective and economical method of preparing a plastic specimen for SEM analysis is to freeze the specimen in liquid nitrogen, which will cause it to become brittle. The brittle specimen will break by bending and provide a fresh surface for analysis.

If a surface for very detailed analysis is needed, mount the specimen in a liquid resin, which hardens, followed by polishing with grit to provide a very smooth and flat surface. Images on a smooth polished surface are more easily resolved for SEM, EDXRA, ESCA, AES, and SIMS analyses.

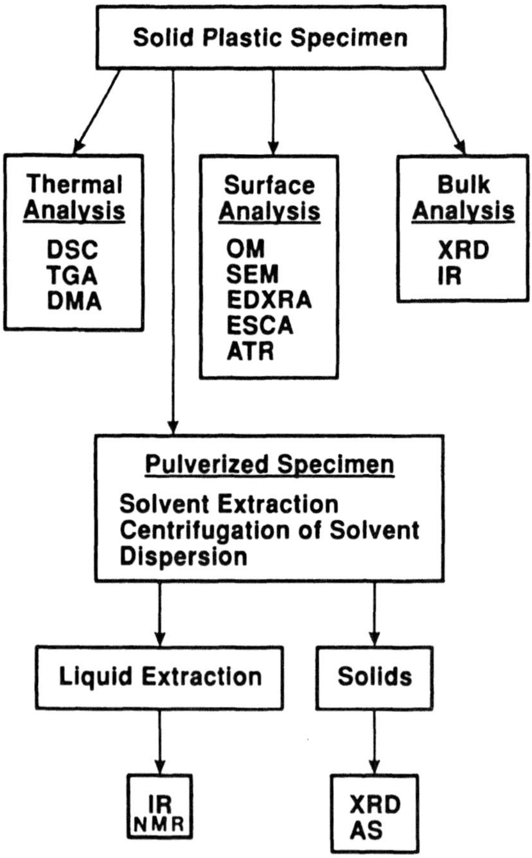

Figure 9.2. Scheme for deformulation of solid plastic specimen.

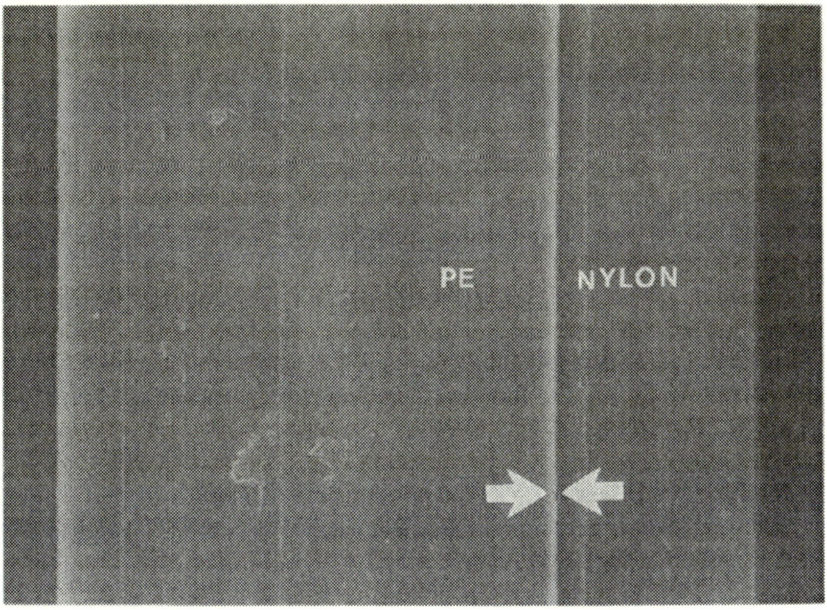

Figure 9.3. SEM micrograph of laminated plastic film.

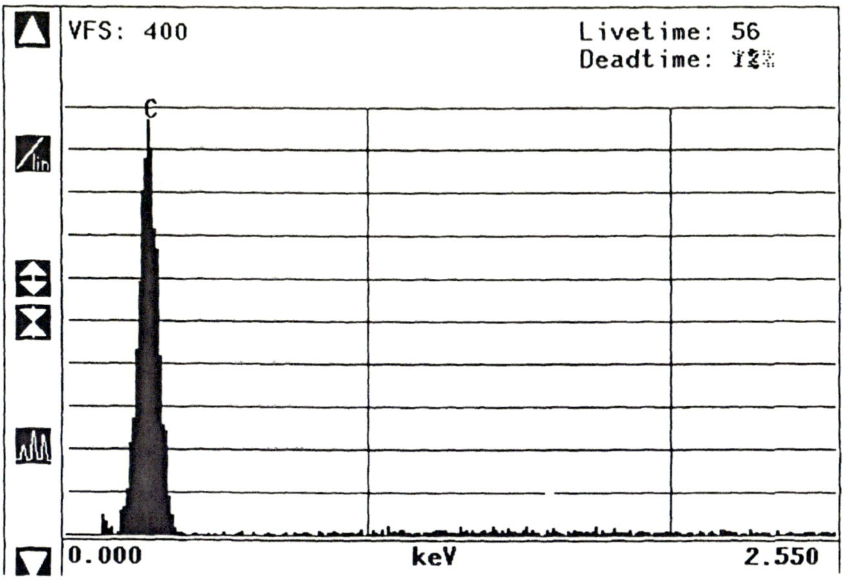

Figure 9.4. EDXRA spectrogram of left side of laminated film.

A detailed scheme for deformulation of solid plastic specimens is shown in Fig. 9.2. Often, the chemical class of the plastic material is identified by IR if there is no interference from heavy loading of pigments or fillers.

Example 1. A plastic film specimen is hardened in liquid nitrogen and broken followed by mounting and polishing. An SEM micrograph of a coextruded plastic film specimen is shown in Fig. 9.3. EDXRA spectrograms of the left and right sides show only carbon on the left side (Fig. 9.4), and carbon, nitrogen, and oxygen on the opposite side (Fig. 9.5).

The films are analyzed using Fourier transform infrared spectroscopy with microscopic and ATR attachments. Infrared spectra of both sides of the materials in Fig. 9.3 are generated without damaging the specimen and are shown in Figs. 9.6 (left) and 9.7 (right). The films are identified as polyethylene and polyamide.

A DSC thermogram (Fig. 9.8) of the composite specimen, consisting of the complete structure in Fig. 9.3, generates melting temperatures that correlate to low-density polyethylene (LDPE) and polyamide (nylon 6,6).

This specimen is a nylon 6/LDPE laminated film. A materials and products search reveals that the LDPE film is a SCLAIRFILM SL-1 (Du Pont Canada) laminating LDPE product, and the polyamide film is a Dartek F101 (Du Pont Canada) laminating film of the nylon 6,6.

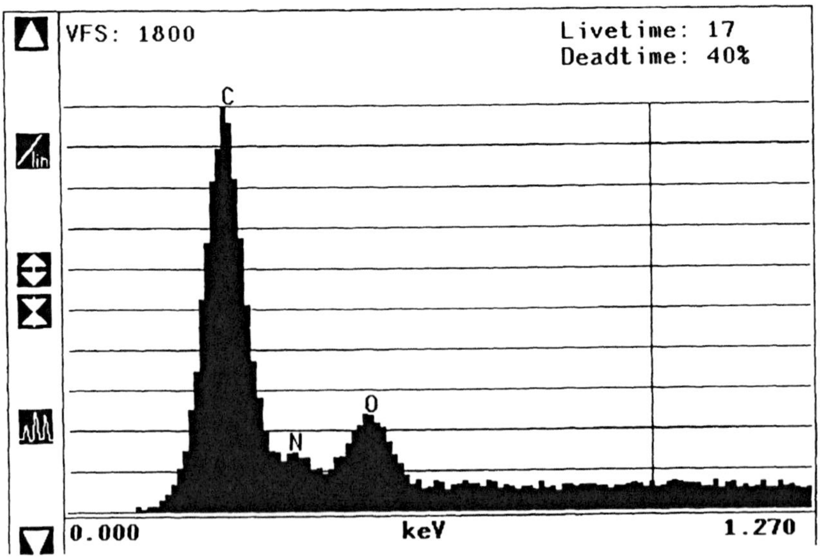

Figure 9.5. EDXRA spectrogram of right side of laminated film.

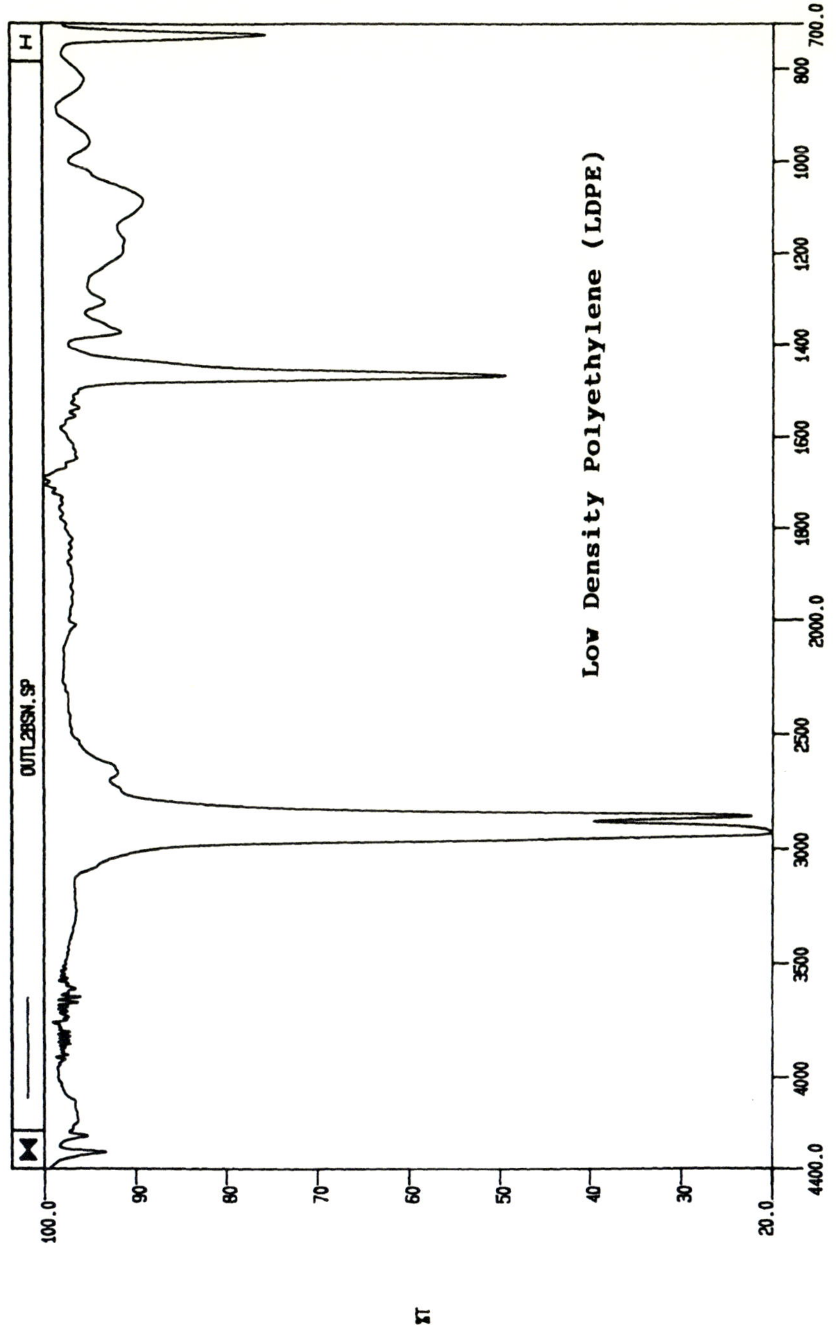

Figure 9.6. IR spectrum of left side of laminated film.

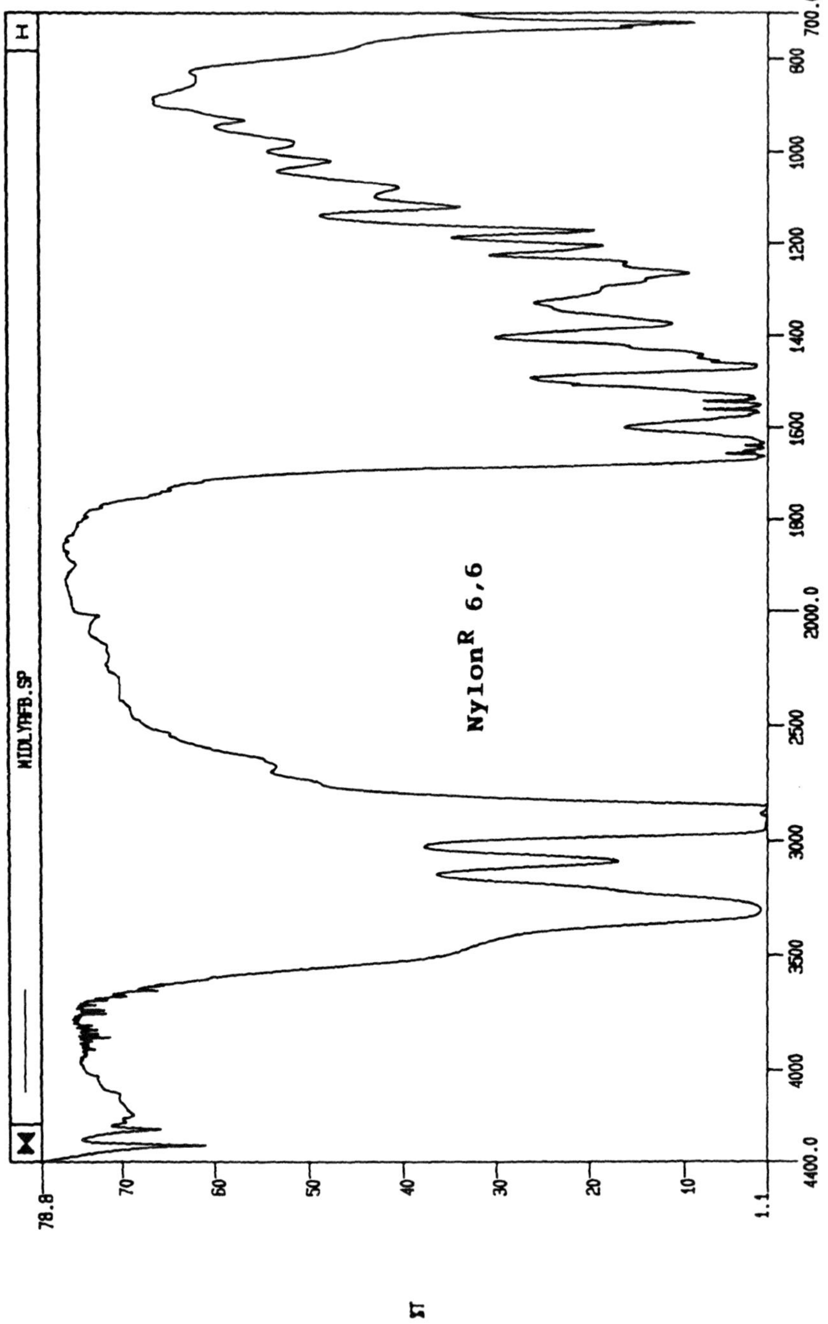

Figure 9.7. IR spectrum of right side of laminated film.

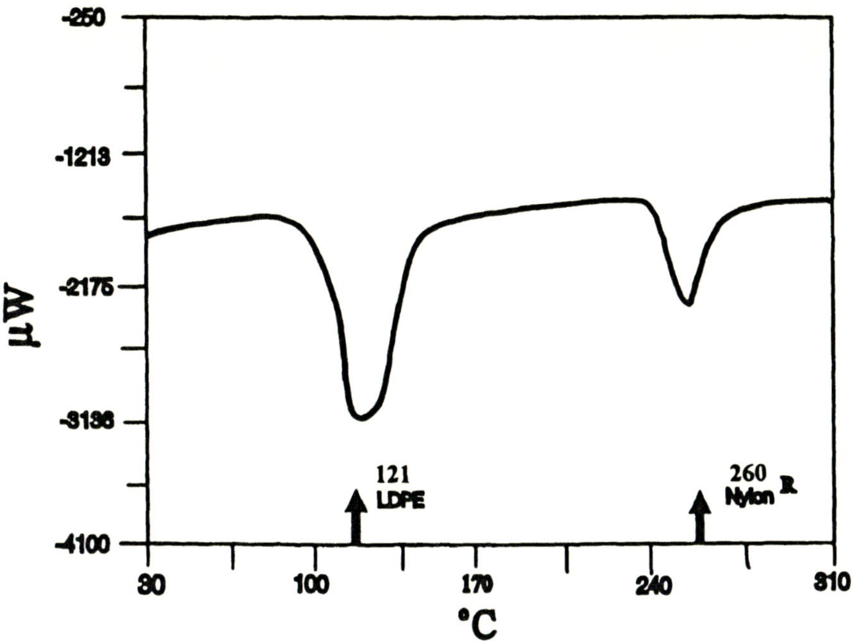

Figure 9.8. DSC thermogram of laminated film.

Carbon-filled plastics and elastomers (rubbers) cause a problem when analyzed by IR spectroscopy. The carbon particles scatter the IR energy. Microscopic IR beams are better for this application as the small beam can focus on a pure resin region of the specimen.

Another method for preparing small pieces of pigmented/filled sample is to dissolve the specimen in solvent followed by separation of solids by centrifugation. The polymer will remain in solution and the solvent is removed by oven drying. If the polymer is difficult to dissolve, fluxing in hot solvent (see Fig. 6.4) will disintegrate the specimen.

Soft elastomeric materials can often be prepared for SEM analysis by freezing in liquid nitrogen and breaking. They can be pulverized to powder by freezing in liquid nitrogen while hammering the specimen.

9.2. LIQUID SPECIMENS

Liquid specimens are polymers dissolved in solvent, in dispersion, or of very low molecular weight. The specimen may contain pigments and additives. A scheme for preparation of specimens for deformulation of polymers in liquid form is shown in Fig. 9.9. If the specimen contains obvious color and turbidity, then it must be prepared for complete deformulation by separating components as shown in Fig. 9.10.

Separation of solids from liquids is followed by separation of solvents from polymer and additives. Eventually, every component is separated and the specimen is completely deformulated.

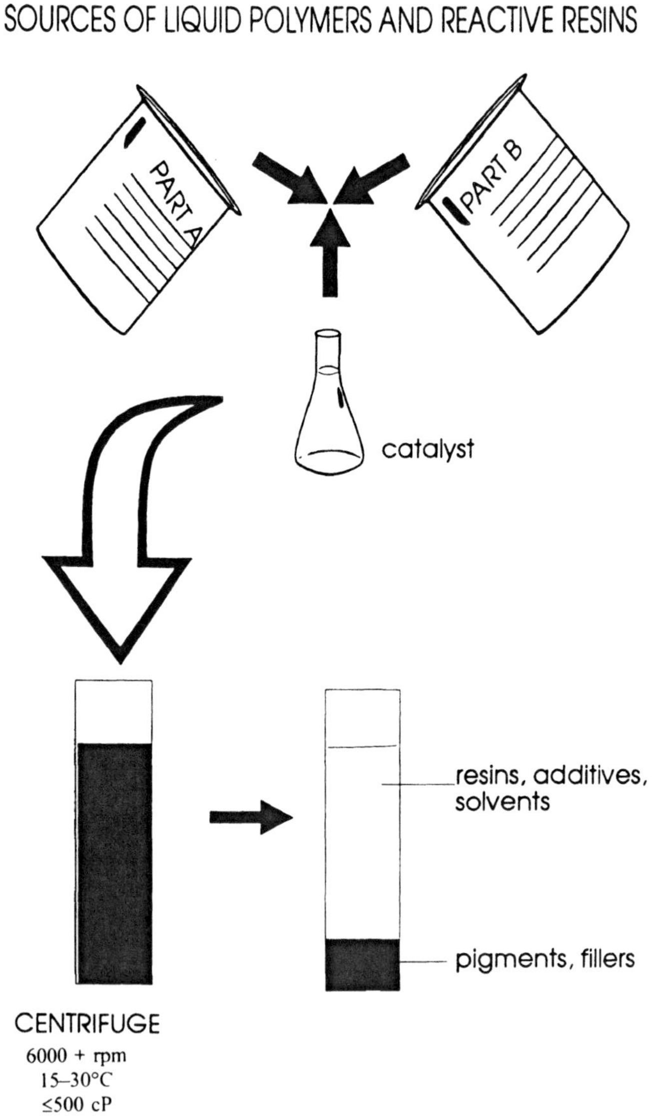

Figure 9.9. Scheme for preparation of liquid plastic specimen for deformulation.

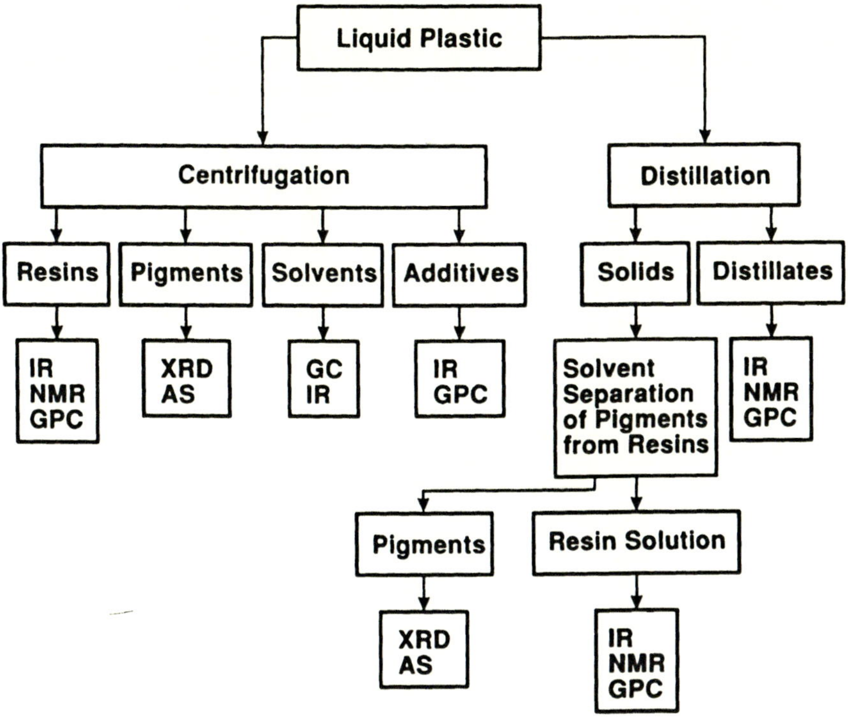

Figure 9.10. Scheme for deformulation of liquid plastic specimen.

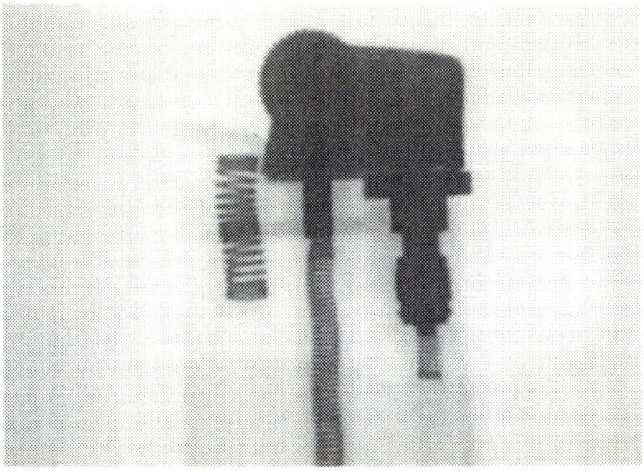

Figure 9.11. X-ray micrograph of a disposable lighter. Dark areas are metal and light areas are plastic.

9.3. NONDESTRUCTIVE EXAMINATION OF PLASTIC PARTS

A useful tool for nondestructively examining plastic parts before chemical analysis is the X-ray microscope (XRM). The XRM can peer into a solid material and answer important questions as to what is in the plastic part and how many different materials comprise the part.

Example 2. The lighter in Fig. 9.11 is an X-ray microscope image of a liquid fuel disposable lighter. The image shows that dense metal parts are molded into the lighter plastic case. Using this information, the deformulation plan may include a cross-sectioning of the lighter to examine all of the parts that are shown in the image.

9.4. REFORMULATION

Generate a table of components versus percent weight. This is the formulation recipe. Acquire materials from suppliers listed in Table 8.2. Formulate the recipe and compare it to the original material. Compare physical properties to confirm a successful reformulation.

10

Adhesives Formulations

10.1. GENERAL

Adhesives unite materials, creating a whole that is greater than the sum of its parts (Skeist and Miron, 1977). Their volume is small compared to the metals, glass, wood, paper, fibers, rubber, and plastics they bond. The "adhesive" bonds "adherends," which are substrates such as glass, metal, plastics, and wood (Dann, 1970).

In a typical adhesive bond, the basic components are:

SUBSTRATE/INTERFACE/**ADHESIVE**/INTERFACE/SUBSTRATE

Adhesives may be classified in many ways including mode of application and setting, chemical composition, cost, suitability for various adherends, and end products. Chemical composition will be the preferred method of classification as the theme of this book is "analysis of adhesives," but other methods related to formulating will be discussed for the reader's information.

10.1.1. Applications

Adhesives must be applied to substrates in a fluid form to wet the surfaces, which requires low viscosity to flow onto the surfaces while eliminating voids. After application to surfaces (adherends), the adhesive must solidify to develop bonding strength. The transition from fluid to solid may be accomplished in the following ways (Skeist and Miron, 1977):

1. Cooling of a thermoplastic. Thermoplastics soften and melt when heated, becoming hard again when cooled. Methods of applying adhesives in this way include hot-melt applicators; dry powders that are heated after application; and extruders.
2. Release of solvent or carrier. Solutions and latices contain the adhesive composition in admixture with water or organic solvents. These liquids

lower the viscosity to permit wetting of the substrate. After wetting has been accomplished, they must be removed.
3. Polymerization. The fluid adhesive is applied to the substrate followed by rapid polymerization to bond the substrates. The reaction-sensitive adhesives fall into two main groups: condensation and addition polymerizations.
4. Pressure-sensitive adhesives. These adhesives are fluid applied and do not undergo a chemical reaction. After wetting the substrates, they remain in the gel state which is a tackiness capable of being removed rather than a permanent bond.

10.1.2. Origin

1. Natural. Starch, dextrins, asphalt, animal and vegetable proteins, natural rubber, and shellac.
2. Semisynthetic. Cellulose nitrate and other cellulosics, polyamides derived from dimer acids, and castor oil-based polyurethanes.
3. Synthetic. Vinyl-type addition polymers: polyvinyl acetate, polyvinyl alcohol, acrylics, unsaturated polyesters, butadiene-acrylonitrile, butadiene-styrene, neoprene, butyl rubber, and polyisobutylene. Polymers formed by condensation and other stepwise mechanisms: epoxies, polyurethanes, polysulfide rubbers, and the reaction of formaldehyde with phenol, resorcinol, urea, and melamine.

10.1.3. Solubility

Adhesives can be categorized by solubility or fusibility of the final adhesive (glue) line.

1. Soluble. Thermoplastics, starch and derivatives, asphalts, some proteins, cellulosics, vinyls, and some acrylics.
2. Insoluble. Thermosets, phenol- and resorcinol-formaldehyde, urea- and melamine-formaldehyde, epoxies, polyurethanes, natural and synthetic rubbers if vulcanizes, anaerobics, and unsaturated polyesters.

10.1.4. Method of Cure or Cross-Linking

Cross-linking usually involves the reaction of two chemical intermediates; examples are:

1. Formaldehyde condensed with phenol and resorcinol
2. Formaldehyde condensed with urea and melamine
3. Isocyanate reacted with polyol to produce polyurethane
4. Epoxide reacted with primary amine or polyamide-amine

5. Unsaturated polyester copolymerized with styrene
6. Sulfur-vulcanized diene rubbers

Cross-linking may also take place among molecules of a single species as follows:

1. Epoxide catalyzed with tertiary amine
2. Dimethacrylate compounded anaerobically so that it will polymerize when air is excluded
3. Peroxide-vulcanized rubbers

Moisture curable adhesives can cross-link when exposed to water and examples follow.

1. Isocyanate prepolymers
2. Silicones
3. Polysulfides
4. Unsaturated polyesters
5. Cyanoacrylates
6. Epoxy resins

10.2. FORMULATIONS OF ADHESIVES BY USE

Widely used adhesive formulations are provided in Tables 10.1–10.34. The reader is referred to Skeist and Miron (1977), manufacturers, suppliers, and others for a more comprehensive list of formulations. A comprehensive list of adhesive terms is contained in Table 10.40.

An excellent source of adhesives formulations and suppliers is *Adhesives Age* published by Communications Channels, Inc., a division of Argus Press Holdings, Inc., P.O. Box 1147, Skokie, Illinois.

11

Adhesives Materials

11.1. INTRODUCTION

This chapter reviews materials commonly used in adhesives products. The major sources of this information were Skeist and Miron (1977) and *Adhesives Age* (1993). Adhesives materials suppliers are shown in Table 11.1.

11.2. SYNTHETIC RESINS

11.2.1. Polyvinyl Acetal

The principal applications for polyvinyl acetal adhesives are glass and metal (Farmer and Jemmott, 1990), but they have excellent adhesion for paper, fibers, and plastics. Monsanto, DuPont, and Union Carbide have been the leading suppliers in the United States of polyvinyl butyral. DuPont supplies safety glass interlayer under the trade name Butacite and Monsanto, Saflex. Union Carbide offers polyvinyl butyral resin as Bakelite. Monsanto produces polyvinyl formal resin under the trade name Formvar.

Polyvinyl acetals are manufactured by reacting one molecule of aldehyde with two molecules of alcohol in the presence of an acid catalyst. Films of polyvinyl acetals are characterized by their high resistance to aliphatic hydrocarbons, mineral, animal, castor, and blown oils.

11.2.2. Polyvinyl Acetate

General-purpose wood glue (household white glue) consists of an emulsion of polyvinyl acetate and polyvinyl alcohol. The excellent adhesion of polyvinyl acetate emulsions to cellulosic and other materials gave rise to an abundance of applications including bookbinding, paper bags, milk cartons, drinking straws, envelopes, folding boxes, and many more. Among the manufacturers of polyvinyl acetate are Air Products and Chemicals, National Starch, Union Carbide, (Jaffe *et al.*, 1990).

Among the main uses for polyvinyl acetate emulsions are interior and exterior flat paints. In the textile industry, polyvinyl acetate emulsions impart durability and strength to finishes. The paper industry uses small-particle-size polyvinyl acetate emulsions as pigment binders for clays in paper and paperboard coatings.

11.2.3. Polyvinyl Alcohol

Polyvinyl alcohol (PVA), a dry solid, is a water-soluble synthetic resin (Jaffe *et al.*, 1990). It is produced by the hydrolysis of polyvinyl acetate. The resins are excellent adhesives and form tough clear films. However, being very hydrophilic, PVA must be protected from moisture. The primary uses for PVA in the United States are in textile and paper sizing, adhesives and emulsion polymerization.

11.2.4. Polyvinyl Butyral

See Section 11.2.1, polyvinyl acetal.

11.2.5. Polyisobutylene and Butyl

Butyl rubber and polyisobutylene are elastomeric polymers used quite widely in adhesives and sealants as primary elastomeric binder and as tactifiers and modifiers (Higgins *et al.*, 1990). Polybutylene is a homopolymer and butyl rubber is copolymer of isobutylene and a small amount of isoprene.

Applications of butyl and polyisobutylene include pressure-sensitive adhesives in automotive and architectural sealants.

11.2.6. Acrylics

Acrylic adhesive polymers, in solvent solution and aqueous emulsion forms, are widely used as thc basis for adhesives for pressure-sensitive tapes, labels, and other decorative and functional pressure-sensitive products (Gehan, 1990).

Acrylic adhesive polymers are synthesized from a wide selection of acrylic and methacrylic ester monomers and with low levels of monomers having pendent functional groups useful for post-cross-linking and/or adhesion uses. Specifically, acrylic adhesives are based mainly on ethyl, butyl, and 2-ethyl hexyl acrylate monomers and small quantities of methyl methacrylate together with other specialty acrylic monomers. Often the acrylic monomers are copolymerized with other vinyl monomers such as vinyl acetate, vinyl chloride, styrene, etc. The synthesis of linear polymers of very high molecular weight is possible because of the high reactivity of the vinyl groups in the monomers. A major manufacturer of acrylic emulsions is Rhom & Haas.

Acrylic adhesives for contact adhesives are used where immediate high bond strength is required. The adhesive is applied to joining surfaces, dried, and bonded together. Cure time can vary from several minutes to hours. Contact adhesives are used to manufacture furniture and countertops of high-pressure plastic and particle board, prefabricated curtain walls, assemblies of cold-roll steel to honeycomb cardboard, and others. Heat and pressure bonding provides other applications such as heat-seal food packaging, vacuum forming operations as automotive door panels, and heat sealing of cellophane to metal foil and metallized polyester film for polypropylene film. The applications for acrylic adhesives are vast.

11.2.7. Anaerobics

Anaerobic adhesives are single-component liquids or pastes that can be stored for prolonged periods of time at room temperature in the presence of oxygen, but harden rapidly to form strong bonds when applied to surfaces that exclude oxygen (air). Oxygen is a free radical scavenger and the curing or reaction proceeds via free radical initiation of the polymerization process. Loctite Sealant Grade A was the first anaerobic sealant (Rooney and Malofsky, 1990). A similar material for nuclear applications was characterized by Gooch (1982). The basic advantages of anaerobic adhesives are fast assembly of surfaces and parts and cost reduction as they are easily applied and form strong bonds rapidly at room temperature. They are a one-component product and require no premixing.

Starting materials for widely used formulations include prepolymers based on polytetramethylene glycol and hydrogenated bisphenol A capped with diisocyanates and hydroxyalkylmethacrylates. Hydroperoxides are added to the formulation to initiate the polymerization and set up the adhesive in the absence of oxygen. Many variations have been made to these formulations including the use of fillers and primers. When the anaerobic adhesive is packaged, an air space is left in the container to block the curing reaction.

Applications of anaerobic adhesives include conveniently locking threaded fasteners, liquid gaskets, porous metal impregnation, and sealing pipe thread.

11.2.8. Cyanoacrylates

The popular one-drop "super glues" are based on cyanoacrylate materials. The cyanoacrylate monomers polymerize or cure when they contact moisture or water. Most surfaces contain microfilms of water which is sufficient to catalyze the reaction. Alkyl cyanoacrylate adhesives are unique among adhesives because they are the only single-component, "instant" bonding adhesives that cure at ambient conditions without required external energy (Coover, *et al.*, 1990). Major producers of these adhesives include Loctite Corporation, National Starch Company, Henkel AKG, Toa Gosei, and Alpha Techno.

Alkyl-2-cyanoacrylate monomers are highly reactive compounds and will polymerize via anionic and/or free radical mechanisms. The anionic reaction route predominates and is catalyzed by small amounts of a weak base such as water. Ultraviolet light and heat can cause polymerization. Acid (Lewis or protonic) stabilizers are employed to prevent premature polymerization.

Applications for cyanoacrylate adhesives include household cementing jobs, bonding weather stripping to automotive bodies, and the repair of flexible PVC side trim strips for automobiles.

11.2.9. Ethylvinyl Alcohol (EVA)

The EVA resins are usually incorporated into hotmelt adhesives, discussed in Section 11.2.10.

11.2.10. Polyolefins

Polyolefin adhesives are primarily of the hotmelt type. The growth of hotmelt adhesives is related to: rapid set time, ease of dispensing, elimination of solvents, elimination of hazardous materials, wide formulating latitude, and others.

A typical ethylene-vinyl acetate-based hotmelt has three components (Eastman and Fullhart, 1990): a polymer, 30–40%; a modifying or tackifying resin, 30–40%; and a petroleum wax, 20–30%. The quantity and relative amount of each material is governed by the performance requirements of the adhesive. The polymer forms the base or strength of the adhesive; the modifier provides surface wetting and tack; and the wax lowers melt viscosity.

Through the 1960s, ethylene and vinyl acetate monomers made ethylene-vinyl acetate resins with 18–40% acetate content. They were developed for a wide variety of uses. Later polyethylene and polypropylene became less expensive and more prevalent in hotmelt adhesives used for packaging, paper substrates, paperboard cartons, and corrugated containers.

Tackifiers are usually hydrocarbon resins, rosin esters, and polyterpenes. Waxes are paraffin-type or very-low-molecular-weight hydrocarbons.

11.2.11. Polyethylene Terephthalate

Polyesters are the reaction product of dibasic acids with polyfunctional hydroxyl-bearing materials. Linear saturated and unsaturated polyester resins have been successful for hotmelt adhesives. Polyethylene terephthalate (PET) has been widely used for fibers and films, but also for hotmelt adhesives.

Polyesters serve in the shoe industry to extend the life of different parts of the shoe. Polyester-amide copolymers have been employed to attach automobile parts.

11.2.12. Nylons

Terpolymers of nylon were developed to decrease melt viscosity of the original homopolymers. Terpolymers include: nylon 6, 6-6, 6-10 (DuPont), nylon 6, 6-6, 12 (Emser Werke), and others (Rossitto, 1990).

Many of these hotmelt adhesives are used in fabric bonding.

11.2.13. Phenolic Resins

In acidic media, phenolics that are formed when the molar ratio of formaldehyde to phenol is greater than one are called resoles. The phenol moieties are terminated with reactive hydroxymethyl groups ($-CH_2OH$), known as methylol groups. In basic media, if the molar ratio of formaldehyde to phenol is less than one, the polymer becomes phenol terminated and is called novolak (Tobiason, 1990).

Applications for phenolic resins are vast. Examples include coated abrasives or sandpapers, abrasion wheels for polishing stone, and foundry applications as molds (Tobiason, 1990).

11.2.14. Amino Resins

Amino resins are prepared by reacting formaldehyde with a compound containing the amino group $-NH_2$ (Updegraff, 1990). The amino compounds most commonly used are urea and melamine which produce urea formaldehyde and melamine formaldehyde resins.

Amino resins are used to bond plywood and particle board, laminated wood beams, parquet flooring, interior flush door, and furniture assembly.

11.2.15. Epoxies

Epoxy resins (Meath, 1990) are reactive with a number of different curing agents and yield a wide variety of products with different cure requirements. Epoxy resins react via an addition mechanism with no by-products. They possess hydroxyl groups along the molecular chains which provide adhesion to many substrates.

The most widely used epoxy resins are based on bisphenol A and epichlorohydrin which are bifunctional with epoxide pendent groups. It is the pendent groups that react with a host of curing agents such as amines and alcohols. Manufacturers of commercial epoxy resins include Dow Chemical (Epon 828), Ciba-Geigy (Araldite 6010), Interez (Epi-Rez), and Reichhold (Epotuf 37-1410).

11.2.16. Polyurethane

The most widely used polyurethane adhesive components (Schollenberger, 1990) continue to be toluene diisocyanate (TDI), diphenylmethane-4,4′-diisocyanate (MDI), polymethylene polyphenyl isocyanate (PAPI), and triphenylmethane triisocyanate (Desmodur R) together with polyester and polyether glycols.

Polyester-based polyurethanes are more frequently used than polyether systems because of their higher cohesive and adhesive properties.

Major uses for polyurethanes include food packaging, footwear, furniture, automotive, and aircraft.

11.3. SYNTHETIC RUBBERS

11.3.1. Styrene-Butadiene Rubber (SBR)

The process of manufacturing SBR consists of three steps: polymerization, monomer recovery, and finishing (Midgley and Rea, 1990). SBRs are produced by addition copolymerization of styrene and butadiene monomers in either an emulsion or a solution polymerization process.

Uses for SBR include general-purpose and specialty construction of adhesives and tape adhesives. Other applications include pressure-sensitive adhesives for labels, surgical tape, masking, protective wrapping, splicing, and so on.

11.3.2. Nitrile Rubber

Nitrile rubbers are broadly defined as copolymers of a diene and a vinyl unsaturated nitrile (Mackey and Weil, 1990). Manufacturers and products include BFGoodrich (Hycar), Uniroyal Chemical (Paracril CJ), and Goodyear (Chemigum, N3). Nitrile rubbers have good oil resistance which is useful for gaskets and cements in contact with oils. Their good elastomeric and polarity properties provide them with good solvent resistance and compatibility with other polar materials.

Nitrile rubbers are used for laminating polymeric films to metals, laminating polypropylene carpet to plywood, and others.

11.3.3. Neoprene

Neoprene (polychloroprene) combines rapid bond strength development with good tack or self-adhesion, and resistance to oils, chemicals, water, heat, sunlight, and ozone. It is widely used in bonding shoe soles, furniture construction, and others.

Neoprene is produced from the chloroprene monomer, 1-chloro-1,3-butadiene, in an emulsion process. The monomer can add in a number of ways and the *trans*-1,4 addition is the most common.

11.3.4. Butyl Rubber

Butyl rubber is a straight-chain hydrocarbon, and a copolymer of isobutylene and a minor amount of isoprene. There are four curing systems for butyl rubber: (1) the quinoids cure, (2) cure with sulfur or sulfur donor groups, (3) resin cure, and (4) the zinc oxide cure for halogenated butyl rubber only.

The compound has good resistance to heat, light, and weathering. Butyl latex is used in packaging adhesives such as tackifying and flexibilizing additives in higher-strength adhesives based on more brittle polymers. It is useful for laminating and seaming adhesives and specialty binders and coatings for both polyethylene and polypropylene. One supplier of butyl latex is Burke-Palmason Chemical Company.

11.3.5. Polysulfide

Polysulfide liquid polymers originally found wide acceptance for applications requiring a flexible, adhering, chemically resistant composition of matter. They were the first liquid polymers cured at room temperature and found applications on aircraft as sealants. These sealants are noncorrosive and do not produce any by-products harmful to aluminum. Other applications include a quick hose repair compound, a sealant for bolted steel tanks, electrical potting compounds, caulks, and wooden flight decks.

Many of the sealants are prepared from Thiokol LP-2, -32, or -31 as the base polysulfide liquid polymer. The preparation of polysulfide liquid polymers (Panek, 1990) involves the reaction of bischloroethyl formal with a sodium polysulfide solution containing emulsifying and nucleating agents. The sulfur is present as a mixture of disulfide and trisulfide. Next, the resulting high-molecular-weight polymer is split into segments that are terminated by mercaptan groups. The average molecular weight is 4000. The cross-linking agent is trichloropropane and the curing agent is 50% lead dioxide, 45% plasticizer, and 5% stearic acid.

11.3.6. Silicone

Silicone resins possess a wide range of properties. They are resistant to extremes of temperature, UV and infrared radiation, and oxidative degradation (Dean, 1990). Silicone elastomers are useful for caulking and sealant compounds, bonding and abhesion (releasing) materials.

The fundamental component of most silicone sealants is the polymeric siloxane, silanol-terminated polydimethylsiloxane. A catalyst is used for cross-linking systems, and moisture is absorbed from the atmosphere for RTV systems.

11.3.7. Reclaimed Rubber

Reclaimed rubber is used for fillers in other adhesives, usually of lower quality.

11.4. LOW-MOLECULAR-WEIGHT RESINS

- Aminoplasts
- Rosin
- Rosin esters
- Polyterpenes
- Petroleum resins
- Coumarone-indene

11.5. NATURAL DERIVED POLYMERS AND RESINS

Natural polymers are usually of plant or animal origin (Gooch, 1980; Sperling, 1983); some examples are:

- Bitumens
- Starch
- Dextrin
- Wheat flour
- Soy flour
- Animal glues

11.5.1. Animal Glues

Animal glues have been used for over a century and were one of the first glues made from natural materials.

- Animal resins. Animal glue is an adhesive of great versatility. This natural polymer is an organic colloid derived from collagen (Brandis, 1990). Animal glue is a protein derived from the hydrolysis of collagen, a principal protein constituent of animal hide, tissue, and bones. Collagen, animal glue, and gelatin are closely related as to protein and chemical composition. Gelatin is considered to be hydrolyzed collagen:

$$C_{102}H_{149}O_{38}N_{31} + H_2O \leftrightarrow C_{102}H_{151}O_{39}N_{31}$$

As a protein, animal glue is essentially composed of polyamides of certain alpha-amino acids. Animal glue is a polydisperse system containing mixtures of similar molecules of widely different molecular weights (20,000 to 250,000 g/mole). Animal glues are soluble in water and insoluble in oils, waxes, organic solvents, and alcohol.

- Fish resins. All such glues or gelatins are derived from collagen, a long-chain protein found mostly in skin and bone. It is insoluble in water, but can be broken down with heat and chemicals (acids or bases) in the presence of water to produce a water-soluble product. The end product can be either a glue or a gelatin depending on the process. The glue would be used for an adhesive. The collagen molecule is made up of varying amounts of 20 different amino acids. Fish skin collagen breaks down more readily than animal skin collagen by heat or enzyme activity.

Properties of fish glue are:

1. Average molecular weight, 30,000 to 60,000 g/mole.
2. End groups on the polypeptide chain are carboxylic, amino, and hydroxyl.
3. Color is light caramel.
4. Odor is mild and indicative of the odorant added.
5. Viscosity is 4000–7000 cP at 70°F.
6. Weight is 9.8 lb/gal.

7. Insoluble in organic solvents.

11.5.2. Casein

Casein is manufactured from skim milk and is a product of the dairy industry. It is a protein that is a natural condensation product of amino acids held together by the amide or peptide bond, –CONH–. The molecule in its native state comprises a great number of different amino acids. Its high molecular weight accounts for its colloidal properties and its value as an adhesive. Hydrolysis destroys the molecule when subjected to strong acid or alkali. Elements found in the ash of the grade of casein used for adhesives include phosphorus, potassium, sodium, and calcium in concentrations of 0.2 to 3%.

Water-resistant casein glue sets to a gel via a slow, chemical reaction, sodium caseinate gradually converting to calcium caseinate. The chemicals are dry-mixed with the casein and shipped to the user. The casein-lime product is readily dispersed in cold water and often used as a common wood glue.

11.5.3. Polyamide and Polyester Resins

Polyamides and polyesters developed for fibers are too high-melting and too fast-setting to be used for adhesives (Rossitto, 1990). Most of the polyamides and polyesters used in hotmelt adhesives are based on copolymers. The most common monomers used for hotmelt polyamides are dibasic acids, amino acids and lactams, and diamines. Polyester amides are made by reacting an aromatic polyester such as PET or PBT with dimer acid. An acid-terminated prepolymer is formed which is then reacted with a diamine to produce blocked polyester-amides, a copolymer.

Applications include continuous lamination of fabric and plastic substrates, toe lasting of shoes, bonding SMC automotive parts, and others.

11.5.4. Natural Rubber

Latex is tapped from the tree *Hevea brasiliensis* and contains about 35% solids (Gazeley, 1990). The rubber particles are removed from the latex and concentrated. It is then processed into rubber products. Natural rubber is similar in composition to the synthetic rubber polyisoprene. Oxidation of the rubber will cause cross-linking or set-up.

Applications of natural rubber adhesives include self-sealing paper envelopes, latex pressure-sensitive adhesives, tile adhesives, vulcanizing latex adhesives, anchor coat for tufted carpets, and many others.

11.6. INORGANIC

Inorganic materials for use in adhesives are categorized as (1) soluble silicates and (2) other inorganic cements such as the insoluble salts in hydraulic and sorel cements, silico-phosphate and other phosphate cements.

The siliceous soluble silicates are characterized by empirical weight ratios of the silica to the alkali content as their compositions are not those of molecular compounds. A solution containing 1.0 mole of Na_2O for each 3.3 moles of SiO_2 will, on a weight basis, have a ratio of 3.22% SiO_2 to 1% Na_2, or as a 3.22 ratio silicate.

11.7. SOLVENTS, PLASTICIZERS, HUMECTANTS, AND WAXES

- Acetone
- Heptane
- Mineral spirits
- Toluene
- Dioctyl phthalate
- Tricresyl phthalate
- Glycerol
- Ethylene glycol
- Paraffin wax

11.8. FILLERS AND SOLID ADDITIVES

- Kaolin
- Bentonite
- Whiting
- Silica
- Zinc oxide
- Magnesium

11.9. CURING AGENTS

- Triethylene tetramine
- Tetraethylene pentamine
- Hexamethylene tetramine
- Phenylene diamine

12

Deformulation of Adhesives

12.1. INTRODUCTION

Adhesives can be pigmented, filled, but most are translucent or transparent. Materials are added to adhesive formulations for the following major purposes:

1. Enhanced adhesion
2. Wetting of substrates
3. Weathering, moisture resistance, etc.
4. Enhanced strength
5. Enhanced curing rate
6. Color

The formulations in Chapter 11 are examples of mixing ingredients to achieve a specific adhesive formulation for one or more applications. Knowing either the type of adhesive or the application gives valuable clues about the other. If neither type nor application is known, it is necessary to start from the beginning and use a proven deformulation scheme.

The following discussion covers methods for the deformulation of solid and liquid adhesive specimens. The methods of analysis are not explained in detail as they were outlined individually in Chapters 2 and 3.

12.2. SOLID SPECIMEN OF ADHESIVE

12.2.1. Surface Analysis

Some common sources of solid adhesive materials are shown in Fig. 12.1. The adhesive material is solid, and may be rubbery, after it sets up on the substrate. Solid specimens of adhesive materials are dried/cured cements, glues, hotmelt adhesives, and others. Taking a representative sample includes scraping off or cutting a

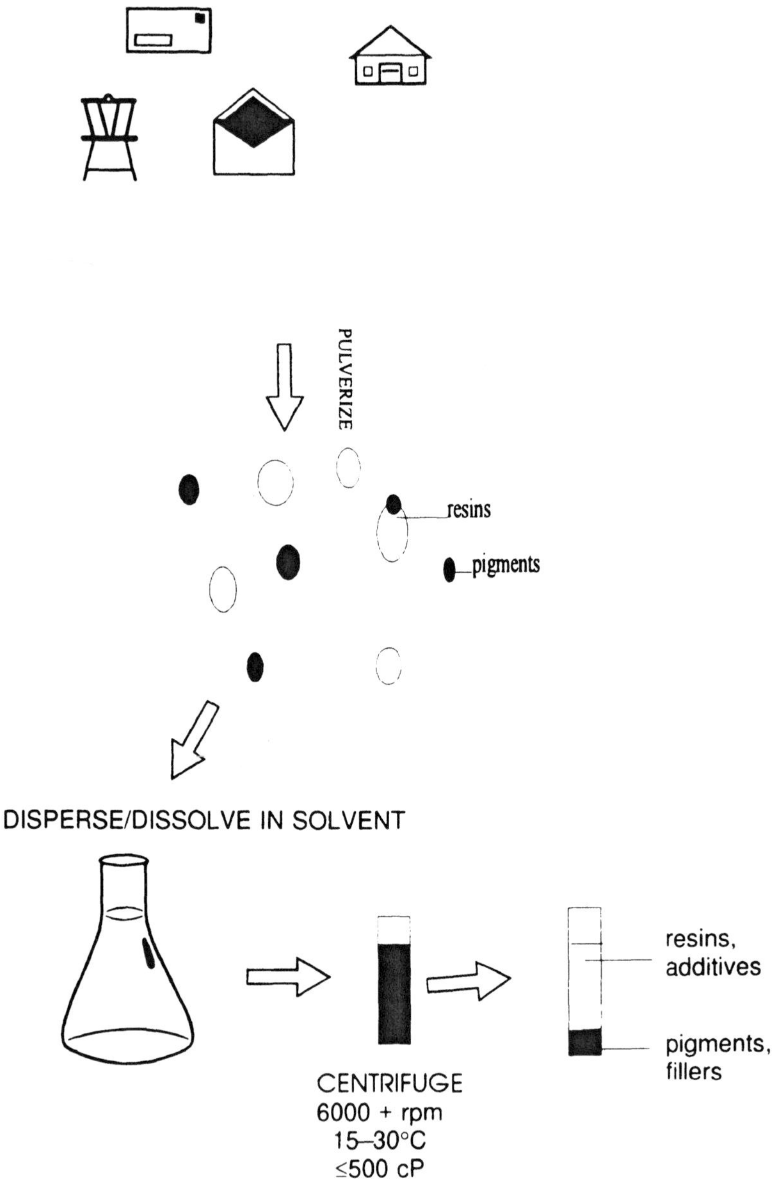

Figure 12.1. Scheme for preparation of solid adhesive specimen for deformulation.

specimen from a substrate. Usually, the solid sample is taken from an application where the adhesive was used.

Preparation of a solid specimen for investigation is illustrated in Fig. 12.1, and the solid specimen is pulverized by freezing with liquid nitrogen followed by hammering. A pulverized specimen will consist of fine particles which are dispersed/dissolved in solvent and may require solvent refluxing (see Fig. 6.4) to separate vehicle from fillers/pigments. The mixture is centrifuged (see Fig. 1.2) to separate the denser pigments/fillers from the resins and solvents. Possibly, the vehicle will separate from the solvent. The oven (105°C, until dry)-dried vehicle and solid particles are analyzed with IR, AS, and XRD.

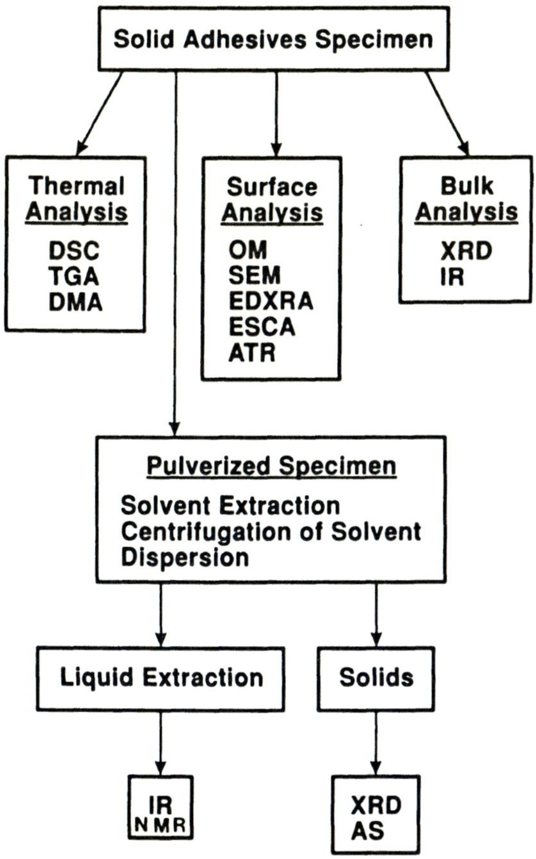

Figure 12.2. Scheme for deformulation of solid adhesive specimen.

A small solid sample can be coated and placed directly in the SEM instrument, but a polished surface specimen in a resin (as acrylic or epoxy) is preferred to enhance the image and resolution.

First, identify the specific application of the adhesive. Next, follow the workable scheme for deformulating a solid adhesive specimen shown in Fig. 12.2. Immediately, observe the specimen with an optical microscope (or similar device) to determine the color, presence of filler, and any other information before proceeding with more extensive and expensive methods. Take a color photograph to document the appearance of natural and magnified images.

Observe the surface of the specimen by SEM so as to characterize fillers and pigments with regard to size, shape, and concentration. Elemental analysis (while in the SEM instrument) will provide identification of elements within the particles and of the vehicle. Other electron microscopic surface examination methods can be employed on the same sample including AES, SIMS, and ESCA, if necessary. ESCA has the capability of chemically analyzing the specimen. ATR infrared spectroscopy can be employed on the surface for development of an IR spectrum for chemical identification.

Bulk analysis is necessary for AS and XRD and the determination of metals, inorganic compounds, etc. Transmission infrared spectroscopy will develop a spectrum from a transparent/translucent solid specimen.

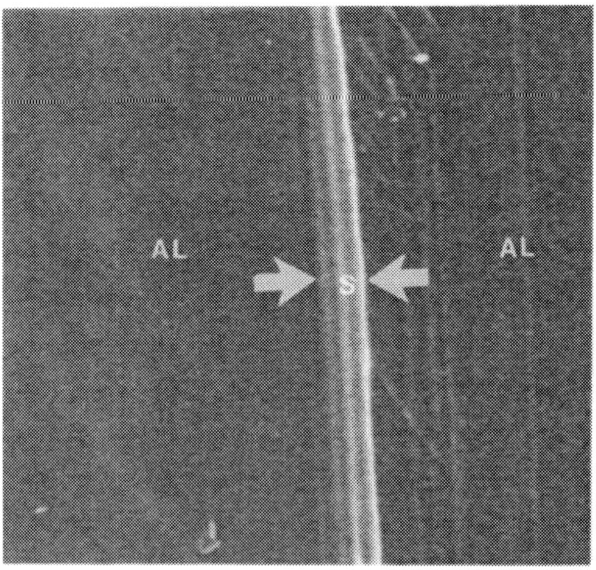

Figure 12.3. SEM micrograph (1000×) of aluminum aircraft panel bonded with polysulfide two-part elastomeric sealant. Sealant layer is highlighted by arrows.

Microscopic infrared examination of the polished specimen before coating for the SEM will provide an infrared spectrum for identification of the matrix or vehicle in the adhesive specimen.

Example. An SEM micrograph (1000×) of a cemented aluminum bond is shown in Fig. 12.3. The very thin cement layer (S) is present between aluminum surfaces (Al). This image was analyzed in an SEM instrument with EDXRA and aluminum was identified in each metal panel, and carbon, oxygen, and sulfur were identified in the cement. Nickel particles were discovered within the adhesive. Microscopic FTIR analysis identified the cement layer as polysulfide. The sample came from a military aircraft, and specifications for this aircraft included conductive polysulfide sealants for fastened aluminum joints/bonds. From a manufacturers' products list, Products Research Corporation manufactures this adhesive.

12.2.2. Bulk Analysis

Elemental analysis of specimens by EDXRA is limited to about 1% concentration by volume. Use AS and ICP for more refined methods of elemental analysis, if necessary. Pigments and fillers are further investigated using XRD and IR.

12.3. LIQUID SPECIMEN OF ADHESIVE

Liquid adhesives will usually be in the form of manufactured products prior to use, and, therefore, in liquid application form. In the case of hotmelt adhesives, the materials are solid prior to use and must be investigated as solid adhesives. A container of manufactured liquid adhesive is shown in Fig. 12.4. The viscosity of the adhesive should be adjusted to 500 cP or below with solvent followed by centrifugation. The volume of each centrifuge tube is 60–100 cm^3, so it may be necessary to fill several tubes. Weigh the tubes to ensure that they counterbalance each other within 0.1 g when centrifuged; large vibrations will develop if they are 1.0 g out of balance. Denser pigments and fillers will sediment to the bottom of the centrifuge tube and polymers/resins form the uppermost layer. Carefully, the layers are separated, oven dried, weighed, and analyzed individually according to the scheme in Fig. 12.5.

Separation by these methods is very convenient and saves time. Chromatography techniques further separate liquid components. The liquid fraction components are resolved and identified by injecting an aliquot into a calibrated HPLC. Also, the volatile liquids are identified by injection into a GC unit.

The molecular weight and distribution is determined by injecting some of the liquid fraction into a GPC.

If a larger specimen is required and solvent is known to be present, a solvent removal method as in Fig. 6.7 is employed. The weighed specimen (up to about 500 g) is weighed again after distillation of all volatile liquids, and each distilled

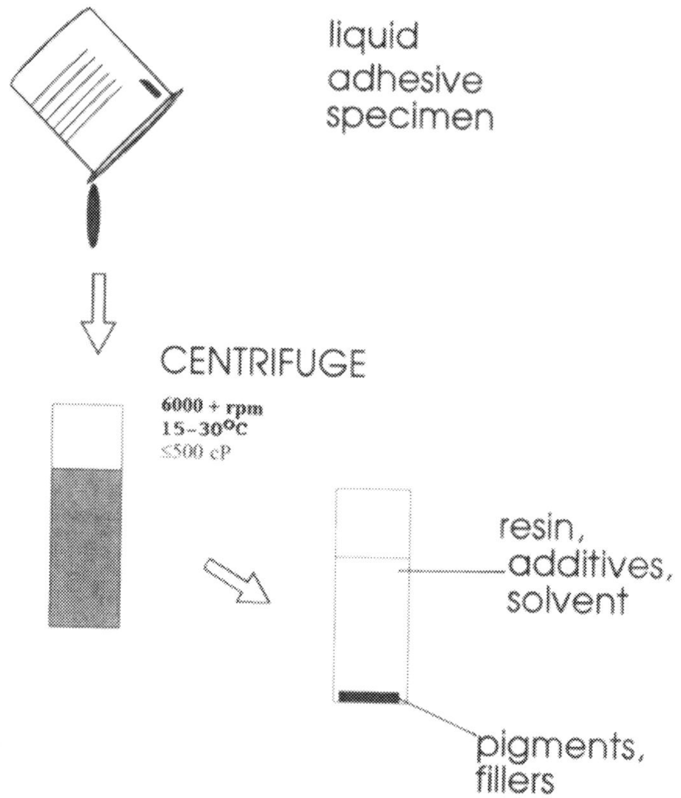

Figure 12.4. Scheme for preparation of liquid adhesive specimen for deformulation.

liquid is gravimetrically/volumetrically measured at 20–25°C. Observing the temperature of each liquid as it distills will determine the boiling temperature, and indicate when to "catch" the next distillate. A transmission IR spectrum can be developed from a liquid cell filled with each solvent.

The qualitative/quantitative results of these analyses will yield a table of components versus percent weight. From these data, reformulation of the original material can be accomplished.

12.4. THERMAL ANALYSIS OF SOLID SPECIMEN

Thermal analysis is listed separately in Fig. 12.2 because it is neither an elemental nor a chemical method of analysis. Logically, it could be placed in the bulk analysis column. Thermal analysis is a destructive method of investigation, and a specimen that can be destroyed must be available. DSC can show the melting temperature T_m, which will indicate whether the adhesive vehicle is thermoplastic

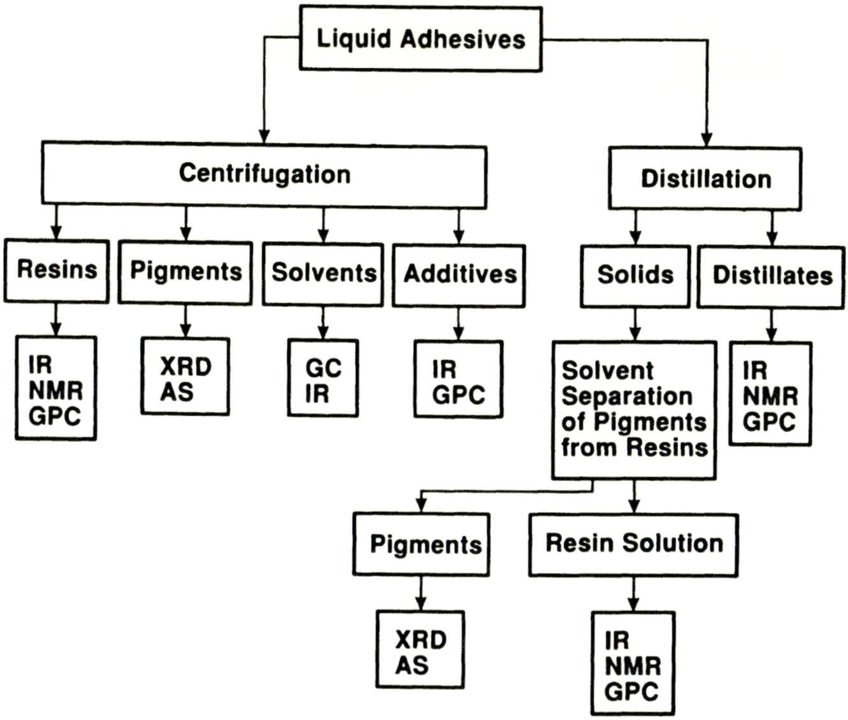

Figure 12.5. Scheme for deformulation of liquid adhesive specimen.

or thermoset. A melting peak will develop if it is thermoplastic (see Fig. 3.20), and if thermoset it will show a glass transition temperature T_g. A decomposition temperature T_d (see Fig. 3.21) will develop in the form of a downward-sloping curve corresponding to a DSC decomposition event (cross-linking and oxidation). The temperature and shape of the TGA decomposition are indicative of classes of polymers and resins, and much can be learned from a TGA curve.

An unfilled and unpigmented curve will show a T_d curve that descends to near 0% (about 0.5% carbon remaining) weight (see Fig. 3.21), but percent weight above zero will show the percent weight of fillers/pigments or other thermally stable solids.

12.5. REFORMULATING FROM DATA

The final test of the deformulation investigation is reformulation using materials identified during the course of the investigation. See Table 11.1 regarding procurement of materials for comparison to existing products.

Deformulating the unknown adhesive specimen by two or more methods is the best way to gain confidence in the results of an investigation.

13

Ink Formulations

13.1. GENERAL

The U.S. Bureau of Census figures (*Printing Ink Handbook*, 1976) indicate that there are approximately 200 ink companies producing inks in about 400 plants throughout the United States.

The National Association of Printing Ink Manufacturers (NAPIM) has been the only national trade association for the printing ink industry since its founding in 1914.

National Association of Printing Ink Manufacturers
777 Terrace Avenue
Hasbrouck Heights, NJ 07604-3110
(201) 288-8454

It consists of printing ink manufacturers engaged in the production and sale of printing inks on the open market in the United States.

The Institute of Paper Science & Technology in Atlanta, Georgia, is the premier facility in the United States for paper research.

Institute of Paper Science & Technology
500 10th Street
Atlanta, GA 30318
(404) 853-9500
Fax: (404) 853-9510

The wide variety of printing applications within the graphic arts requires different types of printing inks suited to the various printing processes, substrates, and end uses as discussed in the *Printing Ink Handbook* (1976).

Some of the many end uses, substrates, and performance needs are listed in Table 13.1. The major printing processes and corresponding inks are:

1. Letterpress
   Heatset
   Quickset
   Rotary
   High-gloss
   Moisture-set
   Water-washable
   News
2. Lithographic
   Web offset
   Sheet offset
   Metal decorating
3. Flexographic
   Solvent
   Water
4. Gravure
   Type A
   Type B
   Type C
   Type T
5. Other printing processes
   Screen printing
   Electrostatic
   Metallic
   Water color
   Cold-set
   Magnetic
   Optical

Practical, but important factors to consider are:

1. Color or colors to be reproduced
2. Printing process to be used
3. Substrate to be printed
4. Processing or converting requirements
5. End-use requirements
6. Cost requirements

For letterpress on coated papers, black inks can be expected to give 150,000 to 200,000 square inches per pound of ink; transparent colors, 125,000 to 175,000 square inches per pound of ink; and opaque colors, 75,000 to 125,000 square inches per pound of ink. The offset process usually gives 50 to 100% more coverage than letterpress.

13.2. LETTERPRESS

In the letterpress process, a plate with raised type is brought into direct contact with the substrate being printed. Gutenberg's revolutionary invention of movable type in about 1450 made possible printing as we know it today. Letterpress inks are viscous tacky systems that usually cure by autoxidation. Some major types of letterpress inks are: rotary ink, quickset ink, heatset ink, high-gloss ink, moisture-set ink, water-washable ink, and news ink.

As mentioned, letterpress inks are viscous, tacky systems. The vehicles are oil or varnish based and contain resins that cure by autoxidation, reaction with oxygen in air. The major exception is news ink, which generally consists of pigment dispersed in mineral oil and drying (or flow-out) by absorption in the paper substrate. Whereas final drying of the ink film is the result of autoxidation of the resin or oil component, initial setting may take place by absorption of ink into the substrate or by evaporation by the application of heat (heatset inks).

Where the letterpress image is being transferred to a rigid surface such as a plastic or two-piece metal container, the image is transferred to a blanket and then to the printed surface. This special form of letterpress printing has become known as "letterset" or "dry offset" because an offset blanket is used, but no water is present in the printing process as is the case in offset lithography.

Rotary inks used today are heatset types although nonheat, slower-drying oil types are also used. Rotary inks are typically used for typography or letterpress printing of books, magazines, and newspapers.

The body of a rotary ink for books is generally fairly fluid and will set up somewhat when not agitated. Book papers are supplied in many different surfaces, and the ink must be formulated to react properly on the surfaces. For example, a smooth hard paper requires a fast-drying ink.

Magazine and catalog inks require a significant degree of nonrub property in this type of rotary ink to sustain folding, handling, and the like.

Heatset ink is usually used in high-speed runs and with good quality. This type of ink requires a vehicle composed of synthetic resins dissolved or dispersed in suitable hydrocarbon solvents. The resins are usually high-melting types with good release at elevated temperatures. The solvent employed has a narrow boiling range with low volatility at room temperature and a fast evaporating rate at elevated temperatures.

Quickset ink types dry by filtration, coagulation, selective absorption, or a combination with autoxidation. The vehicles are special resin–oil combinations that, after the ink has been printed, separate into a solid material which remains on the surface as a dry film, plus an oily material which penetrates into the stock. This rapid separation gives the effect of a quick setting and drying.

High gloss is affected by porosity, degree of sizing, weight and type of paper stock. The more resistant the paper to penetration of the vehicle, the higher is the gloss produced. However, the gloss is primarily dependent on the formulation of the ink. Generally, the smaller the pigment or the more finely dispersed, the higher is the gloss. Modified phenolic and alkyd resins provide satisfactory high-gloss inks. These synthetic resins are often used in conjunction with drying oils to produce vehicles that exhibit minimum penetration and maximum gloss.

Moisture-set inks consist of pigments dispersed in a vehicle composed of a water-insoluble resin dissolved in a water-miscible solvent. When the printing is subjected to steam or a fine mist of water, the water-miscible solvent acts to absorb water and the water-immiscible resin to precipitate and bind the pigment to the paper. Often, humidity in the air is sufficient to set these inks on many substrates. The resins generally employed are maleic or fumaric acid, and modified rosin products that are acidic. Inks of this type are used in printing bread wrappers, milk containers, paper cups, and other applications where rapid printing and immediate handling of the printed matter is important.

Water-washable inks are designed primarily for letterpress printing on kraft paper and corrugated board. The print sets very rapidly to become a water-resistant film. The vehicle consists of a modified rosin soap in glycol solvent.

News ink usually drys by absorption of the ink in the stock. They are used on web presses, and require a very fluid consistency. Black news ink consists primarily of mineral oil and carbon black. Colored news inks are based on colored pigments flushed in mineral oil vehicles.

13.3. LITHOGRAPHIC

In commercial practice most lithographic printing is accomplished via an offset process by transferring the image from the plate to an intermediate roller or blanket and then to the substrate being printed. As most of lithography is accomplished by the offset method, the term *offset* has become synonymous with lithography.

Lithographic inks are viscous inks with varnish systems similar to letterpress varnishes. They differ in that the ink films applied are thinner than letterpress, and pigment content is higher. Also, they must be formulated to run in the presence of water, as water is used to create the nonimage areas of the plate.

In certain limited applications, such as printing of business forms, ink may be transferred directly from the lithographic plate to the printed surface. In this case, the process is known as direct lithography, or "dilitho."

13.3.1. Web Offset Inks

Web offset printing, because of its higher running speeds, requires inks with lower viscosities and tack, but high resistance to emulsification with the fountain solution (water). Web offset inks can be separated into two categories:

1. Nonheatset. Which air dry and heatset, with the assistance of ovens. Nonheatset web offset inks use ink oils which are absorbed into the substrate during the drying process.
2. Heatset. Heatset web offset inks, like heatset letterpress inks, are set by driving off the ink oil in an oven.

13.3.2. Sheet Offset Inks

Most sheet offset inks used today for general commercial printing are quickset letterpress; they set rapidly as the ink oil component penetrates the substrate, and subsequently dry as the vehicle cures by autoxidation. Higher-gloss and more-abrasion-resistant inks, such as those used in carbon printing, are modified with harder resins and often represent a compromise between quicksetting and better abrasion resistance properties. Sheet offset inks are not dried with heat dryers, though some sheet presses do have low-level heat assistance.

Sheetfed offset inks are offered in a broad variety of vehicle systems, which can be categorized as five general classes:

1. Autoxidative. Containing largely natural or synthetic drying oils.
2. Gloss. Drying oils, very hard resins, minimal hydrocarbon solvents.
3. Quickset. Hard soluble resins, hydrocarbon oils and solvents, minimal drying oils and plasticizers.
4. Penetrating. Soluble resins, hydrocarbon oils and solvents, drying and semidrying oils and varnishes.
5. UV curing. Reactive, cross-linking systems that cure by application of ultraviolet radiation.

13.3.3. Metal Decorating Inks

Metal decorating inks are lithographic inks that are specially formulated with synthetic resin varnishes to dry on metal surfaces with high-temperature baking. To decorate formed containers, special offset presses are used which may be either wet or dry offset processes. In either case, the ink systems are similar.

13.4. FLEXOGRAPHIC

Flexographic inks are chemically different from paste inks used for letterpress and lithographic printing. They are low-viscosity inks that dry by solvent evaporation, absorption into the substrate, and decomposition.

There are two main types of flexographic inks: water and solvent. Water inks are used on absorbent paper stocks such as kraft or lightweight paper. Solvent types are used on films such as cellophane, polyethylene, or polypropylene. They may also be used on some paper substrates.

Water-based flexographic inks (Flexography, 1991) are widely used on paper and paperboard including bleached or brown kraft and corrugated. Vehicles for these water-based inks are usually made from ammonia or amine-solubilized protein, casein, shellac-esterified fumarated rosins, acrylic copolymers, or their mixtures. Advantages of water-based inks include good press stability and printability, absence of fire and health hazards, convenience and economy of water as a diluent and for washup. Disadvantages include low gloss and slow drying which limits their use to absorbent stocks.

Solvent-based inks dry mainly by evaporation of volatile solvents which include the lower alcohols together with esters, glycol ethers, and the lower aliphatic hydrocarbons. These solvents are used to dissolve a wide variety of vehicles including nitrocellulose, cellulose ethers and esters, polyamides, acrylics, and modified rosins and ketone resins.

13.5. GRAVURE

The major elements of the gravure process consist of the gravure cylinder on which the image to be reproduced is etched, the impression roller which brings the web of paper, foil, or film into contact with the gravure cylinder, a doctor blade which removes excess ink from the surface of the cylinder, and an ink reservoir in which the cylinder is immersed.

Intaglio printing consists of a process such as gravure and engraving in which the image or design is recessed below the nonimage areas of the engraving, plate, or cylinder. The best end-use example of this process is printing of United States currency.

Gravure or intaglio inks are low-viscosity inks that dry by solvent evaporation. They are very versatile and may be formulated with an exceptionally wide range of resin vehicles. There are four main types of gravure inks. Each has certain specific applications which designate the type of binder and solvent used.

1. Type A is used for publication printing and is the cheaper of the gravure inks.
2. Type B is used for publication printing on better-grade stocks than Type A.
3. Type C is used for various types of packaging.
4. Type T is used for package printing, primarily food cartons.

13.6. OTHER INKS

13.6.1. Screen Printing

This printing system (formerly known as silkscreen) is a stenciling technique in which a heavy film of ink is applied through a mesh screen in the form of a design. Its former name related to the material used to support the stencil.

Variation in mesh size permits control of the thickness of the ink film laid down. Today screens stronger than silk are made of metal mesh and of synthetic fibers.

The surface to be decorated is placed under a stencil and a mass of ink is drawn across the stencil surface by a rubber squeegee. The ink is forced through the open areas of the stencil and deposited on the printed screen.

Screen printing is well suited for the preparation of large posters as the size of the poster is limited only by the ability to make a clean wipe over the screen.

13.6.2. Electrostatic

The basis of the process is an electrically charged conducting stencil that acts as the image-forming master. The stencil is similar to that used in screen process work, and the stencil support has electroconductive properties. Fine mesh is used to obtain high resolution.

13.6.3. Metallic

These inks consist of a suspension of fine metal flakes in vehicles that serve to bind the powders to the surface being printed. The high brilliance and luster characterizing these inks are caused by the "leafing" of metal flakes when they float to the ink surface. Examples of metals are aluminum, bronze, and copper.

13.6.4. Watercolor

These inks are generally employed in the printing of wallpaper, greeting cards, and novelties. Watercolor inks are based on a vehicle composed essentially of gum arabic, dextrin, glycerin, and water. Pigments or dyes can be used as the colorant in this type of ink. Special rollers are required and water is used to wash the press.

13.6.5. Cold-Set

Inks of this type are solid rather than liquid at room temperature. They consist of pigments dispersed in plasticized waxes having melting points ranging from 150 to 200°F (65.6 to 93.3°C).

They are used on presses with fountains that are heated above the melting point of the inks. The inks are melted and maintained in a fluid condition until they are impressed on the relatively cold paper, where they revert almost instantly to their normal solid state. The advantages of these inks are that they do not smudge or "setoff," are almost tack-free when in the fluid state, neither skin in the cans nor dry on the presses, and yield sharper printing results, as they do not penetrate into the pores of the paper.

13.6.6. Magnetic

Magnetic inks are employed in an electronic system for character recognition that is used for sorting and calculating items such as bank checks, business forms, and others. Magnetic inks are made with pigments (e.g., iron oxides) that are magnetized after printing and the printed characters can later be recognized by electronic reading equipment. These are formulated to produce exceptionally high-grade printing.

13.6.7. Optical or Readable

Optical reading equipment has become very useful for reading information on products. The inks used for forming "bar codes" and other reading images are very precisely formulated to provide dense stripes or bars to be used with laser or other reading equipment. Usually, the bar code is a carbon pigment ink producing a dense black stripe when printed.

13.7. INK FORMULATIONS

The ink formulations are shown in Tables 13.2–13.44. Each component has a purpose that is essential for the overall performance of the ink. The vehicle is the binder (e.g., resin, rosin, polymer) which carries the pigment to a surface and dries to produce an image. The vehicle requires a solvent such as alcohol to reduce the viscosity so that it flows easily onto a surface. If the vehicle is water dispersible or soluble, then water is the solvent. The increasingly stringent environmental regulations are moving ink formulations in the direction of water systems. The color of the ink is provided by a pigment or dye. The pigment is usually a solid particle with a color, and the dye is a chemical compound that is soluble in a solvent. Pigments and dyes can be used together to achieve a desired appearance.

Pigments are listed by color name rather than chemical composition as is the custom with formulators. Properties of pigments and dyes differ as the pigments are usually inorganic solid particulates and dyes are soluble organic compounds.

13.8. VARNISHES

Varnishes cover inks to enhance appearance and protection. Varnishes contain no pigments and are formulated for transparency rather than color.

Examples are provided in Tables 13.45 and 13.46.

14

Ink Materials

14.1. GENERAL

The formulations in Chapter 13 contained ingredients used in the manufacture of printing inks which fall into three categories:

1. Liquids such as vehicles
2. Solids such as pigments
3. Supplementary additives such as driers

The raw materials (Leach and Pierce, 1988), chemical description, and sources of materials are provided in Table 14.1.

The vehicle acts as a carrier for the pigment and as a binder to affix the pigment to the printed surface. The nature of the vehicle determines in large measure the tack and flow characteristics of a finished ink.

14.2. VEHICLES

14.2.1. Nondrying Oil Vehicle

Inks printed on soft absorbent papers, such as news and comics inks, dry by the absorption of the vehicle into the paper. The vehicle consists of nondrying, penetrating oils such as petroleum oils, rosin oils, and others, used in combination or modified with various resins to impart suitable tack and flow characteristics.

14.2.2. Drying Oil Vehicle

Autoxidation drying is the type used in most letterpress and offset inks today. It also plays an important role in other types of drying processes by imparting final, thoroughly hard drying after the inks have been initially "set."

Autoxidation generally proceeds in two stages: the absorption of oxygen from air, and the cross-linking or hardening of the vehicle. Only the second stage produces a physical change and development of a hardened film.

Drying oils include, but are not limited to, the following:

1. Linseed oil
2. Cottonseed oil
3. China wood oil
4. Castor oil
5. Perilla oil
6. Soybean oil
7. Petroleum drying oils
8. Fish oil
9. Rosin oil
10. Synthetic drying oils

Linseed oil or litho varnish is the most widely used. Raw linseed oil is not suitable as a printing ink vehicle, and it must be converted by boiling or bodying. Bodying the oil increases the viscosity and adjusts other properties. The temperature used determines the "body" or viscosity of the oil. Linseed oil varnishes have excellent wetting properties for most pigments, and they have good transfer qualities and provide good binding on paper.

14.2.3. Others

Combining oils and synthetic resins can obtain faster and harder drying inks. Chemical modification of oils such as development of an alkyd resin can produce significant improvements in inks.

- Solvent-resin
- Resin oil
- Resin wax
- Water-soluble gum
- Waterborne
- Photoreactive

14.3. SOLVENTS

Solvents are used to thin the vehicle or varnish so that the ink can be applied to form a wet film and transfer to the surface of the substrate. Solvents include water toluene, alcohols, and in waterborne systems, water. Common solvents used in inks are listed in Table 14.2.

14.4. INORGANIC PIGMENTS

Pigments are the solid coloring matter in inks (Leach and Pierce, 1993), but they also determine the specific gravity, opacity or transparency, and resistance to light, heat, and chemicals.

14.4.1. Black Pigments

Black pigments are mostly furnace black and thermal black. Furnace blacks are produced by cracking oil in a continuous furnace and are smaller than thermal blacks. Thermal blacks are made in batch furnaces by cracking natural gas. The primary composition of furnace and thermal blacks is carbon. Mineral blacks are used for special purposes such as magnetic recognition of printed characters.

14.4.2. White Pigments

Opaque pigments reflect light from their surfaces and cover or hide the background on which they are printed. Widely used white pigments are:

1. Titanium dioxide
2. Zinc sulfide
3. Lithopones
4. Zinc oxides

These pigments can be used alone or in combination with other pigments to add opacity or lighten the color.

Transparent pigments do not reflect light at the surface, but transmit light or allow light to pass through the film of ink to be reflected from the surface on which it is printed. Transparent pigments do not hide the background, but allow the background to be seen through the film. Common transparent pigments are:

1. Aluminum hydrate
2. Magnesium carbonate
3. Calcium carbonate
4. Barites
5. Clays

14.4.3. Chrome Yellow

Chrome yellow is produced in a number of shades, from the greenish primrose shade, through the lemon and chrome shades, all the way into the orange. It is generally lead chromate, modified with other lead compounds, especially lead sulfate.

14.4.4. Chrome Green

Chrome green is largely a mixture of chrome yellow with iron blue.

14.4.5. Chrome Orange

Chrome orange and molybdate orange are modified lead compounds similar in structure to chrome yellow. All chrome colors are fast (stable) to light, opaque, and have large specific gravities. Some chrome colors darken on exposure to sulfur compounds in polluted air.

14.4.6. Cadmium (Selenide) Yellows

Oranges and reds are very fast to light and have excellent soap and alkali resistance. They are useful for long exterior exposures where extreme permanency is required, and for soap wrappers where resistance to alkali and soap is necessary.

14.4.7. Cadmium–Mercury Reds

Cadmium–mercury reds range from bright red to deep red, and their properties are similar to those of the older cadmium reds (cadmium selenide).

14.4.8. Vermilion

Vermilion is a red mercury sulfide pigment, heavy in specific gravity, brilliant, and opaque. It is useful where extreme hiding power and resistance to sulfur are important.

14.4.9. Iron Blue

Also made in a number of shades such as milori blue, bronze blue, Prussian blue, and toning blue, iron blue is actually a chemical compound of iron. Iron blues are light in specific gravity, transparent, and permanent to light when used in full strength.

14.4.10. Ultramarine Blue

Ultramarine blue is a mineral pigment, generally transparent, and permanent to light.

14.5. METALLIC PIGMENTS

14.5.1. Silver

Silver is usually aluminum powder.

14.5.2. Gold

Gold is usually a mixture of copper, brass, and other metal flakes to produce varying shades of gold.

14.6. ORGANIC PIGMENTS

Organic pigments are the largest group of pigments used in printing inks.

14.6.1. Yellows

Yellows are primarily yellow lakes, hansa yellows, and diarylide yellows. Yellow lakes are produced from several dyes and pigments of different hues. They are useful when yellow must be printed over darker colors, but not hide or cover them. They are usually transparent.

Lake pigments are usually transparent coloring substances produced from organic dyes by depositing the colors on one of the transparent white materials. They may be considered as dyed transparent white pigments, usually alumina hydrate.

Hansa yellows are strong, permanent, and resistant to many chemicals. They are produced in a variety of hues and are used frequently for strengthening the color of chrome yellows. Diarylide yellows are usually not so lightfast as hansa yellows, but are more transparent. They are used for toning chrome yellows where extreme lightfastness is required.

14.6.2. Oranges

Oranges most commonly used in printing inks are diarylide and pyrazolone orange, a yellow shade orange that combines good fastness properties with tinctorial strength. It is fast to acid, alkali, water, soap, and wax.

14.6.3. Reds

An example of red pigment is naphthol red (or permanent red FRR). This is a strong, bright, clean, yellow shade red, with excellent resistance to acids, alkali, soap, and detergent. It is fairly lightfast.

14.6.4. Blues

An example is PMTA Victoria Blue PMYA brilliant blue. It has a bright reddish blue of high tinctorial strength and purity of hue. It has good lightfastness, and is affected by polar solvents.

14.6.5. Greens

An example of a green pigment is PMTA deep green. This is a bright bluish green of maximum strength with a clean undertone. It has fair lightfastness, and poor resistance to alkali, soap, and strong solvents.

14.6.6. Fluorescents

Powders are created by pulverizing solutions of fluorescent basic or reacted dyes in resins. Fluorescent dyes/pigments have the property of converting short-wavelength radiation into longer wavelengths giving brilliant colors.

14.7. FLUSHED PIGMENTS

When a pigment is manufactured, it is not dried but sold as a paste. The flushed pigment prevents clustering of particles and assists distribution and mixing in the ink formulation.

14.8. DYES

Dyes are used in printing inks because of their optical properties (e.g., transparency, high purity, and color strength). They are distinguished from pigments by their solubility in printing ink vehicles. Dyes are used primarily as toners.

14.9. ADDITIVES

To impart special properties to inks, ingredients such as driers, waxes, lubricants, reducing oils, antioxidants, gums, starches, and surface-active agents are used.

Some may be incorporated directly into the vehicle during cooking. Others may be added during formulating. Others can be added after formulating.

14.9.1. Driers

Driers act as catalysts to speed the autoxidation and drying of the vehicle. Drier are compounds of lead, cobalt, copper, iron, manganese, zinc, zirconium, and other metals. Too much drier causes the vehicle to "skin" and dry on the press. Each vehicle requires a specific drier.

14.9.2. Waxes and Compounds

Waxes are used primarily to prevent setoff and sheet sticking and to improve scuff resistance. The most common waxes are paraffin wax, beeswax, carnauba wax, microcrystalline, ozocerite, and polyethylene. The wax may be cooked directly into the varnish or prepared as a compound and added directly to the ink. Micronized waxes are widely used to shorten (reduce flow) an ink whereas compounds are used to reduce the tack of an ink.

14.9.3. Lubricants and Greases

Cup grease, wool grease, petroleum jelly, and tallow will reduce the tack of an ink and cause it to set quickly. They will also help lubricate the ink so that it distributes and transfers properly. Too much lubrication will cause an ink to become greasy and print poorly.

14.9.4. Reducing Oils and Solvents

These are thin-bodied oils and are used in much the same way as the greases. They aid penetration and rapid setting. High-boiling solvents and thinners may be used in letterpress and lithographic inks to reduce the tack. In flexographic and gravure inks, special care must be taken to use solvents that are compatible with the vehicles used in the inks.

14.9.5. Body Gum and Binding Varnish

Body gum and binding varnish are used to add viscosity to an ink. They pull the ink together and help it to print sharply. In lithographic inks they help to overcome emulsification, improve drying, and prevent chalking of the ink.

14.9.6. Antioxidants or Antiskimming Agents

These agents are sometimes used to reduce excessive drying and skinning on the press. They are very active chemically and should be used with caution. Excessive amounts will prevent the ink from drying on the paper after printing.

14.9.7. Corn Starch

Corn starch and other dry powders such as dry magnesia are used to prevent setoff and to body-up an ink. Too much will cause caking, piling, and fillup.

14.9.8. Surface-Active Agents

These chemicals are used to obtain better wetting and dispersion of pigments. Their use must be controlled as the selection of these materials is critical for each application.

Ink materials and suppliers are listed in Table 14.1.

15

Deformulation of Inks

15.1. INTRODUCTION

Liquid inks are more like paints as they are well pigmented/filled for color and opacity effects. They are usually viscous and contain either water or organic solvents. Therefore, the method of deformulating inks is similar to that of paint and coatings. Of course, inks are not paints and are formulated for printing, drawing, writing, and the like.

15.2. DEFORMULATION OF SOLID INK SPECIMEN

Sources of inks and methods of preparation are shown in Fig. 15.1. These solid specimens are scraped or cut from the substrate before proceeding with the deformulation. The specimen can be cut with a sharp razor blade or frozen in liquid nitrogen and broken to reveal a fresh surface. Also, the specimen can be pulverized to particles, swelled in solvent, and separated by centrifugation.

Figure 15.2 outlines a series of steps for complete deformulation of a solid ink specimen.

The exposed cross-sectional surface of the specimen is examined by optical and electron microscopy to determine the magnified appearance of the specimen. The specific magnification is left to the operator to resolve the image. The shape and size of particles are observed directly, and EDXRA determines the elements present in the particles and vehicle down to an atomic number of 5.

Example. A magnified image of the author's initials written in black pen ink is shown in Fig. 15.3. The section of the "J" marked with an arrowhead is magnified 5000× and 10,000×, and the carbon particles are visible in the micrographs. The bottom-to-top flow of the pen while writing the letter is obvious from the elongated or channeled vehicle and particles. EDXRA of the particles identified them as carbon. Microscopic IR identified the vehicle resin matrix as acrylic resin. DSC thermal analysis of a specimen of the ink generated a glass transition event. A

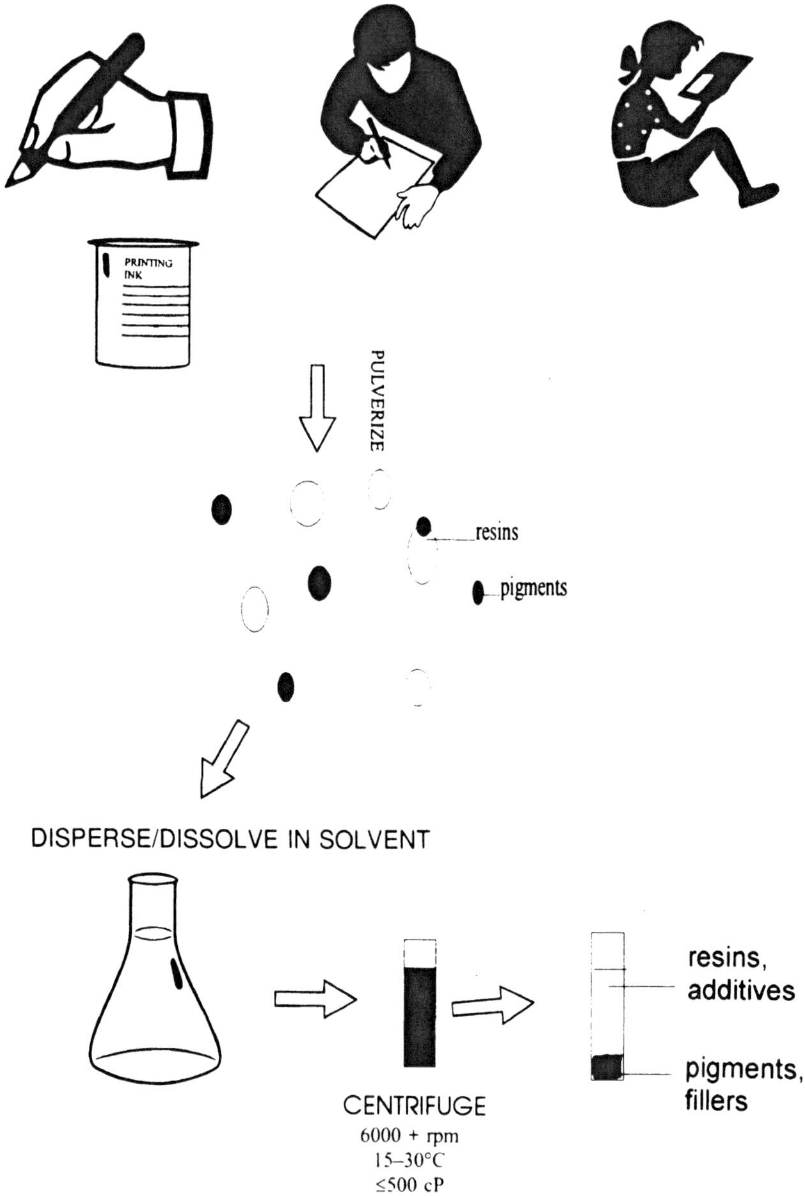

Figure 15.1. Scheme for preparation of solid ink specimen for deformulation.

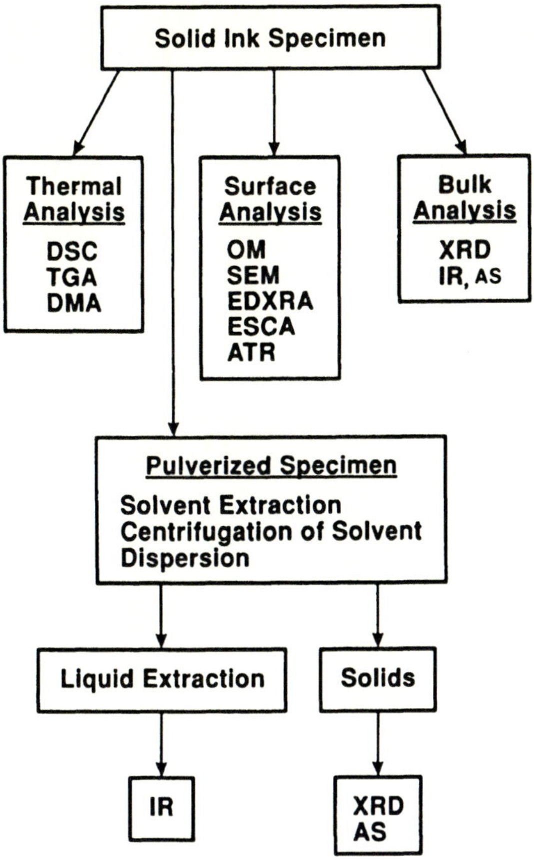

Figure 15.2. Scheme for deformulation of a solid ink specimen.

thermogravimetric analysis showed 26% carbon in the dried film. A percent solids determination on the liquid ink showed 16% of the specimen to be solids.

A bulk specimen can be further deformulated using XRD for inorganic crystalline materials, IR for vehicle chemical identification, and AS for accurate quantitative metals and other elemental identification.

To further refine the specimen and investigation, the specimen is frozen in liquid nitrogen and hammered to pulverize it. The particles are swelled in solvent to soften the ink and separate particles from vehicle. Refluxing the solid specimen in hot solvent (see Fig. 6.4) is a more effective method of swelling the ink. The solvent will extract soluble components (liquid extraction) which are analyzed by HPLC and GPC. The swollen vehicle will either dissolve or form gel particles, either of which are oven dried and analyzed by IR and/or NMR.

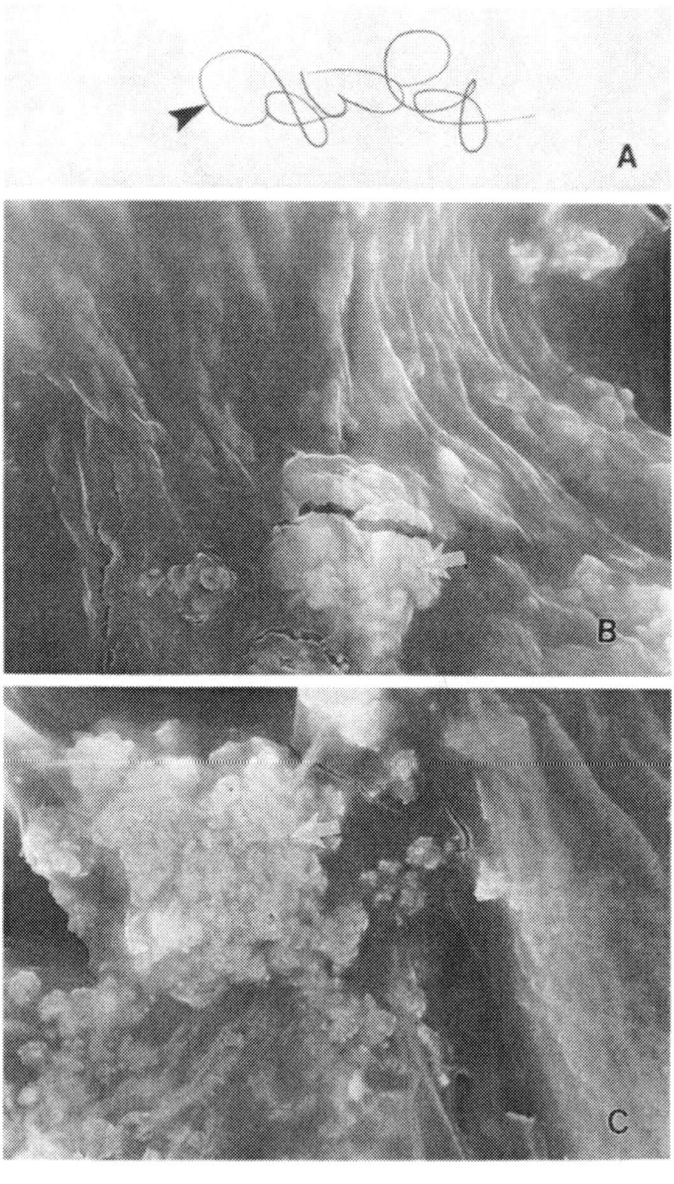

Figure 15.3. SEM micrographs of washable black writing pen ink.

The solids are analyzed by XRD and AS. The EDXRA data give valuable preliminary information about the composition of the specimen, which saves time selecting tools for investigation.

Further investigation of a solid specimen includes AES, SIMS, and especially ESCA for microscopic chemical analysis of surfaces. ESCA provides chemical composition data of vehicle (resins and polymers) and pigments and fillers. However, EDXRA does not detect elements below about 1% in formulated materials (practically speaking), and parts per million concentrations of elements will not be detected. Don't rely too heavily on EDXRA.

15.3. DEFORMULATION OF LIQUID PAINT SPECIMEN

A scheme for the preparation of a liquid ink specimen for deformulation is shown in Fig. 15.4. A liquid ink is ready for centrifugation (see Fig. 1.2) to separate components if the viscosity is 500 cP or less; if not, the viscosity is adjusted with solvent. Weigh each centrifuge tube, then weigh the specimen in the tube. Make sure that the tubes and specimens are within 0.1 g of each other to prevent vibration during centrifugation. Centrifuge several tubes (60–100 cm^3) until the specimen is separated into distinct layers. Remove the layers individually using a pipette for the liquid and a small spatula for the solids. The liquids may be recentrifuged to remove any turbidity and keep the solids from this separation.

Filtration is not recommended except when a centrifuge is not available. Even a low-speed centrifuge is preferable to filtration as the solids will adhere to the filter media and the liquid must be rehandled with much inherent error.

The solids will not be completely free from vehicle, so transfer each layer to a centrifuge tube (60–100 cm^3), add a solvent, and recentrifuge. Pour off the solvent, then oven dry (105°C for 2–3 hours) each layer and weigh prior to following the analytical scheme. Overdrying will cause oxidation and loss of weight, so weigh the solids as soon as constant weight is achieved. Perform an EDXRA evaluation before proceeding to other methods shown in Fig. 15.2. The EDXRA spectrogram will provide an elemental profile (metals, etc.) of the solids which will aid the investigator when performing XRD and AS analyses.

These purification steps yield specimens that will generate reliable data when investigated with analytical instruments. Interferences from cross contamination (vehicle, etc.) will reduce the quality of the data. There is no substitute for good sample preparation, and there cannot be good analytical instrumental analysis unless the sample is adequately prepared.

Weigh each liquid layer, and oven dry an aliquot to drive off volatile liquids such as solvents and leave the higher-molecular-weight vehicle (resins and polymers). Analyzing the vehicle component according to the scheme in Fig. 15.5 will yield valuable GPC and IR data for molecular weight and chemical identification, respectively. It is important to first identify the vehicle to choose a carrier solvent

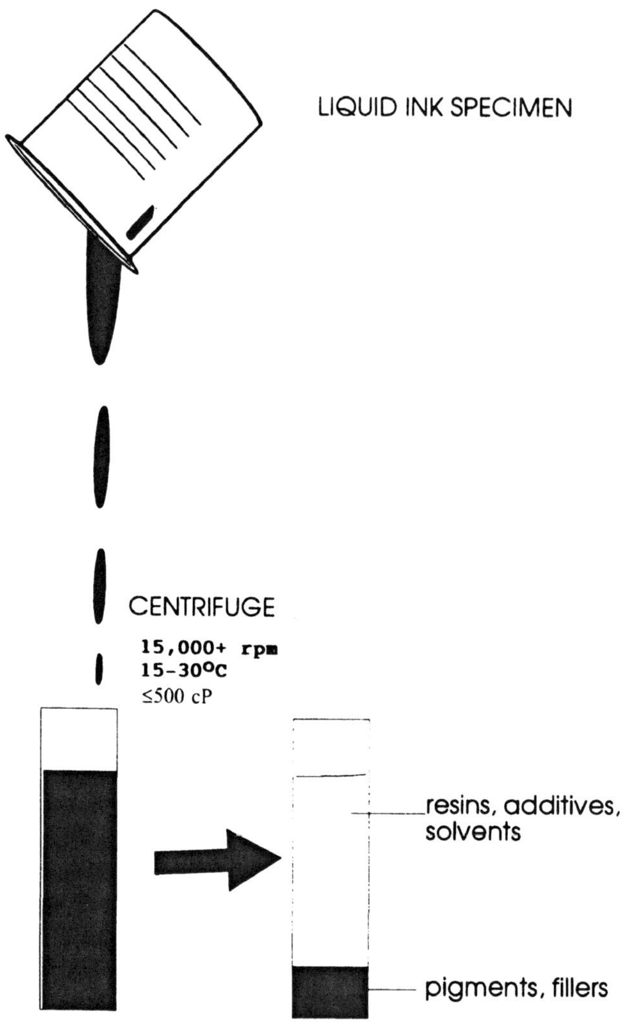

Figure 15.4. Scheme for preparation of liquid ink specimen.

for preparation of the GPC specimen. The GPC specimen must be prepared in the same solvent (same HPLC) as the carrier solvent and filtered before injection into the injection port.

Many ink products such as flexographic inks are water-based because of environmental and health regulations. Therefore, the major liquid component is water, which is easily identified by IR.

The additives (e.g., flow agents and rheology aids) will be present in concentrations of less than 5.0% and particular attention must be given to these compo-

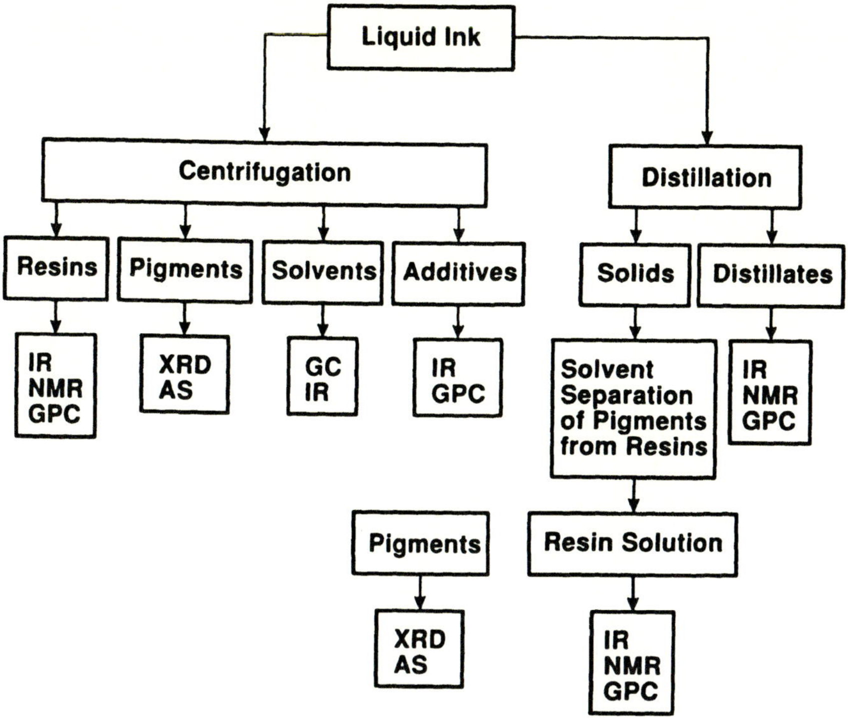

Figure 15.5. Scheme for deformulation of liquid ink specimen.

nents. They are detected by HPLC and will usually be found with the higher-molecular-weight vehicle. If not separated from the vehicle, additives will show a separate, if not convoluted, IR spectrum along with the vehicle. Once the vehicle has been identified, the additional IR absorbance is electronically separated from the total spectrum which is useful for identifying the additives. Identification of additives will correspond to HPLC peaks not associated with those generated by the vehicle.

Catalysts are usually present in concentrations of less than 1.0% which makes them more difficult to identify. They are usually found with the vehicle, but may distill with solvents. If the solvent proves to be water, measure the pH and this will provide a clue to the presence of bases and acids. Parts per million concentrations of metallic ions (and others) in a catalyst are detected by AS (including ICP) which gives information about the total identification.

By this point in the scheme for deformulation, the volatile liquids are all that is left for analysis in the sample. Take an aliquot of the centrifuged liquid component and inject it into a GC or HPLC. An effective method for evaluating the volatile component of the centrifuged liquid component is to inject a head-space vapor specimen into a GC. This consists of heating (about 100°C) a few cubic centimeters of the centrifuged liquid component in a closed vessel to create a vapor of volatile liquids (solvents, water, etc.) at the top of the vessel. A syringe is used to remove a head-space specimen through a rubber septum in the top of the vessel followed by injection in a GC. The GC will separate each solvent, etc. Water is usually an interference in this method. A better method is to distill the centrifuged liquid component and analyze each distillate separately. This assumes that a sufficient quantity of the specimen is available.

The scheme in Fig. 15.5 provides a plan for completely deformulating a liquid ink specimen. If a few hundred grams of original specimen is available, it is weighed, the water or solvent is separated by distillation (see Fig. 6.7) and then measured gravimetrically and volumetrically to determine the amount of solvent. Separation of mixed solvents is accomplished by observing the boiling temperature during distillation and catching each distillate in a separate receiving flask. Each solvent is placed in a liquid IR cell and an IR thermogram is generated. Other methods include GC and HPLC to identify and quantify solvents.

15.4. REFORMULATION

After performing these investigations, prepare a table of components versus percent weight. Acquire materials from the generated table, and reformulate the original recipe. Compare the properties of the new formulation with the original and any published specifications.

References

Adhesives Age. 1977. Communications Channels, Inc., Argus Press Holdings, Skokie, Illinois, pp. 128–45.

Adhesives Age. 1993. Communications Channels, Inc., Argus Press Holdings, Skokie, Illinois, pp. 128–45.

Amelinck, S. 1964. *The Direct Observation of Dislocations*. Academic Press, New York.

———, 1970. *Modern Diffraction and Imaging Techniques in Material Science*. North-Holland, Amsterdam.

American Society for Testing and Materials, 1989. ASTM D 3335, Test method for low concentrations of lead, cadmium and cobalt in paint by atomic absorption spectroscopy, "Annual Book of ASTM Standards," Vol. 6.01, Philadelphia, PA 19103.

ASTM-*Wyandotte Index*. 1963. "Molecular formula list of compounds, names, and references to published infrared spectra." Am. Soc. Testing Materials, spec. tech. publ., 131 (1962), 131-A (1963).

Axel, F. 1992. "Mold release agents." *Modern Plastics Encyclopedia*, p. 177.

Barr, T. L. 1994. *Modern ESCA, The Principles and Practice of X-Ray Photoelectron Spectroscopy*. CRC Press, Boca Raton, Florida, pp. 5, 12, 14.

Bellamy, L. J. 1958. *The Infrared Spectra of Complex Molecules*. 2nd ed. Wiley, New York.

Bertin, E. P. 1970. *Principles and Practice of X-Ray Spectrometric Analysis*. Plenum, New York.

Birks, L. S. 1959. *X-Ray Spectrochemical Analysis*. Wiley–Interscience, New York.

———, 1963. *Electron Probe Microanalysis*. Wiley–Interscience, New York.

Bowie, S. H. U.; and Taylor, K. 1958. A system of ore mineral identification, *Min. Mag.* **99**:265–77, 337–45.

Boyde, A. 1970. "Practical problems and materials in three-dimensional analysis of scanning electron microscopy images." In *Published Proceedings of the Third Annual Scanning Electron Microscopy Symposium*. IIT Research Institute, Chicago, pp. 107–12.

Bragg, W. L. 1933. *The Crystalline State*, Macmillan, New York.

Braksmayer, D. 1992. "Flame retardants." *Modern Plastics Encyclopedia*, p. 164.

Brandis, R. L. 1990. "Animal glue." *Adhesives Handbook*. Van Nostrand–Reinhold, Princeton, pp. 123–29.

Brandrup, J.; and Immergut, E. H. 1975. *Polymer Handbook*. Wiley, New York, pp. IV-1–IV-267.

Bunn, C. W. 1961. *Chemical Crystallography, 2nd ed.*, Oxford University Press, New York.

Cahn, H. L. 1974. "Silicone." *Technology of Paints, Varnishes and Lacquers*. Van Nostrand–Reinhold, Princeton, pp. 223–57.

Cameron, E. N. 1969. *Ore Microscopy*, John Wiley and Sons, New York.

Chang, C. C. 1971. "Auger electron spectroscopy." *Surface Sci.* **25**, 53–9.

Clark, G. L. 1955. *Applied X-rays, 4th ed.*, McGraw-Hill, New York.

Collins, E. A.; Bares, J.; and Billmeyer, F. W. 1973. *Experiments in Polymer Science*. Wiley, New York, pp. 154–67.

Colo, S. M. 1986. "The mechanical properties of polymers." Proceedings of the Pittsburgh Conference, Paper 729.

Colthup, N. B.; Daly, L. H.; and Wiberley, S. E. 1964. *Introduction to Infrared and Raman Spectroscopy*. Academic Press, New York.

Cooke, C. J.; and Duncumb, P. 1969. "Performance analysis of a combined electron microscope and electron probe microanalyzer, EMMA." In *Fifth International Congress on X-Ray Optics and Microanalysis* (Mollenstedt, G.; and Gaukler, K. H., eds.). Springer-Verlag, Berlin, pp. 245–7.

Coover, H. W.; Dreifus, D. W.; O'Connor, J. T. 1990. "Cyanoacrylates adhesives." *Adhesives Handbook*. Van Nostrand–Reinhold, Princeton, pp. 463–70.

Crewe, A. V. 1970. "High-resolution scanning microscopy of biological specimens." *Ber. Bunsenges. Phys. Chem.* **74**, 1181–7.

Cunningham, Davis and Graham. 1986. X-Ray Microscopy. *J. Microsc.* **144**, Pt. 3, December, 261–75.

Dann, J. R. 1970. "Forces involved in the adhesive process." *J. Colloid Interface Sci.* **32**(2), 302–30.

Dean, J. W. 1990. "Silicon adhesives and abhesives." *Adhesives Handbook*. Van Nostrand–Reinhold, Princeton, pp. 522–9.

Dean, J. A.; and Raines, T. C., eds. *Flame Emission and Atomic Absorption Spectrometry*. Vol. 1, *Theory*. 1969. Vol. 2, *Components and Techniques*. 1971. Vol. 3, *Elements and Matrices*. 1974. Dekker, New York.

Deanin, R. D. 1985. "Foamed plastics." in: *Applied Polymer Science*. American Chemical Society. Pp. 469–90.

Dieckmann, D. 1992. "Plasticizers." *Modern Plastics Encyclopedia*, p. 184.

Dotson, S. 1992. "Modifiers." *Modern Plastics Encyclopedia*, p. 175.

Drews, M. J.; Barker, R. H.; Hatcher, J. D. 1985. "Fiber-forming polymers," in: *Applied Polymer Science*. American Chemical Society. Pp. 441–65.

Duncumb, P. 1969. "Recent advances in electron probe microanalysis." *J. Phys. E* **2**, 553–60.

Eastman, E. F.; and Fullhart, L. 1990. "Polyolefin and ethylene copolymer hot-melt adhesives." *Adhesives Handbook*. Van Nostrand–Reinhold, Princeton, pp. 408–20.

Elias, H.-G. 1977. *Macromolecules*. Plenum, New York, pp. 263–79, 338–9, 373–420, 863–5, 927, 933, 939, 980, 1001, 1010.

Ennis, R. S. 1992. "Degradability additives." *Modern Plastics Encyclopedia*, p. 163.

Farmer, D. H.; and Jemmott, B. A. 1990. "Polyvinyl adhesives." *Adhesives Handbook*. Van Nostrand–Reinhold, Princeton, pp. 423–35.

Ferraro, J. R. 1968. *Anal. Chem.* **40**:4, 24A (April).

Fifoot, R. E. 1992. "Fluoroplastics." *Modern Plastics Encyclopedia*, pp. 18–20.

Fisch, M. 1992. "Antioxidants." *Modern Plastics Encyclopedia*, p. 146.

Flick, E. W. 1985. *Printing Ink Formulations*, Noyes Publications, Park Ridge, New Jersey. pp. 169–81.

Fox, D. W.; and Peters, E. N. 1985. "Engineering thermoplastics: Chemistry and technology." *Applied Polymer Science*. American Chemical Society, Washington, D.C., p. 500.

Friedman, S. K. 1992. "Surface active agents." *Modern Plastics Encyclopedia*, p. 196.

Frisch, K. C.; and Kordomenos, P. 1985. "Urethane coatings." *Applied Polymer Science*. American Chemical Society, Washington, D.C., pp. 985–90.

Fry, J. S.; Merriam, C. N.; and Boyd, W. H. 1985. "Chemistry and technology of phenolic resins and coatings." *Applied Polymer Science*. American Chemical Society, Washington, D.C., pp. 1141–55.

Gazeley, K. F. 1990. "Natural rubber adhesives." *Adhesives Handbook*. Van Nostrand–Reinhold, Princeton, pp. 167–80.

Geelan, B. J. 1992. "Foaming agents." *Modern Plastics Encyclopedia*, p. 167.

Gehan, D. R. 1990, "Acylic resins," in: *Adhesives Handbook*, (Skeist, I., ed.), pp. 437–45.

Gianturco, M. 1965. In *Interpretive Spectroscopy* (Freeman, S. K., ed.). Van Nostrand–Reinhold, Princeton, Chap. 2.

Gilfrich, J. V.; and Birks, L. S. 1968. Spectral distribution of x-ray tubes for quantitative x-ray fluorescence analysis, *Anal. Chem.* **40**, 1070–80.

Gooch, J. W. 1980. *Autoxidative Polymerization of Vegetable Oils*. Ph.D. dissertation, University of Southern Mississippi.

———, 1982. "Emulsified oils and alkyds to generate polymers." *Use of Renewable Materials for Coatings and Plastics* (Sperling, L., ed.). Plenum, New York, pp. 303–20.

———, 1993. *Lead-Based Paint Handbook*. Plenum, New York, pp. 37–92.

Gordon, S. 1992. "Colorants." *Modern Plastics Encyclopedia*, p. 154.

Grivet, P. 1965. *Electron Optics*. Peragamon, London. pp. 195–200.

Grove, E. 1971. *Analytical Emission Spectroscopy*. Dekker, New York.

Haine, R.; and Cosslett, V. E. 1961. *The Electron Microscope, the Present State of the Art*. Wiley–Interscience, New York.

Hall, C. E. 1966. *Introduction to Electron Microscopy, 2nd edition*, McGraw-Hill, New York.

Harrick, N. J. 1967. *Internal Reflection Spectroscopy*. Wiley, New York.

Hattori, K. 1992. "Color concentrates." *Modern Plastics Encyclopedia*, pp. 155–8.

Heidenreich, R. D. 1964. *Fundamentals of Transmission Electron Microscopy*. Wiley–Interscience, New York.

Hemsley, D. A. 1984. *The Light Microscopy of Synthetic Polymers*. Oxford University Press, London, pp. 10–24, 26–35.

Henke, B. L.; Newkirk, J. B.; and Mallett, G. R., eds. 1970. *Advances in X-Ray Analysis*. Vol. 13. Plenum, New York.

Herberg, G. 1945. *Molecular Spectra and Molecular Structure*. Vols. 1, 2. Van Nostrand–Reinhold, Princeton.

Hercules, S. H.; and Hercules, D. M. 1974. *Surface Characterization by ESCA*. Plenum, New York, pp. 307–34.

Higgins, J. J.; Jagisch, F. C.; Stucker, N. E. 1990. "Butyl rubber and polyisobutylene," in: *Adhesives Handbook*, (Skeist, I., ed.), pp. 185–99.

Hoffman, F. L. 1927. *Deaths from lead poisoning*, U.S. Dept. of Labor, No. 426.

Howell, P. G. T.; and Boyd, A. 1972. In *Scanning Electron Microscopy* (Johari, O.; and Corvin, I., eds.). IIT Research Institute, Chicago, pp. 233–40.

Hutchins, G. A. 1974. "Electron probe microanalysis," in: *Characterizations of Solid Surfaces* (Kane, P. F.; and Larrabee, G. B., eds.). Plenum, New York, p. 468.

Infrared Spectroscopy—Its Use in the Coatings Industry. 1969. Federation of Societies for Paint Technologies, Philadelphia.

Isings, J. 1961. In *Encyclopedia of Microscopy* (Clark, G. L., ed.). Reinhold, New York, p. 390.

Jacobs, M. H. 1971. "Microstructural studies with a combined electron microscope and electron probe microanalyzer (EMMA-3)." In *Published Proceedings of the Twenty-Fifth Anniversary Meeting EMAG*, Inst. Phys.

Jaffe, H. L.; Rosenblum, F. M.; and Daniels, W. 1990. "Polyvinyl acetate emulsions for adhesives." *Adhesives Handbook*. Van Nostrand–Reinhold, Princeton, pp. 381–5.

Johari, O. 1971. "Total materials characterization with the scanning electron microscope." *Res./Dev.* **22**(7), 12–20.

Johari, O.; and Samuda, A. V. 1974. "Scanning electron microscopy," in: *Characterization of Solid Surfaces* (Kane, P. F.; and Larrabee, G. B., eds.). Chapter 18, Plenum Press, New York.

Jones, S. J.; and Boyde, A. 1970. "Experimental studies on the interpretation of bone surfaces studied with SEM," in: *Proceedings of the Third Scanning Electron Microscopy Symposium*, IIT Research Institute, Chicago, pp. 195–200.

Joseph, M. L. 1986. *Introductory Textile Science*. Holt, Rinehart & Winston, New York, pp. 29, 46, 63, 64–79, 93, 101, 114, 124, 135, 141.

Kamath, V. R. 1992. "Organic peroxides." *Modern Plastics Encyclopedia*, pp. 184–7.

Kane, P. F.; and Larrabee, G. B. 1974. *Characterization of Solid Surfaces*. Plenum, New York.

Kay, D. 1961. *Techniques for Electron Microscopy*. Blackwell Scientific Publications, Oxford.

Klemperer, C. 1953. *Electron Optics*, Cambridge University Press, London.

Krause, A.; Lange, A.; and Ezin, M. 1979. *Plastics Analysis Guide*. Macmillan, New York, pp. 8–10, 17–22.

Leach, R. H.; and Pierce, R. J., eds. 1988. *The Printing Ink Manual*, Blueprint—Chapman & Hall, London, pp. 141–287.

Lenhart, S. J. 1992. "Antimicrobials." *Modern Plastics Encyclopedia*, New York, pp. 14–634.

Lerner, L. R.; and Salzman, M. 1985. "Color pigments." *Applied Polymer Science*. American Chemical Society, Washington, D.C., pp. 1271–90.

Levesque, J. 1992. "Smoke suppressants." *Modern Plastics Encyclopedia*, p. 188.

Liebhafsky, H. A.; Pfeiffer, H. G.; Winslow, E. H.; and Zemana, P. D. 1960. *X-Ray Absorption amd Emission in Analytical Chemistry*. Wiley, New York.

Liebhafsky, H. A.; Pfeiffer, H. G.; and Winslow, 1964. "X-ray methods: absorption diffraction and emission," in: *Treatise on Analytical Chemistry*, Vol. 5, Part I., (Kolthoff, I. M.; and Elving, P. J., eds.). Wiley–Interscience, New York, Chapter 60.

Liebhafsky, H. A.; Pfeiffer, H. G.; Winslow, E. H.; and Zemana, P. D. 1960. *X-ray Absorption and Emission.*

Low, M. J. D. *Anal. Chem.* **41**:6, 97A (May 1969); *J. Chem. Educ.* **47**:A163, A255, A349, A415 (1970).

———, 1970. *J. Chem. Educ.* **47**, A163, A255, A349, A415.

Lupinski, J. H. 1985. "Polymers and the technology of electrical insulation." *Applied Polymer Science*. American Chemical Society, Washington, D.C., p. 524.

McCrone, W. C. 1974. "Light microscopy." *Characterization of Solid Surfaces*. Plenum, New York, pp. 10, 11, 18, 29–30.

MacDonald, N. C. 1971. In *Scanning Electron Microscopy* (Johari, O.; and Corvin, I., eds.). IIT Research Institute, Chicago, pp. 89–96.

Mackey, D. E.; and Weil, C. E. 1990. "Nitrile rubber adhesives." *Adhesives Handbook*. Van Nostrand–Reinhold, Princeton, pp. 206–25.

Martens, C. R. 1974. *Technology of Paints, Varnishes, and Lacquers*. R. E. Krieger, Huntington, New York. Pp. 24–5, 27.

Meath, A. R. 1990. "Epoxy resin adhesives." *Adhesives Handbook*. Van Nostrand–Reinhold, Princeton, pp. 347–60.

Mesch, K. A. 1992. "Lubricants." *Modern Plastics Encyclopedia*, p. 172.

Midgley, C. A.; and Rea, J. B. 1990 "Styrene butadiene rubber adhesives." *Adhesives Handbook*. Van Nostrand–Reinhold, Princeton.

Miller, I. K.; and Zimmerman, J. 1985. "Condensation polymerization and polymerization mechanisms." *Applied Polymer Science*. American Chemical Society, Washington, D.C., pp. 160–2, 167.

Miner, L. H. 1992. "Aramid hybrids." *Modern Plastics Encyclopedia*.

Modern Plastics Encyclopedia. 1992. Hightstown, New Jersey, p. 208.

Monte, S. J. 1992. "Titanates." *Modern Plastics Encyclopedia*, p. 163.

Morrison, R. T.; and Boyd, R. N. 1973. *Organic Chemistry*. Allyn & Bacon, Boston, pp. 405–50.

Nakanishi, K. 1962. *Infrared Absorption Spectroscopy–Practical*, Holden-Day, San Francisco.

Nyquist, R. P.; and Kagel, R. O. 1971. *Infrared Spectra of Inorganic Compounds*. Academic Press, New York.

Osterholtz, F. 1992. "Silanes." *Modern Plastics Encyclopedia*, p. 160.

Panek, J. R. 1990. "Polysulfide sealants and adhesives." *Adhesives Handbook*. Van Nostrand–Reinhold, Princeton, pp. 307–15.

Park, R. A. 1985. "Vinyl resins used in coatings." *Applied Polymer Science*. American Chemical Society, Washington, D.C., pp. 1181–95.

Phillips, F. J. 1992. "Carbon glass hydrids." *Modern Plastics Encyclopedia*, p. 210.

PPG Industries, Inc. 1992. "Glass fibers." *Modern Plastics Encyclopedia*, pp. 212–14.

Prescott, R. 1992. "Carbon fibers." *Modern Plastics Encyclopedia*, pp. 208–10.

Printing Ink Handbook. 1976. The National Association of Printing Ink Manufacturers, New York.

Ringwood, R. 1992. "Stabilizers." *Modern Plastics Encyclopedia*, p. 190.

Robertson, J. M. 1953. *Organic Crystals and Molecules*. Cornell University Press, Ithaca, New York.

Rooney, J. M.; and Malofsky, B. M. 1990. "Anaerobic adhesives." *Adhesives Handbook*. Van Nostrand–Reinhold, Princeton, pp. 381–5.

Rossitto, C. 1990. "Polyester and polyamide high performance hot-melt adhesives." *Adhesives Handbook*. Van Nostrand–Reinhold, Princeton, pp. 478–498.

Roulin-Moloney, A. C. 1989. *Fractography*. Elsevier, New York, pp. 233–86.

Rubin, I. I. 1972. *Injection Molding*. Wiley, New York, pp. 296–7, 323–37.

Rutherford, H. J. 1992. "Fragrances." *Modern Plastics Encyclopedia*, p. 171.

Sadtler Research Laboratories. 1963. *Catalog of Infrared Spectrograms*. Philadelphia.

Shafrin, E. G. 1977. "Critical surface energy of polymers." *Adhesives Handbook*. Van Nostrand–Reinhold, Princeton, pp. 67–8.

Shell Chemical Company, Technical Bulletin-Polyethyene Extruded Film. 1995. Houston, Texas.

Siegbahn, K. 1967. *ESCA, Atomic, Molecular, and Solid State Structure Studied by Means of Electron Spectroscopy*. Almquist & Siks, Uppsala, Sweden.

Siegbaum, K.; Nordling, C.; Johansson, G.; Hedman, J.; Heden, P. F.; Hamrin, K.; Gelius, U.; Bergmark, T.; Werme, L.; Manne, R.; and Baery, Y. 1969. *ESCA Applied to Free Molecules*, North Holland - American Elsevier, Amsterdam, New York.

Silverstein, R. M.; Bassler, G. C.; and Morrill, T. C. 1974. *Spectroscopic Identification of Organic Compounds*. Wiley, New York, pp. 5–19.

Skeist, I. 1990. *Handbook of Adhesives*. Van Nostrand–Reinhold, Princeton.

Skeist, I.; and Miron, J. 1977. *Handbook of Adhesives*. Van Nostrand–Reinhold, Princeton, pp. 3–65, 160–90, 423, 451, 522.

Slade, P. E.; and Jenkins, L. T., eds. 1970. *Techniques and Methods of Polymer Evaluation*. Vol. 1, *Thermal Analysis*. 1966. Vol. 2, *Thermal Characterization Techniques*. 1970. Dekker, New York.

Sloane, H. J. 1971. "The technique of Raman spectroscopy. A state of the art comparison to infrared." *Appl. Spectrosc.* **25**, 430.

Son, P.-N. 1992. "UV stabilizers." *Modern Plastics Encyclopedia*, p. 196.

Sperling, L., ed. 1983. *Use of Renewable Materials for Coatings and Plastics*. Plenum, New York.

Sproull, W. T. 1946. *X-rays in Practice*, McGraw-Hill, New York.

Stevens, V. L.; and Lalk, R. H. 1980. Solvent option for air quality compliance, Water-Borne and Higher-Solids Coatings Symposium, sponsored by University of Southern Mississippi and Southern Society for Coatings Technology, New Orleans, LA, (March 10–12, 1980).

Switzer, G.; Axelrod, J. M.; Lindberg, M. L.; and Larsen, E. S. 1948. Tables of spacings for angle 2θ, Cu Kα, Cu $K\alpha_1$, Cu $K\alpha_2$, Fe Kα, Fe $K\alpha_1$, Fe $K\alpha_2$, Circular 29, Geological Survey, U.S. Department of the Interior, Washington, DC; Tables for conversion of X-ray diffraction angles to interplanar spacings, Publications AMS 10, Government Printing Office, Washington, DC.

Tess, R. W. 1985. "Solvents." *Applied Polymer Science*. American Chemical Society, Washington, D.C., pp. 661–96.

Thomas, G. 1962. *Transmission Electron Microscopy of Metals*. Wiley, New York.

Thomas, L. E. 1971. *Course Notes in Electron Microscopy*, University of Pennsylvania.

Tobiason, F. L. 1990. "Phenolic resin adhesives." *Adhesives Handbook*. Van Nostrand–Reinhold, Princeton, pp. 316–20.

Tobin, M. C. 1971. *Laser Raman Spectroscopy*. Wiley–Interscience, New York.

Uihlein, J. 1992. "Alloys and blends." *Modern Plastics Encyclopedia*, pp. 15–17.

Updegraff, I. N. 1990. "Amino resin adhesives." *Adhesives Handbook*. Van Nostrand–Reinhold, Princeton, pp. 341–46.

Van Drumpt, J. D. 1992. "Antistats." *Modern Plastics Encyclopedia*, p. 150.

VISTA Technical Bulletin - PVC Extruded Pipe. 1995. Houston, Texas.

Washabaugh, F. J. 1992. "Mineral fillers." *Modern Plastics Encyclopedia*, p. 220.

Wasilczyk, G. J. 1992. "Polyurethane foam catalysts." *Modern Plastics Encyclopedia*, p. 187.

Weast, R. C., ed. 1978. *CRC Handbook of Chemistry and Physics*. 59th ed. CRC Press, Boca Raton, Florida, p. F120.

Weismantel, G. E. 1981. *Paint Handbook*. McGraw–Hill, New York, pp. 1-1–1-50, 3-1–3-40.

Wilkes, H. H., Jr. 1972. "A practical guide to internal reflectance spectroscopy." *Am. Lab.* **4**, 11, 42.

Willard, H. H.; Merritt, L. L.; and Dean, J. A., eds. 1974. *Instrumental Methods of Analysis*. Van Nostrand–Reinhold, Princeton, pp. 150–88.

Williams, L. L.; Updegraff, I. H.; and Petropoulos. 1985. "Amino resins." *Applied Polymer Science*, American Chemical Society, Washington, D.C., pp. 1101–15.

Wills, J. 1977. "Oleoresinous adhesives." *Adhesives Handbook*. Van Nostrand–Reinhold, Princeton, pp. 241–49.

Wyckoff, R. W. 1949. *Electron Microscopy, Technique, and Applications*. Wiley–Interscience, New York.

Young, R. D. 1971. "Surface microtopography." *Phys. Today* **24**, 42–8.

Zeitler, E. 1971. In *Scanning Electron Microscopy* (Johari, O.; and Corvin, I., eds.). IIT Research Institute, Chicago, pp. 25–32.

Appendix

Table 1.1. Properties of Materials and Methods of Analysis

Property	Method of Analysis
Color	OM(S/B)
Virtual image and magnification	OM(S)
High topological magnification	EM(S)
Subsurface analysis	AES(S)
Elemental identification	EDXRA(S/B)
Chemical identification	EPM(S), AES(S), ESCA(S), IR(S/B), AS(B), XRD(B), NMR(B), GC(B), HPLC(B)
Crystal form and degree of crystallization	XRD(B), UV(B)
Melting temperature	DSC, DTA(B)
Glass transition temperature	DSC(B)
Decomposition temperature	TGA(B)
Modulus versus temperature	DMA(B)
Coefficient of thermal expansion	TMA(B)
Polymer/resin molecular weight	GPC(B)
Surface energy	G(S)
Viscosity	V(B)
X-ray imaging	XRM(S/B)

Legend: S, surface analysis; B, bulk analysis; OM, optical microscopy; EM, electron microscopy; EDXRA, energy-dispersive X-ray analysis; EPM, electron probe microanalysis; AES, Auger electron spectroscopy; ESCA, electron scanning chemical analysis; IR, infrared spectroscopy; AS, atomic spectroscopy; XRD, X-ray diffraction spectroscopy; GPC, gel permeation chromatography; HPLC, high-performance liquid chromatography; GC, gas chromatography; UV, ultraviolet spectroscopy; NMR, nuclear magnetic resonance; DSC, differential scanning calorimetry; TGA, thermogravimetric analysis; TMA, thermomechanical analysis; DMA, dynamic mechanical analysis; DTA, differential thermal analysis; V, viscosity; XRM, X-ray microscopy; G, goniometer.

Table 2.1. Infrared Absorption Frequencies, Chemical Groups, and Compounds

Bond	Compound Type	Frequency Range (cm^{-1})
–C–H	Alkanes	2850–2960
$-CH_2$	Alkanes	1450
$-CH_3$	Alkanes	1325–1400
–C–H	Alkenes	3020–3080
	Aromatic rings	3000–3100
	Alkynes	3300
–C=C–	Alkenes	1640–1680
–C≡C–	Alkynes	2100–2260
–C=C–	Aromatic rings	1500, 1600
–C–O	Alcohols, ethers, carboxylic acids, esters	1080–1300
–C=O	Aldehydes, ketones, carboxylic acids, esters	1690–1760
–O–H	Monomeric alcohols, phenols	3610–3640
	Hydrogen-bonded alcohols, phenols	3200–3600
	Carboxylic acids	2500–3000
–N–H	Amines	3300–3500
–C≡N	Nitriles	2210–2260
$-NO_2$	Nitro compounds	1515–1560 1345–1385

Sources: Morrison and Boyd (1973), Willard *et al.* (1974).

Table 3.1. ^{1}H-NMR Chemical Shifts and Types of Protons

Proton	Structure (**H**)	Chemical Shifts (δ), ppm
Cyclopropane		0.2
Primary	RCH_3	0.9
Secondary	R_2CH_2	1.3
Tertiary	R_3CH	1.5
Vinylic	C=C–**H**	4.6–5.9
Acetylenic	C≡C–**H**	2–3
Aromatic	Ar–**H**	6–8.5
Benzylic	Ar–C–**H**	2.2–3
Allylic	C=C–CH_3	1.7
Fluorides	**H**C–F	4–4.5
Chlorides	**H**C–Cl	3–4
Bromides	**H**C–Br	2.5–4
Iodides	**H**C–I	2–4
Alcohols	**H**C–OH	3.4–4
Ethers	**H**C–OR	3.3–4
Esters	RCOO–C**H**	3.7–4.1
	HC–COOR	2–2.2

(continued)

Table 3.1. (Continued)

Proton	Structure (**H**)	Chemical Shifts (δ), ppm
Acids	**H**C-COOH	2–2.6
Carbonyl compounds	**H**C-C=O	2–2.7
Aldehydic	RC**H**O	9–10
Hydroxylic	RO**H**	1–5.5
Phenolic	ArO**H**	4–12
Enolic	C=C-O**H**	15–17
Carboxylic	RNH_2	10.5–12
Amino	RNH_2	1–5

Source: Morrison and Boyd (1973).

Note: **H** is the subject proton.

Table 4.1. Paint Formulation and Components

Vehicle	Pigments
Nonvolatile vehicles	
Flame sprayed resins	Opaque
Plasma sprayed resins	Translucent
Solvent-based vehicles	Transparent
Oils	Special-purpose pigments
Resins	
Driers	
Additives	
Lacquer vehicles	
Cellulosics	
Resins	
Plasticizers	
Additives	
Water-based vehicles	
Acrylic	
Polyvinyl acetate	
Styrene-butadiene	
Other polymers and emulsions	
Selected copolymers	
Additives	
Solvents	
Trade sales/maintenance aliphatic solvents, and in some cases aromatics	
Chemical/industrial solvents, including in some cases aromatics	
Lacquer solvents, such as ketones, esters, and acetates	

Source: Weismantel (1981). Reprinted with permission of McGraw–Hill.

Table 4.2. Typical Formulation of a Waterborne Latex-Type Paint

Component	Percent Weight
Opaque pigment	20.0
Extender pigment	15.0
Pigment dispersant	0.1
Protective colloid	1.2
Latex	40.0
Preservative	0.5
Fungicide (optional)	—
Coalescing agent	2.0
Defoamer	0.1
Thickener	0.5
Water	20.6

Table 4.3. Formulation of Vinyl Acetate-Acrylic Latex

Component	Parts by Weight
Deionized water	75.0
Sodium bicarbonate	0.2
Potassium persulfate	0.3
Vinyl acetate	93.0
2-Ethylene acrylate	7.0
Ethyl oxide-propylene oxide block copolymer	5.0

Table 4.4. Formulation for a Semigloss Latex Paint: Interior, Acrylic 27% PVC (White)

Component	Pounds
Propylene glycol	70.0
Dispersant[a]	11.0
Defoamer[b]	2.0
Titanium dioxide-rutile	250.0
Barites	50.0
(Disperse in Cowles mixer, then add the following in the thindown)	
Propylene glycol	100.0
Acrylic latex[c] (46.5%)	492.7
Defoamer[b]	2.0
Butyl Cellosolve (Premix)	13.7
Surfactant[d]	2.0
Water (Premix)	2.0
Preservative[e]	2.6
Fungicide[f] (45%)	0.5
Water and/or hydroxyethyl cellulose[g] (2.5%)	57.8

Gloss, 45%; solids, 47.8%; pigment volume content, 26.8%; viscosity, 75–80 KU; meets Federal Specification TTP-1511A

Notes: [a]Rhom & Haas—Tamol 731
[b]Nopco Chemical Co.—Nopco NDW
[c]Rhom & Haas—Rhoplex AC 490
[d]Rhom & Haas—Triton GR-7
[e]Dow Chemical Co.—Dowicil 75
[f]Rhom & Haas—Skane M-8
[g]Hercules Chemicals, Inc.—Natrosol 250 MR

Table 4.5. Formulation for Exterior House Paint: Acrylic Modified with 13% Alkyd (White)

Component	Pounds
Hydroxyethyl cellulose[a]	85.0
Water	62.5
Dispersant[b] (30%)	10.5
Dispersant[c]	2.5
Potassium tripolyphosphate	1.5
Defoamer[d]	1.0
Ethylene glycol	25.0
Titanium dioxide[e]	237.5
Zinc oxide[f]	50.0
Talc[g]	187.7
(Grind the materials in a Cowles mixer and add the following)	
Acrylic latex[h]	390.8
Long-oil alkyd[i]	30.8
0.5% cobalt, 0.5% of 6% manganese, and 1.4% of 24% lead in alkyd	
Defoamer[d]	1.0
Tributyl phosphate	9.3
Propylene glycol	34.0
Fungicide[j] (45%)	2.0
Ammonium hydroxide (28%)	1.0
Water	65.3

Pigment volume content, 40%; solids, 41%; viscosity, 72–76 KU

Notes: [a]Hercules Chemicals—Natrosol 250 MR
[b]Rohm & Haas—Tamol 850
[c]Rohm & Haas—Triton CF-10
[d]Nopco Chemical—Nopco NZX
[e]E. I. du Pont—Ti Pure R-960
[f]American Zinc Sales Co.—AZ-11
[g]International Talc Co.—Abestine 3X
[h]Rohm & Haas—Rhoplex AC 388
[i]Ashland Chemical Co.—Aroplaz 1271
[j]Rohm & Haas—Skane M-8

Table 4.6. Formulation for Floor Paint: Acrylic Modified with Epoxy (Gray)

Component	Pounds
Dispersing agent[a]	7.5
Dispersing agent[b]	2.0
Defoamer[c]	2.0
Water	80.4
Titanium dioxide[d]	228.6
Lampblack dispersion	30.0
(Grind in Cowles mixer and add the following in the letdown)	
Water	26.1
Propylene glycol	54.6
Preservative[e]	1.0
Acrylic latex[f] (46%)	485.4
Epoxy emulsion[g] (50%)	49.6
6% cobalt drier	0.2
25% lead drier	1.1
Aluminum oxide[h]	24.8
Butyl Cellosolve	24.4
Hydroxyethyl cellulose[i] (3%)	67.8

Solids, 48.2%; gloss, 39%; viscosity, 60–65 KU

Notes: [a]Rohm & Haas—Tamol 731
[b]Rohm & Haas—Triton CF-10
[c]Colloids, Inc.—Colloid 600
[d]E. I. du Pont—Ti Pure R-9OO
[e]Tenneco Chemicals, Inc.—Super Ad-It
[f]Rohm & Haas—Rhoplex AC 61
[g]Ciba Products Co.—Araldite DP-624
[h]Exolon Co.—SD-No.-220 Mesh
[i]Union Carbide Chemical Co.—WP-4400

Table 4.7. Latex Shingle Stain: Vinyl Acrylic (Red)

Component	Pounds
Water	250.0
Hydroxyethyl cellulose[a]	3.0
Dispersant[b]	4.5
Surfactant[c]	3.0
Potassium triphosphate	1.0
Antifoam[d]	1.0
Ethylene glycol	10.0
Preservative	2.0
Titanium dioxide-rutile[e]	25.0
Aluminum silicate[f]	50.0
Zinc oxide	50.0
Silica[g]	25.0
Black oxide[h]	15.0
Red oxide[i]	65.00
(Grind in Cowles mixer and add the following in the letdown)	
Butyl carbitol	15.0
Water	195.0
Vinyl acrylic latex[j] (55%)	305.0
Antifoam[d]	2.0

Notes: [a]Union Carbide Corp.—QP-52,000
[b]Rohm & Haas—Tamol 850
[c]GAF—CO-630 Surfactant
[d]Witco Chemical Co.—Balab 748
[e]E. I. du Pont—Ti Pure R-960
[f]Indusmun—Minex 4
[g]Johns-Manville—Celite 281
[h]Pfizer—BK-5099
[i]Pfizer—RO 7097, Kroma
[j]Union Carbide Corp.—UCAR 366

Table 4.8. Formulation for Water-Based Acrylic Coil Coating Enamel (White)

Component	Pounds
Deionized water	46.4
N,N-Dimethylethanolamine	0.1
Ethylene glycol	3.4
Nonionic surfactant[a]	2.2
Dispersant[b]	7.3
Defoamer[c]	0.5
Titanium dioxide	211.7
(Mix in a Cowles mixer)	
Defoamer[c]	271.7
Deionized water	35.4
N,N-Dimethylethanolamine	5.6
Acrylic-styrene latex[d]	543.4
Deionized water	97.7
Butyl carbitol	39.8
Melamine resin[e]	31.0

Notes: [a]GAF—Igepal CA 630
[b]Rohm & Haas—Tamol 731
[c]Diamond Shamrock Chemical Co.—Foam Master VF
[d]Union Carbide—UCAR 4510
[e]American Cyanamid—Cymel 303

Table 4.9. Formulation for Polyester Coil Coating Enamel (White)

Component	Pounds
Titanium dioxide	282.1
Polyester resin[a] (70% NV)	95.9
Water	145.6
(Pebble mill 18–24 hours)	
Polyester resin[a] (70% NV)	265.3
Hexamethoxy methyl melamine[b]	64.8
Trimethyl propanediol isobutyrate[c]	63.7
Dimethylethanolamine	1.7
Water	159.1

Notes: [a]Ashland Chemical Co.—Arolon 465.WA8.70
[b]American Cyanamid Co.—Cymel 301
[c]Eastman Chemical Co.—Texanol

Table 4.10. Formulation for Clear Baking Varnish for Direct Roll-Coater Application

Component	Pounds
Acrylic latex[a] (43%)	701.9
Deionized water	27.5
N,N-Dimethylethanolamine	5.9
Hexylene glycol	88.4
Melamine resin[b]	44.2
Defoamer[c] (use as needed)	

Notes: [a]Union Carbide—UCAR 4510
[b]American Cyanamid—Cymel 350
[c]Diamond Shamrock—Foam Master VF

Table 4.11. Formulation for Clear Sealer for Wood-Board Coating

Component	Pounds
Acrylic latex[a] (46.5%)	181.1
Water	649.5
Wetting agent[b]	0.1
Butyl Cellosolve[c]	13.7

Notes: [a]Rhom & Haas—Rhoplex AC 73
[b]Rohm & Haas—Triton GR-7M
[c]Union Carbide—Butyl Cellosolve

Table 4.12. Formulation for Alkyd Automobile Refinishing Enamel

Component	Pounds
Rutile titanium dioxide	260
Soya lecithin	2
Modified tall oil benzoate alkyd resin	615
Lead naphthenate	24
Manganese naphthenate	2
Cobalt naphthenate	2
Methyl ethyl ketoxime	6
Guaiacol (18%)	6
Mineral spirits	75
High flash naphtha	27
(Weight per gallon is 9.84 lb/gal)	

Table 4.13. Formulation for Maintenance Primer, Amine Adduct Type

Component	Pounds
A—base component	
Red lead (97%)	729.6
Celite 266 (Johns-Manville Products Co.)	68.8
Abestine 3X (International Talc Co.)	56.6
Aluminum stearate	3.4
Epon 1001 (Shell Chemical Co.)	170.5
Beetle 216-8 (American Cyanamid Co.)	10.4
MIBK	80.4
Ethylene glycol monobutyl ether	9.0
Toluene	89.5
B—curing agent component	
Epon Curing Agent C-111 (Shell Chemical Co.)	88.9
MIBK	80.3
Ethylene glycol monobutyl ether	9.0
Toluene	90.4
Ethyl alcohol	21.2
Mixing ratio of A:B: 1:1	
Total nonvolatiles: 71.1%	
Weight per gallon: 15.1 lb/gal	

Table 4.14. Formulation for Epoxy/Polyamide Brushing Enamel (Gray)

Component	Pounds
A—base component	
Epon 1001-CX-75 (Shell Chemical Co.)	474.8
Beetle 216-8 (American Cyanamid Co.)	16.6
Titanium dioxide-rutile NC	471.1
Talc No. 399 (Whittaker, Clark and Daniels Co.)	47.1
Bentone 27/ethyl alcohol (1/1)	5.7
Lampblack	2.8
Diacetone alcohol	63.1
Heavy aromatic naphtha (KB-90)	125.3
B—curing agent component	
Epon Curing Agent V1-60 (Shell Chemical Co.)	594.0
Heavy aromatic naphtha (KB-9O)	111.0
Ethylene glycol monoethyl ether	56.0
Mixing ratio by volume: 1.0/1.0	
Weight per gallon: 9.8 lb/gal	

Table 4.15. Formulation for Epoxy-Phenolic Baking Enamel (Green)

Component	Pounds
Chrome oxide	84.8
Epon 1007 (Shell Chemical Co.)	207.6
Methylon 75108 (General Electric Co.)	69.3
Silicone Resin SR-82 (General Electric)	4.7
Phosphoric acid (85%)	5.0
n-Butanol	37.6
Cellosolve acetate	240.8
Xylene	240.8
Epoxy resin/phenolic resin mix ratio: 75/25 by weight	
Total nonvolatiles: 41.4%	
Weight/gallon: 8.9 lb/gal	

Table 4.16. Formulation for Soft Lacquer for Nonferrous Metals

Component	Parts by Weight
Solid acrylic resin	89
Thinner: 59% toluene, 25% MIBK, 10% IPA, 6% Pentoxone	356
Nitrocellulose (HB-14-P), 1/2 sec	87
MIBK	213

Table 4.17. Formulation for White Lacquer on Aluminum

Component	Parts by Weight
Grind portion	
Medium hard acrylic solution resin	105
Methyl ethyl ketone	17
Cellosolve	15
Ti Pure R-900 titanium dioxide	40
Letdown portion	
Methyl ethyl ketone	18
Cellosolve	15
Acrylic resin	744
Toluene	197.2
Ethyl alcohol	50
Benzotriazole	4.4
Epoxidized soybean oil	4.4

Table 4.18. Formulation for Clear Aerosol Lacquer

Component	Parts by Weight
Acrylic resin	17.1
Toluene	14.9
Methylene chloride (or acetone)	13.8
MIBK	4.6
Poly-Solv EE Acetate (or high flash naphtha)	3.6
Sanitizer	1.0
Freon-12 Propellant	45.0

Table 4.19. Formulation for Alcohol-Based Spray Lacquer

Component	Parts by Weight
Alcohol-soluble acrylic resin	10
Isopropyl alcohol	25
n-Propyl alcohol	40
Pentoxone	25

Table 4.20. Formulation for Acrylic Concrete Sealer

Component	Parts by Weight
Acrylic resin	28
Santicizer 160 plasticizer	3
Xylene	27
Toluene	42

Table 4.21. Formulation for Acrylic-Butyrate Wood Lacquer (Nonyellowing)

Component	Parts by Weight
Acrylic resin	21.3
Cellosolve acetate butyrate, 1/2 sec	8.5
Santicizer 160 plasticizer	3.0
DC-510 (1000 centistokes) fluid	0.01
Eastman inhibitor DOBP	0.09
Toluene	32.1
Tecsol, 95%	10.0
Ethyl acetate	5.0
Isobutyl acetate	10.0
Methyl isoamyl ketone	10.0

Table 4.22. Formulation for Steel Coating Lacquer

Component	Parts by Weight
Grind portion	
Ti Pure R-610 titanium dioxide	6.17
Carbon black	0.07
Hard methacrylate solution polymer	4.03
Cellosolve acetate	2.53
Letdown portion	
Hard methacrylate solution polymer	11.60
Santicizer 160 butyl benzyl phthalate	3.76
Cellulose acetate butyrate, 1/2 sec (25% solids)	10.03
MEK	21.97
Toluene	21.97

Table 4.23. Formulation for Thermosetting Appliance Enamel

Component	Parts by Weight
Grind portion	
Ti Pure R-900 titanium dioxide	27.4
Carboxyl functional acrylic	18.3
Letdown portion	
Carboxyl functional acrylic	21.8
Epon 1001 (50% solids)	26.7
Xylene	7.8
Cellosolve acetate	2.6
Raybo 3 (antisilk agent for smoothness)	0.06

Table 4.24. Formulation for White Exterior House Paint

Component	Pounds per 100 Gallons
Grind portion	53.6
Water	10.7
Tamol 731 (25%)	2.5
Nopco NZX	1.0
Ethylene glycol	25.0
Pine oil	3.0
Metasol 57 (100%)	1.8
Ti Pure R-610 titanium dioxide	240.0
Ti Pure FF titanium dioxide	10.0
Talc	100.0
Calcium carbonate	110.0
Letdown portion	
Exterior acrylic emulsion	512.0
Water	7.7
Nopco NZX	1.0
Ammonium hydroxide (28%)	2.0

Table 4.25. Formulation of Wash Primers for Steel (MIL-C-15328A)

Component	Parts by Weight
Base grind	
Vinyl butyral resin	7.2
Basic zinc chromate pigment (insoluble)	6.9
Magnesium silicate (talc)	1.0
Lampblack	0.1
Ethyl alcohol (95%)	48.8
Butanol	16.1
Acid diluent	
Phosphoric acid (85%)	3.6
Water	3.2
Ethyl alcohol (95%)	13.1

Table 4.26. Plasticized Vinyl Acetate Emulsion

Component	Parts by Weight
Lacquer phase (82.0%)	
Vinyl acetate	50.0
Tricresyl phosphate	5.0
Toluene	43.5
Oleic acid	1.5
Water phase (18.0%)	
Distilled water	92.0
28% aqua ammonia	

Table 4.27. Formulation for High-Build Chlorinated Rubber Paint (Red)

Component	Percent Weight
Chlorinated rubber	17.0
Chlorinated paraffin (70% Cl)	11.3
(42% Cl)	5.7
Red iron oxide	9.5
Barites	14.1
Modified hydrogenated castor oil (e.g., Thixatrol ST)	1.8
Xylene	40.6

Note: Brush application

Table 4.28. Formulation for Traffic Paint Based on Chlorinated Rubber and Phenolic

Component	Percent Weight
Chlorinated rubber (10 cps)	6.60
Chlorinated paraffin (42% Cl)	3.18
20 gal tung oil varnish (50% N.V.)	18.90
Rutile titanium dioxide	5.15
Titanium calcium pigment (30% TiO_2)	25.70
Abestine	4.64
Celite	7.30
Mica	5.15
Cobalt naphthenate	0.13
Epichlorohydrin	0.20
Mineral spirits	3.78
Toluene	19.27

Per ASTM D-711-55

Table 4.29. Formulation for Heat-Resistant Aluminum Paint

Component	Pounds per 100 Gallons
G-E silicone resin SR-112 (50%)	279.0
Ethyl cellulose solution (5.5%)	126.5
6% manganese naphthenate	2.3
Solvesso 100	178.2
Alcoa aluminum paste #206 or Reynolds #32	310.0

Note: Brush or spray application

Table 4.30. Formulation for Zinc-Dust, Zinc-Oxide Primer

Component	Pounds per 100 Gallons
Asarco #1 zinc dust	312.5
XX-601 zinc oxide	150.0
#1132 graphite	50.0
Diatomaceous silica	43.8
G-E silicone resin SR-112 (50%)	462.5
Solvesso 100[a]	231.3

Note: [a]For spray gun application, xylene may be substituted for the slower solvent.

Table 4.31. Formulation for Heat-Resistant Metal Primer

Component	Pounds per 100 Gallons
Imperial X-883 zinc yellow	215.5
R-C#1094 indian red[a]	161.3
MicroVelva A	161.3
G-E silicone resin SR-120	445.2
70:30 xylene/*n*-butanol	222.2

Note: [a]C. K. Williams #8098 red oxide may be used in place of R-C#1094 on equal weight basis.

Table 4.32. Formulation for High Infrared Reflectance Missile Coating (White)

Component	Pounds per 100 Gallons
Zinc sulfide	650
G-E silicone resin SR-112 (50%)	292
G-E silicone resin SR-82 (60%)	129
Acryloid B-66 (40%)	183
Nuogel AO	9
Xylene	54
(Air dry and bake for complete hardness)	

Table 4.33. Formulation for Heat-Resistant Enamel (Black)

Component	Pounds per 100 Gallons
Ferro F-2302 black	56.7
#1132 graphite	113.4
Micalith G	56.7
7% ethyl cellulose T-200 in toluene	410.1
G-E SC-3900, 20% in *n*-butanol	9.5
G-E silicone resin SR-82 (60%)	94.5
Aroplaz 7323 (60%)	68.0
6% cobalt octoate	0.8
6% manganese octoate	0.5
Xylene	88.2
Weight per gallon: 9.0 lb/gal	
Viscosity: 84 KU	
[Bake for 30 minutes at 204°C (400°F) and age for 16 hours in air; the film withstands 1/8th inch bend and 24-hour immersion in gasoline.]	

Table 4.34. Formulation for Cocoa Brown High-Temperature Baked Appliance Enamel

Component	Pounds per 100 Gallons
TiO_2 RANC	57.5
Ferro F-6112 red brown	86.0
Bentone 11	14.3
G-E silicone resin SR-120 (65%)	689.0
Cymel 301	78.8
Catalyst 1010	5.4
6% manganese naphthenate	7.2
6% iron naphthenate	1.4
70/30 xylene/*n*-butanol	44.8

Weight per gallon: 9.85 lb/gal

Viscosity: 63 KU

[Reduce 5:1 by volume with the solvent blend and spray. Bake for 1 hour at 260°C (500°F). Hardness is 4H.]

Table 4.35. Formulation for Light Brown Electrical Resistor Coating

Component	Pounds
Ferro F-6109 light yellow brown	25.2
Ferro F-6112 red brown	25.2
325-mesh mica	149.2
Antimony oxide KR	28.2
Zinc oxide XX-4	16.3
Santocel CS	8.9
Bentone 38	7.7
Denatured ethyl alcohol (95%)	3.4
G-E silicone resin SR-112 (50%)	180.3
G-E silicone resin SR-125 (50%)	180.3
6% manganese naphthenate	3.0
Xylene	332.5

Weight per gallon: 9.5–9.7 lb/gal

Viscosity: 61–63 KU

Table 4.36. Formulation for Coil or Strip Coating

Component	Parts by Weight
Titanium dioxide (nonchalking)	294
Magnesium silicate (325 mesh)	26
Magnesium silicate (extra fine)	101
Silicone/polyester vehicle (50% nonvolatiles)	553
Hexamethoxymethyl melamine resin	31
Acid catalyst	3
Solvesso 150	150

Table 4.37. Formulation for Interior Appliance Epoxy Powder Coating (White)

Component	Percent Weight
DER 663U[a] (epoxy resin)	47.4
DER 673MF[a] (flow agent in epoxy resin)	10.0
DEH 41[a] (hardener and catalyst)	2.6
Benzoin	0.1
TiO_2 (pigment)	24.9
$BaSO_4$ (filler)	15.0
(Oven cure for 10 minutes at 180°C)	

Note: [a]Dow Chemical Company

Table 4.38. Formulation for Exterior/Interior Epoxy-Polyester Appliance Powder Coating

Component	Percent Weight
DER 662[a] (epoxy resin)	32.0
Uralac P 2980[b] (polyester resin)	34.0
Modaflow III[c] (flow agent)	0.8
Benzoin	0.5
TiO_2 (pigment)	27.7
Talc (pigment)	5.0
(Oven cure for 8 minutes at 180°C)	

Notes: [a]Dow Chemical Company
[b]DSM Resins
[c]Monsanto

Table 4.39. Formulation for Low-Gloss Epoxy-Polyester Powder Coating

Component	Pecent Weight
Araldite GT 6084[a] (epoxy resin)	39.7
Uralac 2450[b] (polyester resin)	14.8
B55[c] (hardener)	5.5
Flow agent	0.5
Benzoin	0.1
TiO_2 (pigment)	30.0
$CaCO_3$ (pigment)	9.4
(Oven cure for 20 minutes at 200°C, gloss at 60° is 40%)	

Notes: [a]Ciba-Geigy
[b]DSM Resins
[c]Huls

Table 4.40. Formulation for Polyester-Polyurethane Powder Coating

Component	Percent Weight
Uralac P2115[a] (polyester resin)	46.6
B1065[b] (blocked IPDI)	11.9
Flow agent	0.5
Benzoin	1.0
TiO_2 (pigment)	30.0
$BaSO_4$ (pigment)	10.0
(Oven cure for 15 minutes at 200°C)	

Notes: [a]DSM Resins
[b]Huls

Table 4.41. Formulation for Polyester-Hydroxyalkyl Amide System Powder Coating

Component	Percent Weight
Grilesta V76-12[a] (TMA free polyester)	56.0
Primid XL 552[b] (beta-HAA)	3.0
Flow agent	0.8
Benzoin	0.2
TiO_2 (pigment)	40.0
(Oven cure for 5 minutes at 200°C)	

Notes: [a]EMS
[b]Ciba-Geigy

Table 4.42. Formulation for Epoxy/Phenolic Pipe Coating Powder Coating

Component	Percent Weight
DER 642U[a]	42.5
DER 672U[a]	6.5
DEH 81[a]	21.0
Iron oxide red	13.0
$BaSO_4$	16.5
Aerosil R972[b]	0.5
(Cure by residual heat curing from preheating of pipe, 220–240°C)	

Notes: [a]Dow Chemical Company
[b]Degussa

Table 4.43. Formulation for Epoxy/Phenolic Chemical-Resistant Powder Coating

Component	Percent Weight
Araldite GT 7203[a] (epoxy resin)	57.5
Corlan 100[b] (phenolic novolac)	12.0
Benzoin	0.1
2-Methyl imidazole (catalyst)	0.1
$BaSO_4$ (pigment)	15.9
Iron oxide red (pigment)	14.2
Aerosil R972[c]	0.2

Notes: [a]Ciba-Geigy
[b]Isovolta
[c]Degussa

Table 5.1. List of Paint Materials, Descriptions, and Suppliers

Material	Description	Manufacturer
Abestine 3X	Talc	International Talc Co.
Acryloid resins	Resins	Rhom & Haas
Additives	General additives	Troy Chemical Co.
Aerosil R972	Additive	Degussa Co.
Alcoa Aluminum pastes	Aluminum pigments	Aluminum Company
Amsco Solvents	Solvents, thinners	Amsco Co.
Araldite	Epoxy emulsion	Ciba-Geigy Co.
Araldite GT 6084	Epoxy resin	Ciba-Geigy Co.
Aroclor resins	Resins	Monsanto Co.
Arolon 465.WA.8.70	Polyester resin	Ashland Chemical Co
Aroplaz resins	Resins	Archer Daniels Midland Co.
Aroplaz 1271	Long-oil alkyd resin	Ashland Chemical Co.
AZ-11	Zinc oxide	American Zinc Sales Co.
Balab 748	Antifoaming agent	Witco Chemical Co.
Bentone 11, 38	Pigments	National Lead Co.
BK-5099	Black oxide	Pfizer Corp.
Butyl Cellosolve	Butyl Cellosolve	Union Carbide Corp.
B55	Hardener for resins	Huls Co.
B1065	Blocked isophthalic diisocyanate	Huls Co.
Catalyst 1010	Catalysts	Cytec, Inc.
Celite 281	Silica	Johns-Manville Co.
Corlan 100	Phenolic novolac	Isovolta Co.
Cymel 301	Hexamethoxy methyl amine	Cytec, Inc.
Cymel 303	Melamine resin	Cytec, Inc.
Cymel 350	Melamine resin	Cytec, Inc.
DER 662	Epoxy resin	Dow Chemical Co.
DER 663U	Epoxy resin	Dow Chemical Co.
DER 673MF	Flow agent in epoxy resin	Dow Chemical Co.
Dowicil 75	Preservative	Dow Chemical Co.
Drying agents		Davison Chemical Co., Minerals and Chemicals Philipp Corp.
Epon Resins	Epoxy resins	Shell Chemical Co.
Ethyl cellulose	Thickener	Hercules Powder Co.
Ferro colors	Colored pigments	Ferro Corp.
GAF-CO-630	Surfactant	GAF Corp.
G-E Silicone	Silicone resins	General Electric Co., Silicone Products Division
G-E SR-82	Silicon resin	General Electric Co.
G-E SR-112 (50%)	Silicone resin	General Electric Co.
G-E SR-125	Silicone resin	General Electric Co.

(*continued*)

Table 5.1. (Continued)

Material	Description	Manufacturer
#1132 Graphite	Black pigments	Joseph Dixon Crucible Co.
Imperial Color	Colored pigments	Imperial Color Div., Hercules Powder Co.
Mica, 325 mesh	Mica pigments, etc.	English Mica Co.
Micalith G	Mica pigments, etc.	English Mica Co.
MicroVelva A	Pigments	Carbola Chemical Div., International Talc Co.
Minex 4	Aluminum silicate	Indusmun Co.
Modaflow	Flowing agent	Monsanto Co.
Nuogel AO	Additive	Nuodex Products Div., Tenneco Chemicals
Nuosperse 657	Dispersing agent	Nuodex Products Div., Tenneco Chemicals
Pliolite Resins	Resins	The Goodyear Tire & Rubber Company, Chemical Division
Q panels	Metal test panels	The Q Panel Company
R-C iron oxides	Colored pigments	Reichard-Coulston Co.
Rhoplex AC 61	Acrylic latex	Rhom & Haas
Rhoplex AC 73	Acrylic latex	Rhom & Haas
Rhoplex AC 388	Acrylic latex	Rhom & Haas
Rhoplex AC 490	Acrylic latex	Rhom & Haas
Santocel CS	Additives	Monsanto Co.
Solvesso solvents	Solvents	Humble Oil & Refining Co.
Tamol 731	Dispersing agent	Rhom & Haas
Tamol 850	Dispersing agent	Rhom & Haas
Titanium dioxide	White pigments	Titanium Pigments Div., National Lead Co.
Triton CF-10	Trimethyl propanediol isobutyrate	Eastman Chemical Co.
Uralac 2450	Polyester resin	DSM Resins
Uralac P2115	Polyester resin	DSM Resins
Uralac P 2980	Polyester resin	DSM Resins
Zinc Dust #1	Metallic pigments	American Smelting & Refining Co.
Zinc oxide	White pigments	The New Jersey Zinc Co.
Priming pigment agents		National Lead Co., Mineral Pigments Corp., Holland-Succo Color Co.
Modaflow III	Flow agent	Monsanto Co.
Natrosol 250 MR	Water and/or hydroxy ethyl cellulose	Hercules Chemicals Inc.
Nopco NDW	Defoamer	Nopco Chemical Co.
Colloid 600	Defoamer	Colloids, Inc.
Nopco NZX	Defoamer	

(continued)

Table 5.1. (Continued)

Material	Description	Manufacturer
Foam Master VF	Defoamer	Diamond Shamrock Chemical Co.
Nonionic surfactant	Igepal CA-630	GAF Corp.
QP-52,000 Resins and Oils	Hydroxyethyl cellulose	Union Carbide Corp. Allied Chemical Corp., Archer Daniels Midland Co., The Baker Castor Oil Co., Hercules Powder Co., Marbon Chemical Division, Borg-Warner Corp, Neville Chemical Co., Shell Chemical Co., and others
RO 7097, Kroma	Red oxide	Pfizer Corp.
SD-No.-200 Mesh	Aluminum oxide	Exolon Co.
Super Ad-It	Aluminum oxide	Tenneco Chemicals, Inc.
Skane M-8	Fungicide	Rhom & Haas
Triton GR-7	Surfactant	Rhom & Haas
Ti Pure R-960	Titanium dioxide	E. I. du Pont
Ti Pure R-900	Titanium dioxide	E. I. du Pont
UCAR 4510	Acrylic-styrene latex	Union Carbide Corp.
WP-4400	Hydroxyethyl cellulose (3%)	Union Carbide Corp.

Note: Raw materials and producers can be found in *Chemical Week*, Buyers Guide Issues.

Table 7.1. Formulation for Typical Polystyrene Injection Molded Part

Component	Percent Weight
Resin: polystyrene (or other)	97.0
Dye: organic color pigment	2.0
Color enhancer: titanium dioxide	0.5

Source: Du Pont Technical Bulletin (1995).

Table 7.2. Formulation for Typical Thermoset Injection Molded Parts

Component	Percent Weight
Thermoset resin	98.00
Heat curing catalyst	0.05
Dye/pigment	1.95

Table 7.3. Typical Formulation for Polyester Fibers

Component	Percent Weight
Step 1. Spin poly(ethylene terephthalate) fiber	100
Step 2. Add color with disperse dye with chemical auxiliary, sulfonated lignins	(Variable)

Source: M. J. Drews (1985).

Table 7.4. Typical Formulation for Transparent Polyethylene Extruded Film

Component	Percent Weight
Polyethylene resin (Shell Chemical Co.)	99.0
Antistat agent	0.5
Lubricant	0.5

Source: Shell Chemical Company Technical Bulletin (1995).

Table 7.5. Formulation for Typical Flexible Urethane Foam

Component	Parts by Weight
Polyol (trifunctional) MWA = 3000	100
Toluene diisocyanate	46
Organotin catalyst	0.4
Silicone surfactant	1.0
Tertiary amine catalyst	0.2
Water	3.6
Monofluorotrichloromethane	0–15
Density (lb/ft^3)	1.4
Tensile strength (lb/ft^2)	14.0
Elongation (%)	220
Tear strength (lb/in)	2.2
Indent load deflection (lb)	
25% deflection	30
65% deflection	57

Source: R. D. Deanin (1985).

Table 7.6. Formulation for Typical Rigid Urethane Foam

Component	Parts by Weight
Polyether polyol (Hydroxyl No. 460)	100
N,N,N′,N′-Tetrakis (2-hydroxypropyl) ethylenediamine	8
Triethylene diamine	0.3
N,N-Dimethyl ethanolamine	0.5
Dibutyl tin dilaurate	0.02
Silicone surfactant	1.5
Trichlorofluoromethane	38
Toluene diisocyanate	107
NCO/OH ratio	1.03
Feed temperature (°F)	80
Mold temperature (°F)	125
Tack-free time (sec)	150
Density (lb/ft^3)	95
Compression modulus ($lb/in.^2$)	60
Flexural modulus ($lb/in.^2$)	900
Shear strength ($lb/in.^2$)	200

Source: R. D. Deanin (1985).

Table 7.7. Formulation for Typical PVC Gel or Plastisol

Component	Percent Weight
PVC resin (General Electric)	65.0
Plasticizer	35.0
Optional: color tint	

Table 7.8. Formulation for Typical Extruded PVC Pipe

Component	Percent Weight
PVC resin	88
Plasticizers: Butyl benzyl phthalate or di-2-ethyl hexyl phthalate	9
Organic dye/pigment	2
Optional:	
Heat stabilizer	
UV stabilizer	

Source: VISTA Technical Bulletin (1995).

Table 8.1. Melting and Glass Transition Temperatures of Some Plastic Materials

Plastic Material	Melting Temperature (°C)	Glass Transition Temperature (°C)
Acetal	175	−85
Acrylic	160	70, 105
Acrylonitrile-butadiene-styrene	190	
Cellulose acetate butyrate		50
Cellulose acetate proprionate		39
Cellulose triacetate	306	70
Chlorinated polyether	181	
Ethyl cellulose		43
Nylon 6	225	50
Nylon 6,6	260	50
Nylon 6,10	213–220	40
Nylon 11	182–194	46
Nylon 12	179	37
Polycarbonate	225	152
Polychlorotrifluoroethylene	220	35–45
Polyethylene	110–141	−125, −20
Polyfluorinated ethylene propylene		11
Polypropylene	172–176	−5, 45
Polystyrene	235	81–100
Polytetrafluoroethylene	330	−113, 20
Polyvinyl chloride	200	70–80
Polyvinylidine chloride	210	−17
Polyvinylidine fluoride	171–210	−39

Sources: Modern Plastics Encyclopedia (1992), I. I. Rubin (1972).

Table 8.2. Plastics Materials and Suppliers

Material	Supplier
Acetal	BASF Corp., Plastic Materials
	Du Pont Canada Inc.
	Du Pont Co., Polymer Products Dept.
	Hoechst Celanese Corp., Engineering Plastics Div.
	ICI Advanced Materials
Acrylamide	Cytec Corp.
Acrylic	Amco Plastic Materials Inc.
	Anerson Developement Co.
	Du Pont Canada Inc.
	Du Pont Co., Polymer Products Dept.
	ICI Resins US
	Reichhold Chemicals, Inc., Emulsion Polymers Div.
	Rhone-Poulenc Inc.
	Rohm and Haas Co.
	Westinghouse Electric Corp., Electrical Materials Div.
Acrylonitrile-butadiene-styrene	Accurate Compounding, Inc.
	Amco Plastic Materials Inc.
	Ashland Chemical Co.
	BASF Corp., Plastic Materials
	Dow Chemical U.S.A.
	GE Co., GE Plastics
	Grace, W.R., & Co., Organic Chemicals Div.
	ICI Advanced Materials
	Monsanto Co.
Acrylonitrile-chlorinated PE-styrene	Fleet Plastics Corp.
	Plastic Compounders of Mass., Inc.
Acrylonitrile-styrene-acrylic (ASA)	Amco Plastic Materials Inc.
	BASF Corp., Plastic Materials
	GE Co., GE Plastics
	Plastic Compounders of Mass., Inc.
Adhesion promoters	Advance Process Supply Co.
	Air Products and Chemicals, Inc.
	Dow Corning Corp.
	Du Pont Co., Du Pont Chemicals
	Exxon Chemical Americas, Polymers Group
	BF Goodrich Adhesive Systems Div.
	Loctite Corp., Industrial Products Group
	Morton International, Inc.
	National Industrial Chemical Co.
	Schering Berlin Polymers Inc.
	Unitex Chemical Corp.
	Westinghouse Electric Corp., Electrical Materials Div.

(continued)

Table 8.2. (Continued)

Material	Supplier
Alkyd	Advance Coatings Co.
	Cosmic Plastics, Inc.
	George, P. D., Co.
	Heller, H., & Co., Inc.
	National Industrial Chemical Co.
	Plastics Engineering Co.
	Resyn Corp.
	Rhone-Poulenc Inc.
	Rich Plastic Products, Inc.
	Sterling Group
	Westinghouse Electric Corp., Electrical Materials Div.
Allyl	Auburn Plastic Engineering, Div. Plastic Warehousing Corp.
	Cosmic Plastics, Inc.
	GCA Chemical Corp.
	Heller, H., & Co., Inc.
	Polytech Industries
	Rogers Corp.
Antiblocking and flatting agents	Advanced Compounding, Div., Blessings Corp.
	Davison Chemical Div., W. R. Grace & Co.
	Degussa Corp., Aerosil and Imported Pigment Products Div.
	Dow Corning Corp.
	GE Silicones
	Plastics Color Chip, Div. of PMC Inc.
	Quantum Chemical Corp., USI Div.
	Spectrum Color, Inc.
	Unipol Consultants
	Whittaker, Clark & Daniels, Inc.
	Zeelan Industries, Inc.
Antifogging agents	Advanced Compounding Div., Blessing Corp.
	Canada Colors & Chemicals, Ltd.
	Henkel Corp.
	Humko Chemical Div., Witco Corp.
	ICI Americas Inc.
	Polyvel, Inc.
	Unichema North America
Antimicrobials	Buckman Laboratories, Inc.
	Canada Colors & Chemicals, Ltd.
	Dow Chemical U.S.A.
	Ferro Corp., Bedford Chemical Div.
	Huls America Inc.
	ICI Americas Inc.
	Morton International, Industrial Chemicals & Additives
	Napp Chemical Co.
	Plastics & Chemicals, Inc.

(continued)

Table 8.2. (Continued)

Material	Supplier
Antioxidants	Akzo Chemicals Inc.
	Atochem North America
	Canada Colors & Chemicals Ltd.
	Ciba-Geigy Corp.
	DuPont Co., DuPont Chemicals
	Ethyl Corp., Chemicals Group
	Ferro Corp., Bedford Chemical Div.
	BF Goodrich Co., Specialty Polymers & Chemicals Div.
	Goodyear Tire & Rubber Co., Chemical Div.
	Grace, W.R., & Co., Organic Chemicals Div.
	Hoechst Celanese Corp., Polymer Additives
	Mobay Corp.
	Monsanto Co.
	Morton International, Industrial Chemicals & Additives
	Plastics Color Chip, Div. of PMC Inc.
	Quantum Chemical Corp., USI Div.
	Uniroyal Chemical Co., Inc.
Antistats	Akzo Chemicals, Inc.
	Argus Div., Witco Corp.
	Canada Colors & Chemicals, Ltd.
	Ferro Industrial Products Ltd.
	General Color & Chemical Co., Inc.
	ICI Americas Inc.
	National Industrial Chemical Co.
	Plastics Color Chip, Div. of PMC Inc.
	Quantum Chemical Corp., USI Div.
	Schering Berlin Polymers Inc.
Aramid fiber reinforcements	Chemfab, Chemical Fabrics Corp.
	Creative Coatings Corp.
	Hexcel Corp., Trevarno Div.
	North American Textiles
Bismaleimide	Ciba-Geigy Corp., Plastics Div.
	GCA Chemical Corp.
	Polyply Inc.
	Shell Chemical Co.
	Unipol Consultants
Blown film	
1. Ethylene-vinyl acetate (EVA)	Allied Plastics Supply Corp.—1,2,3
2. Polyethylene, high-density (HDPE)	Exxon Chemical Co., Polymers—1,3
3. Polyethylene, low-density (LDPE or LLDPE)	Gaska Tape, Inc.—4
4. Polyvinyl chloride	Goodyear Tire & Rubber Co., Films Div.—3,4

(*continued*)

Table 8.2. (Continued)

Material	Supplier
Brighteners	Allied Color Industries, Inc. Mobay Corp. Sandoz Chemicals Corp.
Bulk molding compounds (BMC)	Colortech Inc. Ferro Industrial Products Ltd. ICI Polyurethanes Group Jet Moulding Compounds Ltd.
Calendered film and sheet 1. Polyvinyl chloride & copolymers, flexible 2. Polyvinyl chloride & copolymers, rigid 3. Polyvinylidene chloride	Allied Plastics Supply Corp.—1,2 Commercial Plastics and Supply Corp.—1,2,3
Carbon blacks and graphite	Akzo Fortril Fibers, Inc. BASF Structural Materials, Inc.
Carbon fibers	Cabot Corp. FRP Supply, Div. of Ashland Chemical, Inc. Hercules, Inc.
Catalysts and promoters	Ethyl Corp., Chemicals Group Ferro Corp., Bedford Chemical Div. Huls America Inc. Morton International, Industrial Chemicals & Additives Reichhold Chemicals, Inc. Schering Berlin Polymers Inc.
Cellulosics	Advance Resins Corp. Dow Chemical U.S.A. Eastman Chemical Products, Inc. Plastic Compounders of Mass., Inc. Plastic Extruders, Inc.
Clarifiers	Allied Color Industries, Inc. Mitsui Plastics, Inc.
Coextrusions	Acutech Plastics, Inc. Allied Plastics Supply Corp. Dow Chemical U.S.A. Mearl Corp. Reynolds Metals Co. Vulcan Products Inc.
Colorants 1. Concentrates 2. Dyes 3. Fluorescent 4. Liquid	Accurate Color Inc.—1–8 Akrochem Corp.—1,7 BASF Corp.—2,3,7 Cabot Corp., Special Blacks Div.—7

(continued)

Table 8.2. (Continued)

Material	Supplier
5. Luminescent	Carolina Color Corp.—1,3–7
6. Metallic	CDI Dispersions—1,4
7. Pigments	Chromatics Inc.—1–4
8. Pearlescent	Colortech Inc.—1
	DSM Engineering Plastics—1
	EM Industries Inc.—6–8
	Hoechst Celanese Corp., Colorants & Surfactants Div.
	ICI Advanced Materials
Corrugated sheet and tubing	Allied Plastics Supply Corp.
	Fiber Glass Plastic, Inc.
	Piedmont Plastics, Inc.
Coupling agents	
Silanes	Akzo Chemicals Inc.
	Degussa Corp., Aerosil and Imported Pigment Products Div.
	Dow Chemical Co.
	Ferro Corp., Filled & Reinforced Plastics Div.
	Plastics & Chemicals, Inc.
Titanates	Akzo Chemicals Inc.
	Unipol Consultants
Cross-linking agents	Air Products and Chemicals, Inc.
	Akzo Chemicals Inc.
	Atochem North America, Organic Peroxides Div.
	Dow Corning Corp.
	Quantum Chemical Corp., USI Div.
Emulsifiers	Ashland Chemical, Inc.
	Henkel Corp.
	ICI Americas Inc.
Epoxy	Abatron, Inc.
	Acme Div., Allied Products Corp.
	Ciba-Geigy Corp., Plastics Div.
	DAP Inc.
	Dow Chemical U.S.A.
	Huls America Inc.
	ICI Composites Inc., Fiberite Molding Materials
	Reichhold Chemicals, Inc.
	Rhone-Poulenc Inc.
	Shell Chemical Co.
Ethylene-acid copolymer	Dow Chemical U.S.A.
	Du Pont Canada Inc.
	Reichhold Chemicals, Inc.
	Vinmar Inc.

(*continued*)

Table 8.2. (Continued)

Material	Supplier
Ethylene-ethyl acrylate	Azdel Inc., Southfield, MI
	Modern Dispersions Inc.
	Union Carbide Chemicals and Plastic Co., Inc., Polyolefins Div.
Ethylene-methyl acrylate	Amco Plastic Materials Inc.
	Chevron Chemical Co., Olefin & Derivatives
	Exxon Chemical Americas, Polymers Group
	Exxon Chemical Co., Polymers Group
	Heller, H., & Co., Inc.
	Modern Dispersions Inc.
	Triad Plastics, Inc.
	Vinmar Inc.
Ethylene-vinyl acetate	Ashland Chemical Inc., Thermoplastic Services Dept.
	Chevron Chemical Co., Olefin & Derivatives
	Du Pont Canada Inc.
	Du Pont Co., Polymer Products Dept.
	Mobay Corp.
	Mobil Polymers U.S. Inc.
	Reichhold Chemicals, Inc., Emulsion Polymers Div.
Ethylene-vinyl acrylate	Colonial Rubber Works, Inc.
	Exxon Chemical Americas, Polymers Group
	Heller, H., & Co., Inc.
Ethylene-vinyl alcohol (EVOH)	Heller, H., & Co., Inc.
	Morton International, Inc.
Fillers, glass	Abrasive Machine & Supply Co.
	Advanced Compounding, Div. Blessings Corp.
	3M Co., Engineered Materials, Industrial Specialties Div.
	Zeelan Industries, Inc.
Fillers, metallic	Bakaert Corp.
	Potters Industries, Inc.
Fillers, mineral	
1. Barium	Advanced Compounding, Div. Blessings Corp.—11
2. Calcium carbonate	Alcan Chemicals, Div. Alcan Aluminun—4
3. Clays	Colortech Inc.—2,6,9,10
4. Hydrated alumina	Degussa Corp., Aerosil and Imported Pigment Products Div.
5. Magnesiums	Englehard Corp., Specialty Minerals and Color Group—3,4,9
6. Mica	Georgia Marble Co.—2,4
7. Perlite	Heller H., & Co.—2
8. Quartz	Huber, J. M., Corp., Calcium Carbonate Div.—1,2
9. Silica	ICD Group, Inc., Chemicals Div.—5,9
10. Talc	ICI resins US—2
11. Wollastonite	Mearl Corp.—6,9

(*continued*)

Table 8.2. (Continued)

Material	Supplier
	Mountain Minerals Co. Ltd.—1,9
	New England Resins & Pigments Corp.—1–6,9,10
	Pfizer Minerals, Specialty Minerals Group—2,10
	Plastics Color Chip, Div. of PMC Inc.
	Thiele Kaolin Co.
	Unimin Specialty Minerials—8,9
	United States Gypsum Co., anhydrous & dihydrate calcium sulfate fillers—3,10,11
Fillers, organic	American Wood Fibers
	Composition Materals of America, Inc.
	Heller, H., & Co., Inc.
	ICD Group Inc., Chemicals Div.
	International Filler Corp.
	Shamokin Filler Co., Inc.
	Westinghouse Electric Corp., Electrical Materials Div.
	Wilner Wood Products Co.
Flame retardants	Akzo Chemicals Inc.
	Alcan Chemicals, Div. Alcan Aluminum
	Ampacet Corp.
	BASF Corp., Urethanes
	Ethyl Corp., Chemicals Group
	Ferro Corp., Bedford Chemical Div.
	General Color & Chemical Co., Inc.
	Hoechst Celanese Corp., Polymer Additives
	Morton International, Industrial Chemicals & Additives
	PPG Industries Inc., Chemical Div.
Fluoroplastics	
1. Ethylene-chlorotrifluoroethylene (ECTFE)	Atochem North America, Inc.—5,8
	Cadillac Plastic & Chemical Co.—3,6
2. Ethylene-tetrafluoroethylene (ETFE)	Chapman Associates, Inc.—2–6,8
	Chemical Coatings & Engineering Co.—6
	Deer Polymer Corp.—1–8
3. Fluorinated ethylene propylene (FEP)	Du Pont Canada Inc.—2–4,6
	Du Pont Co., Polymer Products Dept.—2–4
4. Perfluoroalkoxy (PFA)	Fluoro-Plastics, Inc.—3,4,6
5. Polychlorotrifluoroethylene (PCTFE)	
6. Polytetrafluoroethylene (PTFE)	
7. Polyvinyl fluoride (PVF)	
8. Polyvinylidene fluoride (PVDF)	
Foaming agents	
Chemical	Atochem North American, Organic Peroxides Div.

(continued)

Table 8.2. (Continued)

Material	Supplier
	DSM Engineering Plastics
	Du Pont Co., Du Pont Chemicals
	ICI Americas, Inc.
Physical	Expancel/BNobel Industries
	National Industrial Chemical Co.
Glass fiber reinforcements	
1. Chopped strand	Advance Coatings Co.
2. Fabrics	Allied Signal Inc., Fluroglas
3. Filaments and staple	Ferro Corp., Filled and Reinforced Plastics Div.
4. Flakes	Fiber Glass Industries, Inc.
5. Mats (chopped strand; continuous; finishing)	Hexcel
6. Milled fibers	Hexcell Corp., Trevarno Div.
	Manville Sales Corp., Mats, Fiber & Reinforcements Div.
7. Roving	PPG Industries, Inc./Fiber Glass Products
Heat distortion modifiers	Advanced Compounding, Div. Blessings Corp.
	GE Specialty Chemicals
	Unipol Consultants
Impact modifiers	Amoco Chemical Co.
	Atochem North America, Inc.
Ionomer	Ampacet Corp.
	Deer Polymer Corp.
	Du Pont Canada Inc.
	Du Pont Co., Polymer Products Dept.
	Exxon Chemical Americas, Polymers Group
	Exxon Chemical Co., Polymers Flex-O-Glass, Inc.
	Heller, H., & Co., Inc.
	Modern Dispersions Inc.
	Schulman, A., Inc.
	World Plastic Extruders, Inc.
Ketone-based resins	BASF Corp., Plastic Materials
	ICI Advanced Materials
Lubricants (additive)	Accurate Color Inc.
	Advanced Compounding, Div. Blessings Corp.
	Akzo Chemicals Inc.
	Allied Signal Inc., A-C Performance Additives
	Canada Colors & Chemicals, Ltd.
	Daniel Products
	Deer Polymer Corp.
	DSM Engineering Plastics
	GE Silicones
	GE Specialty Chemicals
	Henkel Corp., Plastics Additives

(continued)

Table 8.2. (Continued)

Material	Supplier
	Hercules Inc. ICI Advanced Materials Morton International, Industrial Chemicals & Additives Shell Chemical Co. Witco Corp., Organics Div.
Melamine	BASF Corp., Urethanes Commercial Plastics and Supply Corp. ICI Composites Inc., Fiberite Molding Materials Reichhold Ltd.
Metallizing agents	Delaware Metallizing Associates, Inc. Electro-Kinetic Systems, Inc. United State Bronze Powders, Inc.
Methacrylate-butadiene-styrene (MBS)	Fleet Plastics Corp. Mitsui Plastics, Inc. Polymerland, Inc.
Mica flake reinforcements	Eagle Quality Products Ferro Corp., Filled & Reinforced Plastics Div. KMG Minerals, Inc. Polycom Huntsman, Inc.
Mold release agents	Advanced Compounding, Div. Blessings Corp. Akzo Chemicals Inc. Ampacet Corp. Axel Plastics Research Laboratories, Inc. Polymerland, Inc. Union Camp Corp.
Natural fiber reinforcements	Akrochem Corp. Gelman, Herman A., Co. James River Corp., Solka-Floc Div. Monsanto Co. Nitrile Goodyear Tire & Rubber Co., Chemical Div. Reichhold Chemical, Inc., Emulsion Polymers Div.
Nucleating agents	Allied Signal Inc., A-C Performance Additives ICI Americas Inc. Polycom Huntsman, Inc. Spectrum Colors
Nylons	Akzo Engineering Plastics, Inc. Allied Signal Inc., Engineered Plastics Amco Plastic Materials Inc. BASF Corp., Plastic Materials Cadillac Plastic & Chemical Co. Deer Polymer Corp. DSM Engineering Plastics North America DSM Rim Nylon

(continued)

Table 8.2. (Continued)

Material	Supplier
	Du Pont Co., Polymer Products Dept.
	General Polymers Div., Ashland Chemical, Inc.
	Hoechst Celanese Corp., Engineering Plastics Div.
	Mobay Corp.
	Monsanto Co.
Peroxides, organic	Advance Coatings Co.
	Akrochem Corp.
	Akzo Chemicals Inc.
	Degussa Corp., Aerosil and Imported Pigment Products Div.
	Du Pont Co., Du Pont Chemicals
	Hercules Inc.
	Reichhold Chemicals, Inc.
Phenolic	American Resin & Chemical Corp.
	Ashland Chemical, Inc., Specialty Polymers & Adhesives Div.
	Commercial Plastics and Supply Corp.
	Georgia-Pacific
	ICI Composites Inc., Fiberite Molding Materials
	Reichhold Ltd.
	Westinghouse Electric Corp., Electrical Materials Div.
Plasticizers	Advance Coatings Co.
	Akzo Chemicals Inc.
	Atochem North America
	BASF Corp., Plasticizers
	Ethyl Corp., Chemicals Group
	Ferro Corp., Bedford Chemical Div.
	FMC Corp., Chemical Products Group
	Huls America Inc.
Polyamide-imide	Amoco Performance Products Inc.
	ICI Advanced Materials
	ICI Composites Inc., Fiberite Molding Materials
Polyarylamide	Solvay Polymers, Inc., Performance Polymers
Polyarylate	Amoco Performance Products Inc.
	Canada Colors & Chemicals, Ltd.
	Du Pont Co., Polymer Products Dept.
	General Polymers Div., Ashland Chemical, Inc.
	Hoechst Celanese Corp., Engineering Plastics Div.
	Polymer Corp.
Polyaryl ether	Delta Polymers Co.
	Unipol Consultants
Polybutadiene	Goodyear Tire & Rubber Co., Chemical Div.
	Polymerland, Inc.
	Reichhold Chemicals, Inc., Emulsion Polymers Div.

(continued)

Table 8.2. (Continued)

Material	Supplier
Polybutylene	Fleet Plastics Corp. Huls America Inc. Shell Chemical Co.
Polycarbonate	Amco Plastic Materials, Inc. Cadillac Plastic & Chemical Co. Dow Chemical U.S.A. GE Co., GE Plastics General Polymers Div., Ashland Chemical, Inc. Mobay Corp.
Polyester, thermoplastic 1. Liquid crystal polymer 2. Polybutylene terephthalate (PBT) 3. Polyethylene terephthalate (PET)—Engineering grades 4. Polyethylene terephthalate (PET)—Standard grades	 Advance Resins Corp.—1–4 Akzo Engineering Plastics, Inc.—2,3 Amoco Performance Products Inc.—1 BASF Corp., Plastic Materials—2 Cadillac Plastic & Chemical Co.—3,4 GE Co., GE Plastics—2 General Polymers Div., Ashland Chemical, Inc.—2,3,4 Hoechst Celanese Corp., Engineering Plastics Div.—1–3 ICI Advanced Materials—1,2
Polyester, thermoset 1. Aromatic 2. Unsaturated	 Advance Coatings Co.—2 Amoco Chemical Co.—1 Ashland Chemical, Inc.—2 FRP Supply Div., Ashland Chemical, Inc.—2 ICI Composites Inc., Fiberite Molding Materiais—2 Plastics Engineering Co.
Polyetherimide	Commercial Plastics & Supply Corp. ICI Advanced Materials Westinghouse Electric Corp., Electrical Materials Div.
Polyethylene 1. High-density (HDPE) 2. High-molecular-weight, high-density (HMW-HDPE) 3. Linear low-density (LLDPE) 4. Low-density (LDPE) 5. Ultrahigh-molecular-weight (UHMWPE) 6. Ultralow-density (ULDPE)	 Allied Signal Inc., A-C Performance Additives—5 Chevron Chemical Co., Olefin & Deriviates—1–4 Dow Chemical U.S.A.—1,3,4,6 DuPont Co., Polymer Products Dept—4 Phillips 66 Co., Phillips Plastics Resins—1–3
Polyimide, thermoplastic	Allied Signal Inc., Engineered Plastics
Polyimide, thermoset	Ciba-Geigy Corp., Plastics Div. Epoxy Technology, Inc. ICI Composites Inc., Fiberite Molding Materials

(continued)

Table 8.2. (Continued)

Material	Supplier
	Unipol Consultants
Polyisobutylene	National Industrial Chemical Co.
	Unipol Consultants
Polymethylpentene	Phillips 66 Co.
	Plastics Service Inc.
	Polymerland, Inc.
Polyphenylene oxide, modified	Advance Resins Corp.
	Huls America Inc.
	Polymerland, Inc.
	Westover Color & Chemical Co.
Polyphenylene sulfide	Advance Resins Corp.
	Ferro Corp., Engineering Thermoplastics Div.
	Hoechst Celanese Corp., Engineering Plastics Div.
	Mobay Corp.
	Polymerland, Inc.
Polypropylene	Advance Resins Corp.
	Amco Plastic Materials Inc.
	ARCO Chemical Co.
	BASF Corp., Plastic Foams
	Commercial Plastics & Supply Corp.
	Eastman Chemical Products, Inc.
	Exxon Chemical Americas, Polymers Group
	ICI Advanced Materials
	Phillips 66 Co., Phillips Plastics Resins
Polystyrene	Advance Resins Corp.
	American Polymers Inc.
	Amoco Chemical Co.
	ARCO Chemical Co.
	BASF Corp., Plastic Foams
	Canada Colors & Chemicals, Ltd.
	Commercial Plastics and Supply Corp.
	Fina Oil & Chemical Co., ICI Advanced Materials
Polyurethane, thermoset	
1. For flexible foam	BASF Corp., Urethanes—1–8
2. For rigid urethane foam	Dow Chemical U.S.A., Thermoset Applications—1–8
3. For rigid isocyanurate foam	ICI Polyurethanes Group—1–4,6–8
4. For cast microcellular foam	
5. For RIM urethane elastomers	
6. For RIM polyurea elastomers	

(continued)

Table 8.2. (Continued)

Material	Supplier
7. For RIM structural foam	
8. For RIM thinwall engineering pads	
Polyvinyl acetate	Cadillac Plastic & Chemical Co.
	Commercial Plastics and Supply Corp.
	Heller, H., & Co., Inc.
	National Casein Co.
	National Starch and Chemical Co.
	Reichhold Chemicals, Inc., Emulsion Polymers Div.
	Wacker Chemicals (USA), Inc.
Polyvinyl alcohol	Air Products and Chemicals, Inc.
	Du Pont Canada Inc.
	Heller, H., & Co., Inc.
Polyvinyl butyral	Du Pont Canada Inc.
	Hafner Industries, Inc.
	Heller, H., & Co., Inc.
	Wacker Chemicals (USA), Inc.
Polyvinyl chloride (PVC)	
1. Chlorinated	BF Goodrich Co., Geon Vinyl Div.—1–3
2. Dispersion	Bordon Chemicals—2,3
3. Suspension and others	
Polyvinylidene chloride	Chemical Coatings & Engineering Co.
	Grace, W.R., & Co., Organic Chemicals Div.
	Heller, H., & Co., Inc.
	Sattler, H., Plastics Co. Inc.
Processing aids	Advanced Compounding, Div. Blessings Corp.
	Canada Colors & Chemicals, Ltd.
	General Color & Chemical Co., Inc.
	GE Specialty Chemicals
	Hoechst Celanese Corp., Polymer Additives
Sheet molding compounds	ICI Composites Inc., Fiberite Molding Materials
	Reichhold Chemicals, Inc.
Silicone	Commercial Plastics and Supply Corp.
	DAP Inc.
	Dow Corning Corp.
	GE Silicones
	Huls America Inc.
	Mobay Corp.
Slip agents	Akzo Chemicals Inc.
	Axel Plastics Research Laboratories, Inc.
	Canada Colors & Chemicals, Ltd.

(*continued*)

Table 8.2. (Continued)

Material	Supplier
	General Color & Chemical Co., Inc. Witco Corp., Organics Div.
Smoke suppressants	Advanced Compounding, Div. Blessings Corp. Harwick Chemical Corp. Morton International, Industrial Chemicals & Additives Whittaker, Clark & Daniels, Inc.
Stabilizers	Akzo Chemicals Inc. Ferro Corp., Bedford Chemical Div. BF Goodrich Co., Specialty Polymers & Chemicals Div. Hoechst Celanese Corp., Polymer Additives Huls America Inc. ICI Americas Inc. Morton International, Uniroyal Chemical Co., Inc.
Styrene-acrylonitrile (SAN)	Advance Resins Corp. BASF Corp., Plastic Materials Commercial Plastics and Supply Dow Chemical U.S.A. Ferro Corp., Engineering Thermoplastic Div.
Styrene-butadiene	Advance Resins Corp. Dow Chemical U.S.A. Firestone Synthetic Rubber & Latex Co. Goodyear Tire & Rubber Co., Chemical Div. Grace, W.R., & Co., Organic Chemicals Div. Phillips 66 Co., Phillips Plastics Resins Reichhold Chemicals, Inc., Emulsion Polymers Div. Schulman, A., Inc.
Styrene-maleic anhydride	ARCO Chemical Co. General Polymers Div., Ashland Chemical, Inc. Monsanto Co.
Sulfone polymers 1. Polyarylsulfone 2. Polyethersulfone 3. Polyphenyleulfone 4. Polysulfone	 Advance Resins Corp.—1,2,3,4 Amoco Performance Products Inc.—1,3,4 BASF Corp., Plastic Materials—2,4
Surface-active agents	Grace, W.R., & Co., Organic Chemicals Div. Hexcel ICI Americas Inc. Witco Corp., Organics Div.
Synthetic fiber reinforcements	Carborundum, The, Co., Fibers Div. International Filler Corp.
Thermoplastic elastomers 1. Alloys	 Atochem North America, Inc.—4,5

(continued)

Table 8.2. (Continued)

Material	Supplier
2. Engineering	BASF Corp., Urethanes—4
3. Olefinic	Colonial Rubber Co.—1–3,5
4. Polyurethane	Witcoi Corp.—4,6
5. Styrenic	
6. Polyester	
Thermoplastic molding compounds, reinforced	Deer Polymer Corp.
	Ferro Corp., Horizon Polymers Division
	BF Goodrich Co., Specialty Polymers & Chemicals Div.
	Hoechst Celanese Corp., Engineering Plastics Div.
	ICI Advanced Materials
Thixotropic thickeners	Allied Signal Inc., A-C Performance Additives
	Cabot Corp., Cab-O-Sil Div.
	Degussa Corp., Aerosil and Imported Pigment Products Div.
	Engelhard Corp.
	Lubrizol Petroleum Chemicals Co.
	New England Resins & Pigments Corp.
	Unipol Consultants
	Wacker Chemicals (USA), Inc.
UV absorbers	Argus Div., Witco Corp.
	Canada Colors & Chemicals, Ltd.
	General Color & Chemical Co., Inc.
	ICI Americas Inc.
	Plastics Color Chip, Div. of PMC Inc.
	Zinc Corp. of America
Viscosity depressants	Allied Signal Inc., A-C Performance Additives
	Axel Plastics Research Laboratories, Inc.

Source: Modern Plastics Encyclopedia '95, P.O. Box 602, Hightstown, NJ 08520-9955.

Table 10.1. Formulation for Typical Soybean Interior Plywood Adhesive

Component	Pounds
Water at 60–70°F	175
Untoasted soybean flour	97
Pine oil or equivalent defoamer	3
(Mix 2 min or until smooth)	
Hydrated lime	12
Water at 60–70°F	24
(Mix 1 min)	
50% sodium hydroxide solution	14
(Mix 1 min)	
"N" Brand sodium silicate (Philadelphia Quartz Co.)	25
(Mix 1 min)	
Carbon disulfide in	$1\frac{1}{4}$
Carbon tetrachloride	$\frac{1}{2}$
Flake pentachlorophenol	$4\frac{1}{4}$
(Mix 10 min)	

Table 10.2. Formulation for Typical Blood Plywood Adhesive

Component	Pounds
Water at 145°F	200
Soluble dried beef blood	80
Fir wood flour	18
Pine oil or equivalent defoamer	2
(Mix 10 min)	
Cold water	350
Pine oil or equivalent defoamer	2
(Mix 2 min)	
Hydrated lime in	7
Water at 65–70°F	14
(Mix 2 min)	
"N" Brand sodium silicate solution (Philadelphia Quartz Co.)	35
(Mix 5 min)	

Table 10.3. Formulation of Typical Amylose Starch Adhesive for Corrugated Boards

Component	Parts
Carrier starch (A)	
Water	1192
HAS	424
Borax	6
(Bring to 130°F and add the following with stirring)	
NaOH	36.6
Water	47.5
Raw starch suspension (B)	
Water (at 85°F)	3480
Corn starch	1600
Borax	281
Thermosetting resin	91.2

Table 10.4. Formulation of Typical Cellulose Heat-Seal Adhesive for Packaging

Component	Parts by Weight
Cellulose nitrate (11.4% nitrogen)	43.3
Ester gum	30.4
Dicyclohexyl phthalate	29.3
Hydrogenated castor oil phthalate	10.5
Crystalline paraffin wax (60°C melting point)	3.5
Ethyl acetate	547.0
Ethyl alcohol	20.0
Toluene	289.0

Table 10.5. Formulations of Typical Latex Rubber Adhesives

Component	Pounds	
Self-adhesive envelopes		
60% natural latex	100	
10% potassium hydroxide solution	0.2	
50% aqueous dispersion of zinc diethyldithiocarbamate	0.5	
Floor tile adhesive A		
60% natural latex	100	
Methyl cellulose (added as 5% solution)	at least 5	
Clay	150	
Black reclaim dispersion	50	
Tackifying resin dispersion	30–80	
High-boiling-point naphtha (added as emulsion)	10	
Food jar sealing compound		
60% natural latex (ammonia preserved)	100	
Clay (as 50% dispersion stabilized with food-grade surfactant)	200	
General-purpose pure gum adhesive		
60% natural latex	100	
Zinc diethyldithiocarbamate (50% dispersion)	2	
Ammonium caseinate 10% solution	10 (solution)	
Tufted carpet adhesive and backing		
	Primary Backing	Secondary Backing
Natural latex (high ammonia)	100	100
Stabilizer/wetting agent	1.5	1.0
Thiourea (added as 10% solution)	1.0	1.0
Antioxidant	1.0	1.0
Water	to give 75% solids	
Whiting (added as slurry before thickener)	400	250
Polyacrylate thickener added as 5% solution)	0.2	0.3

Table 10.6. Formulation for Pressure-Sensitive Adhesives

Component	Pounds
PIB-based PSA for removable label stock	
Vistanex L-120	100
Hercolyn	35
Escorez 1315	45
Polybutene H-100	70
Irganox 1010	0.5
Solvent (e.g., heptane)	to coatable viscosity
PSA for vinyl floor tile	
Exxon Butyl 268	100
Vistanex LM-MS	20
Terpene phenolic resin such as Schenectady SP-567	70
Solvent	to coatable viscosity

Table 10.7. Formulation of Butyl Rubber-Based Caulking Compound

Component	Percent Weight
Exxon Butyl 065, 50% in Mineral Spirits	20.50
Vistanex LM-MS	2.05
Isostearic acid	0.51
International fiber talc	30.75
Atomite whiting	20.50
Rutile titanium dioxide	2.56
Schenectady SP-553 Resin	3.60
Polybutene H-300	10.25
Blown Soya Oil, Z_3	1.54
Cobalt naphthalenate drier, 6%	0.05
Cab-O-Sil	2.05
Mineral spirits	5.64

Table 10.8. Formulation for Rope-Hotmelt Rubber-Based Adhesive

Component	Percent Weight
Exxon Butyl 268	20
Beta-Pinene Resin (mp 115°C)	20
EVA (Elvax 250)	20
Low-molecular-weight polyethylene (12,000 Da)	20
Low-molecular-weight polyethylene (20,000 Da)	19
Antioxidant	1

Table 10.9. Formulation of Oil-Resistant Nitrile Rubber Adhesive

Component	Pounds
Recipe A—black curing	
Nitrile rubber	100
Zinc oxide	5
Sulfur[a]	3
EPC Black[b]	50
"AgeRite" Resin D	5
Coumarone-indene resin[c]	25
Refined coal tar	25
Recipe B—nonblack curing	
Nitrile rubber	100
Stearic acid	0.5
Zinc oxide	10
Sulfur[d]	2
Calcium silicate[e]	100
Titanium dioxide	25
Coumarone-indene resin[f]	10
Dibutyl phthalate	10
Accelerator "808"[g]	1.5

Notes: [a]Blackbird
[b]"Wyex"
[c]"Picco"
[d]"Spider"
[e]"Silene" EF
[f]"Picco" 10
[g]DuPont

Table 10.10. Formulation of Styrene-Butadiene Rubber (SBR) for Tire Treads

Component	Pounds
High Mooney SBR (150 ML-4)	100
Koresin	40
Petroleum softener (Sundex 53)	10
HAF carbon black (Philblack 0)	60
Zinc oxide	5
BLE	1.0
Santocure	1.2
DPG	0.3

Table 10.11. Formulation of Styrene-Butadiene Rubber (SBR) Liquid Applied Sealant

Component	Percent Weight
SBR (25% styrene)	12.0
Polymerized rosin	19.0
Methyl ester of hydrogenated rosin	2.0
Aromatic plasticizer	2.0
Soft clay	17.0
Fibrous talc	10.0
Toluene	26.0
Xylene	12.0

Table 10.12. Formulation of Hotmelt Adhesive Based on S-I-S Thermoplastic Rubber

	Parts by Weight		
Component	Two Components (parts)	Three Components (parts)	Four Components (parts)
S-I-S (Kraton 1107 Rubber)	100	100	100
Midblock resin (WingTack 95)	100	100	100
Plasticizing oil (Shellflex 371)	—	40	40
Endblock resin (Cumar LX-509)	—	—	60
Stabilizer (zinc dibutyldithiocarbamate)	5	5	5
Total	205	245	305
Shear adhesion failure temp. (°F)	210	188	220
Rolling ball tack (PSTC-6) (cm)	5.9	0.6	1.8
Probe tack (g)	1300	700	1100
180° peel adhesion (PSTC-1) (pli)	5.3	2.5	3.7
Melt viscosity at 350°F (cP)	200,000	30,000	40,000
Holding power to kraft paper (min)	>2800	5	150
Thermoplastic rubber content (wt %)	49	38	33

Table 10.13. Formulation of Pressure-Sensitive Adhesive Based on S-B-S Thermoplastic Rubber

Component	Pounds
Composition (wt. parts)	
S-B-S (Kraton 1101 Rubber)	100
Midblock resin (Super Sta-Tac 80)	200
Stabilizer	1
Properties	
Rolling ball tack (PSTC-6) (in.)	10
Probe tack (g)	1700
180° peel adhesion (PSTC-1) (pli)	7.6
Shear adhesion failure temp. (°F)	180
Thermoplastic rubber content (wt %)	33
Endblock/midblock ratio	10/90

Table 10.14. Formulation of Contact Assembly Adhesive Based on S-B-S Thermoplastic Rubber

Component	Parts by Weight
S-B-S (Kraton 1101 Rubber)	100
Endblock resin (Picco N-100)	37.5
Midblock resin (Pentalyn H)	37.5
Stabilizer (Antioxidant 330)	0.6

Table 10.15. Formulation for Acrylic Emulsion Ceramic Tile Adhesive

Component	Parts by Weight
Emulsion E-1997 (49% solids)	210.0
Propylene glycol	10.0
Water	70.0
Tamol 731	5.0
Urea	30.0
Defoamer	1.0
Duramite calcium carbonate	500.0
Acramine clear concentrate NS2R	14.0

Table 10.16. Formulation for Neoprene Adhesives

Component	Parts by Weight	
	Decorative Laminates	General-Purpose Industrial Adhesive
Neoprene AC[a]	100	100
Magnesium oxide	5	5
Zinc oxide	2	2
Antioxidant	1	1
Heat-reactive tertiary butyl phenolic resin[b]	—	20
Hexane	275	277
Acetone	215	138
Methyl ethyl ketone	—	138
Toluene	122	138
% solids	20	20

Notes: [a]Mooney viscosity grade used depends on viscosity and performance requirements.
[b]Reacted with magnesium oxide—amount of which is included under magnesium oxide.

Table 10.17. Formulation for Simple Acrylic Engineering Adhesive

Component	Parts by Weight
Part 1	
Methyl methacrylate	85.0
Polymethyl methacrylate	15.0
N,N-Dimethyl aniline	0.5
Part 2	
Benzoyl peroxide	0.5
(Mix 1 and 2 for a shelf life of 1/3 hour)	

Table 10.18. Formulation for Polysulfide Adhesives and Primers

	Parts by Weight					
	Part A					
Component	A	B	C	D	E	F
ILP-2	100	100	100	100	—	100
LP-32	—	—	—	—	100	—
SRF No. 3 black	30	—	—	—	10	—
Stearic acid	1.0	1.0	—	1.0	1.0	1.0
Durez 10693	5.0		—	—	—	—
Calcene TM	—	25.0	—	—	—	—
Titanox RA-50	—	10.0	10.0	—	—	—
Lithopone	—	30.0	90.0	50.0	—	—
Kenflex A	—	—	—	15.0	—	—
Sterling MT	—	—	—	—	10.0	—
Sulfur	—	—	—	—	0.15	—
Thermax	—	—	—	—	—	100
Santicizer E-15	—	—	—	—	—	50
Santicizer 141	—	—	40	—	—	—
Santicizer 261	—	25	—	—	—	—
Methylon AP108	—	5	—	—	—	—
	Part B					
C-15	15	15	15	15	—	—
"Accelerator" C-9	—	—	—	—	13.8	—
"Accelerator" PbO_2	—	—	—	—	—	13.5
Dibutylphthalate	—	—	—	—	—	11.0
Stearic acid	—	—	—	—	—	0.5
Recommended use	Aircraft sealant	Building sealant	Casting compound	Potting compound	Deck seal	For MIL-C-15705A

Table 10.19. Formulation of Polysulfide Adhesives and Primers

Component		Parts by Weight	
Primer Formulations for Use with Polysulfide Sealants			
Primer A		Primer B	
Silane A-187	3.85	Parlon S-10	20.0
Tyzor TPT	1.15	Toluene	30.0
Isopropanol	45.00	Silane 4523	2.5
Primer C			
Parlon S-125	20.0		
Aroclor 1254	6.0		
Aroclor 1260	6.0		
Primer D		Primer E	
Parlon S-10	25.0	Toluene	80.0
Marbon CB-60	25.0	Butyl Cellosolve	5.0
Cellosolve acetate	17.5	Butanol	5.0
Toluene	17.5	Silane A-187	10.0
Aroclor 1242	15.0		

Table 10.20. Formulation for Cold-Pressed Medium Abrasive, General-Purpose Phenolic Adhesives

	Parts by Weight	
Component	#1	#2
Aluminum oxide (#54 grain)	1050	1050
Powdered phenolic two-step resin	130	150
Liquid phenolic one-step resin	20	—
Furfural/cresylic acid, 3/2	—	18
Cold pressed density (g/cm^3)	2.64	2.64

Table 10.21. Formulation for Amino Resin Corrugating for Wood

Component	Parts by Weight
Carrier portion	
Unmodified corn starch	100
Water	325
Caustic soda	15
Secondary portion	
Urea	100
Paraformaldehyde	50
Unmodified corn starch	500
Water (100–110°F)	1585

The carrier portion is heated under agitation with live steam at 160°F for 15 min. About 310 parts of water is added and the mixture is cooled to 118°F and added to the secondary portion. The adhesive thus obtained has a gelatinization range of 146–148°F.

Table 10.22. Formulation for General-Purpose Epoxy Adhesive

Component	Parts by Weight
1. Epoxy resin	100 parts
Versamid 115 or equivalent	70 parts
Filler or reinforcement	as desired
2. Epoxy resin	100 parts
Versamid 115	35 parts
DMP-30	5 parts
Filler or reinforcement	as desired
Formula 2 is a faster-curing adhesive than formula 1, but is not as flexible	
3. Epoxy resin	100 parts
Lancast A	70 parts
Filler or reinforcement	as desired

Formula 3 can also be accelerated with tertiary amines. These formulations are two-component, room-temperature-curing adhesives, which have limited pot life after resin and hardener have been mixed. Filler or reinforcement is added to either resin or hardener before these key ingredients are brought together. Cure can be accelerated by heat.

Table 10.23. Formulation for One-Component Epoxy Adhesive

Component	Parts by Weight
Epoxy resin	100
Bentone 34	25
Alumina	25
Dicyandiamide	6
Cure: 1 to 1.5 hr at 350°F	

Shear strength for Al–Al: 2600 psi at room temperature

Table 10.24. Formulation for Quick-Cure Epoxy Adhesive

Component	Parts by Weight
Component I	
Epoxy resin (eq wt = 190–210)	100
Silica flour (Imsil A-10)	60
Carbon black	0.1
Asbestos	3
Component II	
Dion 3-800LC (polymercaptan)	75
Polyamide (Dion Modifier 38)	12
Dion EH-30 (tertiary amine)	8
Silica flour (Imsil A-10)	50
Titanium dioxide	10
Asbestos	4
Gel time: 8 min at 75°F	
Shear strength for Al–Al: 2270 psi at 75°F	

Table 10.25. Formulation for Polyurethane Adhesive for Cementing Neoprene and SBR Rubbers to Nylon and Dacron

Component	Parts by Weight
"Hylene MP" dispersion (40%)	27.5
Neoprene latex Type 635	173.0
Zinc oxide dispersion (50%)	15.0
Zalba emulsion (50%)[a]	6.0

Note: [a]A hindered phenolic antioxidant—du Pont Elastomer Chemicals Dept.

Table 10.26. Formulation for Polyvinyl Acetal Adhesive

Component	Parts by Weight
One component:	
Polyvinyl butyral	100
Phenolic resin	150
Epoxy resin	100
Aluminum powder	200
Isopropyl acetate	200
95% isopropyl alcohol	100
Two component	
Epoxy resin	100
Phenolic resin	100
Methyl ethyl ketone	200

Table 10.27. Formulation for Ethylene Copolymer-Based Hotmelt Adhesive Used for Bookbinding

Component	Parts by Weight
Elvax 260 EVA[a]	30–40
Rosin ester tackifier, R&B	25–45
F. R. paraffin wax, mp 100–105°C	15–30
White microcrystalline wax,[b] mp 82.2–87.8°C	5–10
Ethyl 330 antioxidant[c]	0.5

Notes: [a]DuPont Company
[b]Bareco Div. Petrolite Corporation
[c]Ethyl Corporation

Table 10.28. Formulation for Stryene Block Copolymer for Bookbinding

Component	Parts by Weight
Elvax 260[a]	20–35
Kraton 1107[b]	15–35
Foral 105[c]	20–40
Shellflex 371[b]	5–10
Microcrystalline wax, mp 76.7–87.8°C	10–15
Antioxidant (Irganox 1010)[d]	0.25

Notes: [a]DuPont Company
[b]Shell Chemical Company
[c]Hercules, Inc.
[d]Ciba-Geigy Corporation

Table 10.29. Formulation of Ethyl Vinyl Acetate Pressure-Sensitive Adhesive

Component	Parts by Weight
EVA copolymer(s)	35–50
Plasticizer	0–20
Tackifier(s)	30–50
Filler	0–5
	Total 100
Antioxidant	0.1–0.5

Table 10.30. Formulation for Cyanoacrylate Adhesive

Component	Percent Weight
Alkyl 2-cyanoacrylates	(not applicable)
Catalyzed by water or alcohol	(trace quantity)

Table 10.31. Formulation for Polyethyleneimine Adhesive for Tire Cords

Component	Parts by Weight
Vinyl pyridine latex (Pliocord LVP-4668)[a]	100
VPX-500[b]	17

Notes: [a]Goodyear
[b]DuPont

Table 10.32. Formulation of Urethane Anaerobic Adhesive

Component	Percent Weight
Basic components	
Polymerizable alcohol, e.g., β-hydroxyethyl methacrylate	
Toluene diisocyanate, or isocyanate-terminated urethane prepolymer	
Organic hydroperoxide, e.g., cumene hydroperoxide	
Example formulation	
Estane Resin 5703F2	11.25
Geon Resin 202	3.75
Tetrahydrofuran	85.00

The solution is applied onto vinyl shoe sole and leather upper component and air dried for 1.0 min under 20 psi (gauge) pressure.

Table 10.33. Polymers for High-Temperature Adhesive Formulations

Base Resin	Maximum 100 Hours Use Temperature (°C)
Polyimide	316
Polybenzimidizole	316
Polyquinoxaline	316
Polyphenylquinoxaline	316
Polyarylsulfone	260
Norbornene-terminated imide	260
Acetylene-terminated phenylquinoxaline	260
Polyarylene ether	232
Modified epoxy phenolic	232

Source: Skeist (1990)

Note: Examples of high-temperature adhesive products are Pyralin (DuPont) and Skybond.

Table 10.34. Formulation for One-Component RTV Silicone Adhesive

Component	Percent Weight
Polymeric silicone (silane-terminated polydimethylsiloxane) (2000–150,000 cP)	approx. 90
Cross-linking component (reactive polyfunctional silane such as tri-tetrafunctional silane, methyltriacetoxysilane)	
Catalyst (tin soaps, alkyl carboxylates)	approx. <2
Curing: When exposed to atmospheric water (or vapor) the cross-linker reacts with the polymeric silicone, and the by-product from curing is acetic acid which volatilizes from the hardening sealant/adhesive.	

Note: Percent weight values are approximated because the the exact weight percent values are dependent on the specific application, hardness, etc.

Table 11.1. Adhesive Materials and Suppliers

Resin	Manufacturer
Polymers	
Butadiene styrene rubber	Adhesive Products, Inc.
	Copolymer Rubber & Chemical Corp.
	Dow Chemical Co.
	Firestone Synthetic Rubber & Latex Co., Div. of Goodyear Tire & Rubber Co., Chemical Division
	Morton International, Inc.
	Shell Chemical Co., A Div. of Shell Oil Co.
Butyl rubber	Adhesives & Chemicals Inc.
	A-Line Products Corp.
	BLH Electronics
	Burton Rubber Processing, Inc.
	Burke-Palmason Chemical Co.
	Exxon Chemical Americas
	National Chemicals Co.
	NiChem, Inc.
	Polysar Rubber Div., Miles Inc.
	TACC International Corp.
	U.S. Rubber Reclaiming Inc.
	R. T. Vanderbilt Co., Inc.
Chlorinated rubber	Aceto Corporation
	Adhesives & Chemicals Inc.
	Bayer AG
	CHEMCENTRAL Corporation
	Copolymer Rubber & Chemical Corp.
	Exxon Chemical Americas
	Goldsmith & Eggleton, Inc.
	Kraft Chemical Co.
	National Chemicals Co.
	NiChem, Inc.
	Pierce & Stevens Corp.

(continued)

Table 11.1. (Continued)

Resin	Manufacturer
Chlorinated rubber	TACC International Corp.
	Uniroyal Chemical Company
	R. T. Vanderbilt Co., Inc.
Fluoropolymers	E. I. du Pont de Nemours & Co., Inc. Polymer Products Dept.
	ICI Americas Inc. Specialty Chemicals Division Americas
	Reichhold Chemicals, Inc.
Natural rubber	Adhesive Products, Inc.
	Akrochem Corporation
	American Writing Ink. Co.
	H. A. Astlett & Co. Inc.
	Firestone Synthetic Rubber & Latex Co., Div. of Bridgestone/Firestone, Inc.
	Goldsmith & Eggleton, Inc.
	Guthrie Latex, Inc.
	TACC International Corp.
	Testworth Laboratories, Inc.
Polybutadiene rubber	Ameripol Synpol Corporation
	Bayer AG
	Firestone Synthetic Rubber & Latex Co., Div. of Bridgestone/Firestone, Inc.
	Goldsmith & Eggleton, Inc.
	Goodyear Tire & Rubber Co., Chemical Division
	NiChem, Inc.
	Polysar Rubber Div., Miles Inc.
	Ricon Resins, Inc.
	TACC International Corp.
	R. T. Vanderbilt Co., Inc.
	Western Reserve Chemical
Polychloroprene	Bayer AG
	Burton Rubber Processing, Inc.
	CHEMCENTRAL Corporation
	E. I. du Pont de Nemours & Co., Inc. Polymer Products Dept.
	Goldsmith & Eggleton, Inc.
	Harwick Chemical Corp.
	Miles Inc.
	Morton International, Inc.
	TACC International Corp.
	R. T. Vanderbilt Co., Inc.
Polyisobutylene	Adhesive Products, Inc.
	A-Line Products Corp.
	BASF Corp.
	Burton Rubber Processing, Inc.
	Carlisle Syntec Systems
	Thermo-Cote, Inc.
	R. T. Vanderbilt Co., Inc.

(*continued*)

Table 11.1. (Continued)

Resin	Manufacturer
Polyisoprene	H. A. Astlett & Co. Inc. Burton Rubber Processing, Inc. Carlisle Syntec Systems R. H. Carlson Company, Inc. Goldsmith & Eggleton, Inc. Goodyear Tire & Rubber Co., Chemical Division Hardman, Div. of Harcros Chemicals Inc. Morton International, Inc. R. T. Vanderbilt Co., Inc.
Polysulfide	Burton Rubber Processing, Inc. Courtaulds Aerospace, Inc. Lu-Sol Corp. Morton International, Inc.
Polyurethane	A1 Technology, Inc. A-Line Products Corp. American Cyanamid Co., Cytec Industries BASF Corp. Bayer AG Courtaulds Aerospace, Inc. Dow Chemical Co. Engineered Materials Systems, Inc. Henkel Corporation ICI Polyurethanes Polyurethane Specialties Co., Inc. Reichhold Chemicals, Inc. Sanncor Industries Inc.
Reclaimed rubber	H. A. Astlett & Co., Inc. Burton Rubber Processing, Inc. U.S. Rubber Reclaiming Inc. Western Reserve Chemical
Silicone rubber	A1 Technology, Inc. Accumetric/Meter-Mix Inc. Bayer AG Dow Corning Corporation Engineer Materials Systems, Inc. Laur Silicone Rubber Compounding, Inc. Loctite Corporation PPG Industries Inc. Rhone Poulenc Inc. Seegott Inc. Tandem Products Wacker Silicones
Miscellaneous polymers	A1 Technology, Inc. (UV cured) Aceto Corporation (polyethyleneimine) Adhesive Products, Inc. (polyvinyl acetates, ethylene vinyl acetates, acrylic pressure sensitives)

(*continued*)

Table 11.1. (Continued)

Resin	Manufacturer
Miscellaneous polymers	Bayer AG (polyester and polyether polyols; ethylene/vinyl acetate copolymer)
	Bostik (polyesters, saturated; polyamides)
	Burton Rubber Processing, Inc. (elastomeric or plastic compounds in slab, strip, or diced form)
	CHEMCENTRAL Corporation (chlorinated polyolefins)
	Courtaulds Aerospace, Inc. (mercaptan-terminated polyether urethane OH & SH-terminated polythioether)
	Crowley Chemical Co. (amorphous polypropylene)
	Crowley Tar Products Co., Inc. (amorphous polypropylene)
	Crusader Chemical Co., Inc. (proprietary)
	Dexco Polymers (styrenic block polymers)
	Dow (EAA-Dow Adhesive Film)
	E. I. du Pont de Nemours & Co., Inc., Polymer Products Dept. (chlorinated polyolefins)
	Exxon Chemical Americas (chlorobutyl, bromobutyl)
	Gencorp Polymer Products (vinyl pyridine latex, butadiene styrene carboxy latices)
	GE Specialty Chemicals (acrylonitrile butadiene styrene)
	Heveatex Corp. (aqueous polymer emulsions and coatings)
	Housmex Inc. (reprocessed rubber)
	IGI Baychem International, Inc. (APP)
	King Industries, Inc. (polyester) Lu-Sol Corp. (anaerobic cyanoacrylate)
	Miles Inc. (polyester, polyethers, ethylene-vinyl acetate)
	Moore & Munger Marketing Inc. (high melt or synthetic waxes)
	National Starch & Chemical Company (resin emulsions, acrylic, vinyl acetate, ethylene-vinyl acetate styrene-arylate)
	Neville Chemical Co. (coumarone-indene petroleum hydrocarbon)
	NiChem, Inc. (polyisobutyl ether)
	Olin Corp. Specialty & Organics Dept. (specialty isocyanates, polyester polyols)
	Revertex Americas (liquid polybutadiene)
	Shell Chemical Co., A Div. of Shell Oil Co. (polybutylene; thermoplastic elastomers)
	Sigma Plastronics, Inc. (epoxy, hydrocarbon)
	3M (fluorinated) Union Carbide Corporation, Solvents & Coatings Materials Div. (caprolactone polyols for polyurethanes)
Fillers	American Gilsonite
	R. E. Carroll, Inc.
	Crowley Chemical Co.
	Crowley Tar Products Co., Inc.
	Van Waters & Rogers Inc.
Potassium silicate	Aremco Products, Inc.
	Ashland Chemical Inc., Sub. Ashland Oil, Inc.
	Hoechst Canada Inc., Industrial Division-Chemicals
	Joseph Turner & Co.

(*continued*)

Table 11.1. (Continued)

Resin	Manufacturer
Sodium silicate	Aremco Products, Inc.
	Ashland Chemical Inc., Sub. Ashland Oil, Inc.
	CHEMCENTRAL Corporation
	E. I. du Pont de Nemours & Co., Inc., Polymer Products Dept.
	Harcros Chemicals Inc.
	Hoechst Canada Inc., Industrial Division-Chemicals
	Occidental Chemical Corp., Corporate Marketing Dept.
	The PQ Corporation (PA)
Miscellaneous minerals	Alcoa (aluminum trihydrate)
	Aluchem Inc. (alumina trihydrate, calcium carbonate)
	American Gilsonite (resin)
	CDI Dispersions (dispersions)
	Flanagan Associates Incorporated
	Georgia Marble Co., Industrial Sales (calcium carbonate)
	Limestone Products Corp. (calcium carbonates)
	Mintec (mica and quartz fillers)
	Moore & Munger Marketing Inc. (microcrystalline or paraffin waxes)
	National Lime and Stone Co. (dolomitic limestone dust)
	Piqua Minerals (calcium carbonate)
	SCM Chemicals, Inc. (micronized silica gel)
	Shamokin Filler Co., Inc. (anthracite mineral filler)
	Spartan Minerals Corporation (aluminum silicate, mica)
	Superior Graphite Co. (graphite)
	Superior Materials Inc. (aluminum silicate, mica, talc, clay, calcium carbonate)
	3M (roofing granules)
	R. T. Vanderbilt Co., Inc. (talc, wollastonite, pyrophyllite, kaolin clay)
	Vista Chemical Company (catapal and dispal alumina)
Protein-based	
Animal	Adhesive Products, Inc.
	Borden Packaging & Industrial Products
	Thomas W. Dunn Corp.
Blood albumin	Adhesive Products, Inc.
Casein	Adhesive Products, Inc.
	American Casein Company
	Borden Packaging & Industrial Products
	Erie Foods International, Inc.
	Harwick Chemical Corp.
	Kraft Chemical Co.
	Victor Najda, Inc.
	National Casein Company
	Ultra Additives, Inc.
Fish	Adhesive Products, Inc.
Shellac	Colony Import & Export Corp.

(*continued*)

Table 11.1. (Continued)

Resin	Manufacturer
Shellac	National Chemicals Co.
	NiChem, Inc.
Soybean	Adhesive Products, Inc.
	National Casein Company
	Protein Technologies International Inc., Polymer Products
Miscellaneous protein-based	American Casein Company (casein protein polymers)
	BASF Corp. (ammonium chloride)
	Thomas W. Dunn Corp. (thermoplastic, water base)
	Faesy & Besthoff Inc. (animal bone meal, blood meal)
	Guthrie Latex, Inc. (palm oil)
Thermoplastic Resins	
Acrylic	Air Products and Chemicals, Inc.
	Allied Colloids Inc.
	Apple Adhesives, Inc.
	H. A. Astlett & Co. Inc.
	Axel Plastics Research Laboratories, Inc.
	BASF Corp.
	Basic Adhesives, Inc.
	Caswell & Co. Ltd.
	CHEMCENTRAL Corporation
	Degussa Corp.
	Dexter Automotive Materials
	Flanagan Associates Incorporated
	Franklin International, Polymer Products Div.
	Hardman, Div. of Harcros Chemicals Inc.
	Loctite Corporation
	Lu-Sol Corp.
	Merquinsa
	Morton International, Inc.
	National Starch & Chemical Company
	Reichhold Chemicals, Inc.
	Resinall Corp.
	Rohm and Haas Co.
	Seegott Inc.
	StanChem, Inc.
	Super Glue Corporation
	TACC International Corp.
	Tandem Products
	Thermo-Cote, Inc.
	3M
	Union Carbide Corporation, Solvents & Coatings Materials Div.
	Union Carbide Corporation, UCAR Emulsion Systems
	Utility Development Corp.
	Zeneca Resins
Cellulose	Bayer AG
	BLH Electronics

(*continued*)

Table 11.1. (Continued)

Resin	Manufacturer
Cellulose	Dow
	Eastman Chemical Co.
	Miles Inc.
	Pierce & Stevens Corp.
	Seal-Peel, Inc.
	Thermo-Cote, Inc.
Polyamide	Adhesive Technologies, Inc.
	Aremco Products, Inc.
	Axel Plastics Research Laboratories, Inc.
	BASF Corp.
	Bayer AG
	Bostik
	R. H. Carlson Company, Inc.
	Caswell & Co. Ltd.
	Dexter Automotive Materials
	DSM Engineering Plastics, Inc.
	EMS-American Grilon, Inc.
	Henkel Corporation
	Miles Inc.
	Pacific Coast Polymers
	RIT-Chem Co., Inc.
	Schering Berlin Polymers Inc.
	TACC International Corp.
	3M
	Union Camp Corporation, Chemical Products Div.
Polyolefin	Adhesive Products, Inc.
	Amoco Chemical Company
	Bayer AG
	Bostik
	Caswell & Co. Ltd.
	Dexter Automotive Materials
	Dow
	DSM Engineering Plastics, Inc.
	Eastman Chemical Co.
	Exxon Chemical Americas
	Hercules Incorporated
	Hoechst Canada Inc., Industrial Division-Chemicals
	R. T. Vanderbilt Co., Inc.
Polystyrene	Amoco Chemical Company
	BASF Corp.
	Dow Chemical Co.
	DSM Engineering Plastics, Inc.
	Innovative Formulations Corp.
	Knight Industrial Supplies, Inc.
Polyvinyl acetate	Adhesive Products, Inc.
	Adhesives & Chemicals Inc.

(continued)

Table 11.1. (Continued)

Resin	Manufacturer
Polyvinyl acetate	Adhesive Technologies, Inc.
	Air Products and Chemicals, Inc.
	Basic Adhesives, Inc.
	Borden Packaging & Industrial Products
	Caswell & Co. Ltd.
	Franklin International, Polymer Products Div.
	Hoechst Canada Inc., Industrial Division-Chemicals
	Jowat Corp.
	Knight Industrial Supplies, Inc.
	Morton International, Inc.
	National Casein Company
	National Starch & Chemical Company
	Pacific Coast Polymers
	Para-Chem Southern, Inc.
	Pierce & Stevens Corp.
	Rohm and Haas Co.
	Southern Resin, Inc.
	StanChem, Inc.
	TACC International Corp.
	Ultra Additives, Inc.
	Union Carbide Corporation
	Union Carbide Corporation, Solvents & Coatings Materials Div.
	Union Carbide Corporation, UCAR Emulsion Systems
	Utility Development Corp.
Polyvinyl alcohol	Adhesive Products, Inc.
	Adhesives & Chemicals Inc.
	Air Products and Chemicals, Inc.
	Caswell & Co. Ltd.
	CHEMCENTRAL Corporation
	E. I. du Pont de Nemours & Co., Inc., Polymer Products Dept.
	Hoechst Canada Inc., Industrial Division-Chemicals
	Kimall Trading Company, Equipment & Chemical Div.
	Knight Industrial Supplies, Inc.
	National Casein Company
	Pacific Coast Polymers
	Perry Chemical Corp.
	Southern Resin, Inc.
	StanChem, Inc.
	TACC International Corp.
	Wego Chemical & Mineral Corp.
Polyvinyl chloride	Borden Packaging & Industrial Products
	Caswell & Co. Ltd.
	Dow
	Goodyear Tire & Rubber Co., Chemical Division
	Hoechst Canada Inc., Industrial Division-Chemicals
	Mar Chem Corp.
	National Casein Company

(*continued*)

Table 11.1. (Continued)

Resin	Manufacturer
Polyvinyl chloride	Occidental Chemical Corp., Corporate Marketing Dept.
	Pacific Coast Polymers
	Pierce & Stevens Corp.
	TACC International Corp.
	Utility Development Corp.
	Vista Chemical Company
Miscellaneous thermoplastic resins	Acheson Colloids Co., Div. of Acheson Industries, Inc. (tetrafluoroethylene)
	Air Products and Chemicals, Inc. (vinyl acetate-ethylene copolymers, ethylene-vinyl chloride copolymers)
	Akrochem Corporation (hydrocarbon/tackifying resins)
	Allied Signal, Inc. (low-molecular-weight polyethylene & polyamide copolymers)
	American Gilsonite (gilsonite hydrocarbon resin)
	Apple Adhesives, Inc. (cyanoacrylate adhesive)
	Arizona Chemical Div., International Paper (hydrocarbon, terpene, rosin, and hybrid resins)
	AT Plastics Inc. (AT polymers) (ethylene vinyl acetate copolymers, low-density polyethylene)
	BASF Corp. (polyvinylidene chloride, polyvinyl ether vinyl chloride, vinyl isobutyl ether copolymers)
	Bayer AG (polycarbonate)
	Bostik (polyester, polyurethane)
	CHEMCENTRAL Corporation (ethylene/vinyl acetate copolymers)
	Dexter Automotive Materials (ethylene vinyl acetate)
	Dover Chemical Corp., a Sub. of I.C.C. Industries (70% chlorinated paraffin)
	DSM Engineering Plastics, Inc. (SAN ABS polycarbonate polypropylene, polyethylene, acetal polyurethane, polysulfone—all fiberglass reinforced)
	E. I. du Pont de Nemours & Co., Inc. Polymer Products Dept. (EVA—ehylene vinyl acetate)
	Eastman Chemical Co. (thermoplastic polyesters)
	EMS-American Grilon, Inc. (polyester)
	Exxon Chemical Americas (polypropylene; low-density, high-density, and linear low-density polyethylene; EVA, EMA)
	GAF Chemicals Corporation (PVP/vinyl acetate, PVP/styrene)
	GE Specialty Chemicals (PPE—polyphenylene ether)
	Goodyear Tire & Rubber Co., Chemical Division (polyester copolyester)
	The C. P. Hall Company (hydrocarbon)
	Henkel Corporation (plastic nylon polyamide)
	Hercules Incorporated (polyterpene; styrene polymers & copolymers; rosin-derived esters; modified rosins; petroleum hydrocarbons)
	Heveatex Corp. (aqueous acrylic & PVC coatings)
	The Humphrey Chemical Co. Inc.–CAMBREX Fine Chemicals Group (alkenyl succinic anhydrides)

(continued)

Table 11.1. (Continued)

Resin	Manufacturer
Miscellaneous thermoplastic resins	Jowat Corp. (EVA)
	Lawter International, Inc. (phenols, esters, hydrocarbon, poly ketones)
	Les Derives Resiniques Et Terpeniques (rosin derivatives terpene phenolic resins, terpene resins)
	Miles Inc. (polycarbonate thermoplastic polyesters, polyurethane)
	National Casein Company (hot melt adhesives, polyvinyl cross-link)
	National Starch & Chemical Company (ethylene-vinyl acetate emulsions)
	Natrochem, Inc. (coumarone indene)
	Neville Chemical Co. (coumarone-indene hydrocarbon)
	Pacific Coast Polymers (EVA)
	Permuthane, Inc. (polyurethane)
	Polysat Inc.
	Polyurethane Corp. of America (polyurethane)
	Polyurethane Specialties Co., Inc. (polyurethane)
	Quantum Chemical Corp., USI Div. (EVA, VAE copolymers, low-molecular-weight PE)
	Reichhold Chemicals, Inc. (terpene-rosin esters, terpene phenolics)
	RIT-Chem Co., Inc. (aromatic hydrocarbons)
	Sekisui-Iko Co., Ltd . (ethyl cyanoacrylate methyl cyanoacrylate)
	Soluol Chemical Co. Inc. (polyurethane)
	Superior Materials, Inc. (gilsonite)
	Union Carbide Corporation, Solvents & Coatings Materials Div. (phenoxy, PVA-PVC copolymers)
	Western Reserve Chemical (phenolic)
Thermosetting resins	
Alkyd polyester	Arakawa Chemical (USA) Inc.
	Bayer AG
	Hoechst Canada Inc., Industrial Division-Chemicals
	Insulating Materials, Inc.
	King Industries, Inc.
	Lawter International, Inc.
	Lu-Sol Corp.
	Miles Inc.
	NiChem, Inc.
	Pacer Technology
	Reichhold Chemicals, Inc.
	TACC International Corp.
Epoxy	A1 Technology, Inc.
	Abatron, Inc.
	Adhesives & Chemicals Inc.
	Air Products & Chemicals, Inc.
	American Cyanamid Co., Cytec Industries
	Apple Adhesives, Inc.

(*continued*)

Table 11.1. (Continued)

Resin	Manufacturer
Epoxy	Aremco Products, Inc.
	Ashland Chemical Inc., Sub. Ashland Oil, Inc.
	Bayer AG
	BLH Electronics
	CHEMCENTRAL Corporation
	Ciba Corporation, Furane Aerospace Products
	Conap, Inc.
	Courtaulds Aerospace, Inc.
	Dexter Automotive Materials
	Dow Chemical Co.
	Henkel Corporation
	Heresite Protective Coatings Inc.
	Hoechst Canada Inc., Industrial Division-Chemicals
	Raybestos Products Co.
	Reichhold Chemicals, Inc.
	Schering Berlin Polymers Inc.
	Seegott Inc.
	Shell Chemical Co., A Div. of Shell Oil Co.
	Union Carbide Corporation, Solvents & Coatings Materials Div.
	Utility Development Corp.
Furan	Cardolite Corporation
	Georgia-Pacific, Chemical Div.
	Wego Chemical & Material Corp.
	Western Reserve Chemical
Phenolic	Akrochem Corporation (Two Step)
	Arakawa Chemical (USA) Inc.
	Aremco Products, Inc.
	BLH Electronics
	Borden Packaging & Industrial Products
	CHEMCENTRAL Corporation
	GE Company
	Georgia-Pacific, Chemical Div.
	Hardman, Div. of Harcros Chemicals Inc.
	Heresite Protective Coatings Inc.
	Hoechst Canada Inc., Industrial Division-Chemicals
	PMC Specialties Group, Inc.
	Raschig Corp.
	Raybestros Products Co.
	Schenectady International, Inc.
	Seegott Inc.
	Wego Chemical & Mineral Corp.
	Western Reserve Chemical
Polyamide	American Cyanamid Co., Cytec Industries
	Arakawa Chemical (USA) Inc.
	Aremco Products, Inc.
	Borden Packaging & Industrial Products
	Georgia-Pacific, Chemical Div.

(continued)

Table 11.1. (Continued)

Resin	Manufacturer
Polyamide	Hardman, Div. of Harcros Chemicals Inc.
	Henkel Corporation
	Jowat Corp.
	Laminating Technology Inc.
	Lawter International, Inc.
	Lu-Sol Corp.
	Miller-Stephenson Chemical Co.
	NiChem, Inc.
	Pacific Anchor Chemical Div. of Air Products & Chemicals
	Pam Fastening Technology Inc.
	Reichhold Chemicals, Inc.
	RIT-Chem Co., Inc.
	Schering Berlin Polymers Inc.
	Sigma Plastronics, Inc.
	TACC International Corp.
	TRA-CON, Inc.
Polyanhydride	Arakawa Chemical (USA) Inc.
	Castall, Incorporated
	Hardman, Div. of Harcros Chemicals Inc.
	Lu-Sol Corp.
	Sigma Plastronics, Inc.
Polyimide	American Cyanamid Co., Cytec Industries
	Aremco Products, Inc.
	BLH Electronics
	Ciba
	Engineered Materials Systems, Inc.
	Mavidon Corporation
	Poly Organix, Inc.
Resorcinol	Borden Packaging & Industrial Products
	Georgia-Pacific, Chemical Div.
	Hoechst Canada Inc., Industrial Division-Chemicals
	National Casein Company
	Schenectady International, Inc.
Silicone	Accumetric/Meter-Mix Inc.
	Ashland Chemical Inc., Sub. Ashland Oil, Inc.
	Bayer AG
	John H. Calo Co.
	R. H. Carlson Company, Inc.
	Castall, Incorporated
	CHEMCENTRAL Corporation
	Dow Corning Corporation
	Loctite Corporation
	Lu-Sol Corp.
	McKesson Chemical Co.
	Miles Inc.
	Rhone Poulenc Inc.

(*continued*)

Table 11.1. (Continued)

Resin	Manufacturer
Urea	Acheson Colloids Co., Div. of Acheson Industries, Inc.
	American Cyanamid Co., Cytec Industries
	Borden Packaging & Industrial Products
	Georgia-Pacific, Chemical Div.
	National Casein Company
	NiChem, Inc.
	Sentry/Custom Services Corp.
	Southern Resin, Inc.
	Wego Chemical & Mineral Corp.
Miscellaneous thermosetting resins	A1 Technology, Inc. (film adhesives)
	American Cyanamid Co., Cytec Industries (polyurethane)
	Ashland Chemical Inc., Sub. Ashland Oil, Inc.
	Ciba (bismaleimide)
	Dow (vinyl ester)
	Dymax Corp. (urethane, polyester)
	Epoxy Coatings Co. (water-based epoxy systems; UV curable)
	Georgia-Pacific, Chemical Div. (melamine-formaldehyde)
	Goodyear Tire & Rubber Co., Chemical Division (polyester copolymers)
	Heveatex Corp. (Resorcinol-formaldehyde latex compounds)
	IMPCO, Inc. (styrene-free polyester)
	Kemstar Corp. (aramid Kevlar resins)
	King Industries, Inc. (polyurethane)
	Lu-Sol Corp. (cyanoacrylate)
	Morton International, Inc. (polysulfide, epoxy)
	National Casein Company (polyurethane, 2 part)
	Permuthane, Inc. (polyurethanes) Poly Organix, Inc. (bismaleimides)
	Polyurethane Corp. of America (polyurethane)
	Polyurethane Specialties Co., Inc. (polyurethane)
	Reichhold Chemicals, Inc. (polyester, epoxy, phenolic)
	Sartomer Co. Inc. (photo initiators)
	Super Glue Corporation (cyanoacrylate adhesives)
	TACC International Corp. (urethane)
Vegetable	
Dextrin	Adhesive Products, Inc.
	American Maize Products Co.
	Avebe America, Inc.
	Borden Packaging & Industrial Products
	Caswell & Co. Ltd.
	Corn Products, a unit of CPC International, Inc.
	Knight Industrial Supplies, Inc.
	Kraft Chemical Co.
	National Casein Company
	National Starch & Chemical Company

(*continued*)

Table 11.1. (Continued)

Resin	Manufacturer
Natural gums (arable, karaya, tragacanth)	Adhesive Products,Inc. Ashland Chemical Inc., Sub. Ashland Oil, Inc. Avebe America, Inc. Colony Import & Export Corp. Hercules Incorporated Kraft Chemical Co. TIC Gums Incorporated
Soybean	Adhesive Products, Inc. Ashland Chemical Inc., Sub. Ashland Oil, Inc. Kraft Chemical Co. Protein Technologies International, Inc., Polymer Products Werner G. Smith, Inc.
Starch (corn, tapioca, wheat, potato, sage)	Adhesive Products, Inc. American Maize Products Co. (Corn) Avebe America, Inc. Borden Packaging & Industrial Products Chemstar Products Co. Corn Products, a unit of CPC International, Inc. Knight Industrial Supplies, Inc. Kraft Chemical Co. National Casein Company National Starch & Chemical Company Wood Rosin Adhesive Products, Inc. John H. Calo Co. CHEMCENTRAL Corporation Flanagan Associates Incorporated Harwick Chemical Corp. Hercules Incorporated Kraft Chemical Co. Reichhold Chemicals, Inc.
Miscellaneous vegetable	American Maize Products Co. (corn syrup glucose) Arizona Chemical Div. International Paper (tall oil rosin) Chemstar Products Co. (water-soluble starch derivatives) Composition Materials Co., Inc. (wood flour, walnut shell flour, pecan shell flour, rice hull flour) Georgia-Pacific, Chemical Div. (tall oil rosin) Hercules Incorporated (terpene resins) Ligno Tech USA (calcium and sodium lignosulfurates) Pacific Anchor Chemical Div. of Air Products & Chemicals (walnut, safflower, and linseed oils) Protein Technologies International, Inc. Technichem, Inc. (tall oil rosin) Union Camp Corporation, Chemical Products Div. (tall oil rosin)
Miscellaneous bases	
Ceramics	Aremco Products, Inc.

(*continued*)

Table 11.1. (Continued)

Resin	Manufacturer
Ceramics	Basic Adhesives, Inc.
	BLH Electronics
	Carborundum
	Miles Inc.
	Pacer Technology
	Tandem Products
	3M
Enamels	Miles Inc.
	National Chemicals Co.
	NiChem, Inc.
	Sanncor Industries Inc.
	Schenectady International, Inc.
	StanChem, Inc.
Lacquer	National Chemicals Co.
	National Starch & Chemical Company
	NiChem, Inc.
	Pierce & Stevens Corp.
	Polyurethane Corp. of America
	Polyurethane Specialties Co., Inc.
	P.S.H. Industries, Inc.
	Sanncor Industries, Inc.
	Sentry/Custom Services Corp.
	StanChem, Inc.
Varnishes	Conap, Inc.
	Dow Corning Corporation
	Heresite Protective Coatings Inc.
	Mavidon Corporation
	National Chemicals Co.
	NiChem, Inc.
	Pierce & Stevens Corp.
Other bases	A-Aroma Tech, Inc. (odorants)
	Air Products and Chemicals, Inc. (miscellaneous polymers)
	American Casein Company (casein protein polymers)
	American Cyanamid Co., Cytec Industries (primers, primers-solvent base and foaming additives)
	Borden, HPPG Div. (cyanoacrylate adhesives, wood and leather glue, anaerobic sealants)
	Dynamold, Inc. (high-heat-resistant epoxy potting compound, adhesive; epoxy-based moldable shim materials)
	GAF Chemicals Corporation (N-vinyl-2-pyrrolidone copolymers)
	W. L. Gore & Assoc. Inc. (fluoropolymer etching services)
	Insulating Materials, Inc.
	Kenrich Petrochemicals, Inc. (dispersions)
	Morton International (cyano acrylates)
	Natrochem, Inc. (rosin oils)
	Pacer Technology (cyanoacrylate)
	Polyurethane Corp. of America (polyurethane)
	Polyurethane Specialties Co., Inc. (latices, polyurethane)

(continued)

Table 11.1. (Continued)

Resin	Manufacturer
Other bases	Reichhold Chemicals, Inc. (ethylene vinyl acetate)
	Sentry/Custom Services Corp. (water-based polyurethanes)
	Werner G. Smith, Inc. (waxes, coupling agents, blown fish and soybean oils)
	Superior Graphite Co. (graphite)
	Superior Materials Inc. (hydrocarbon resins)
	UCB Radcure, Inc. (UV curable)
	Ultra Additives, Inc. (hydrocarbon emulsions)
	Union Carbide Corporation, UCAR Emulsion Systems (latices-acrylics, PVA)
	Vanguard Chemical International, Inc. (nitrocellulose solutions, nitrocellulose)

Source: *Adhesives Age* (1993).

Table 13.1. Printing Process and Drying System

Printing Process	Drying System	Vehicle
Letterpress, news	Absorption	Nondrying oil
Letterpress, offset	Oxidation	Drying oil
Letterpress, offset	Quick-setting	Resin oil
Letterpress, letterset	Precipitation	Glycol-resin
Letterpress	Cold-setting	Resin wax
Gravure, flexographic	Evaporation	Solvent resin

Table 13.2. Formulation of Acrylic Black Ink

Component	Percent Weight
Elftex 8 Carbon Black	13.0
Huber 80 Kaolin Pigment	6.0
MP-22 Wax	1.0
Colloid 675 Defoamer	1.0
Isopropyl alcohol	3.0
Gro-Rez 2050 Acrylic Resin Solution	35.0
Ammonia (28%)	0.5
Water	39.5
Transaid 1280 Polymeric Material	1.0

Source: Grow Polymer, technical data sheet, starting formulation.

Table 13.3. Formulation of Acrylic Foil Ink

Component	Percent Weight
Blue pigment	20.0
MPP-123 Polyethylene Wax	0.5
Isopropyl alcohol	6.0
Grocryl 6057 Modified Acrylic Copolymer	40.0
Water	32.2
Ammonia (28%)	1.3

Source: Grow Polymer, technical data sheet, starting formulation.

Table 13.4. Formulation for Acrylic-Polyethylene Ink

Component	Percent Weight
Flexiverse Dispersion	40.0
Gro-Rez 2020 Acrylic Resin Solution	49.0
Growax 35 Polyethylene Emulsion	5.0
Defoamer	0.2
Transaid 1280 Polymeric Material	1.0
Water	4.8

Source: Grow Polymer, technical data sheet, starting formulation.

Table 13.5. Formulation for Acrylic-Wax Ink

Component	Percent Weight
Red Lake C Acroverse Chip	16.45
Water	15.05
Isopropyl alcohol	2.10
Ammonia (28%)	0.70
Morpholine	0.70
Grocryl 6057 Modified Acrylic Copolymer	57.00
Growax 35 Polyethylene Emulsion	4.00
Isopropyl alcohol	4.00
(Color: red flexo/foil ink)	

Source: Grow Polymer, technical data sheet, starting formulation.

Table 13.6. Formulation for Varnish Ink

Component	Percent Weight
Filtrez 5001 Varnish	56.5
Water	18.0
Antifoam	0.5
Barium lithol red pigment	25.0
(Color: red varnish ink, water-type barium lithol red base)	

Source: FRP, technical bulletin, suggestion formulation.

Table 13.7. Formulation for Acrylic Metallic Ink

Component	Percent Weight
Aluminum metallic powder	15.0
Joncryl 1535 Acrylic Mixing Vehicle	85.0
(Color: aluminum, 43 seconds #3-Zahn)	

Source: S. C. Johnson & Son, Inc., graphic art information, JONCRYL 1535, suggested formula.

Table 13.8. Formulation for Metallic Ink, Acrylic/Vinyl/Resin

Component	Percent Weight
Vinyl resin	9.75
Acrylic resin	9.75
Modified rosin	3.25
Powdered Polywax	1.95
Methyl ethyl ketone	20.15
Toluene	20.15
Obron XM-18 Pigment or Obron XM-18G Pigment	35.00
(Color: aluminum leaf)	

Source: Obron, technical bulletin, Obron Introduces Glittering Gravures, suggested formulation.

Table 13.9. Formulation for Alkali-Resistant Acrylic Ink

Component	Percent Weight
Joncryl 537 Acrylic Emulsion Polymer	90.0
Butyl Cellosolve solvent	7.0
Carbitol solvent	2.0
Aromatic 150 solvent	1.0
(Ink vehicle only or alkali and detergent resistance with good adhesion to polystyrene and vinyl films)	

Source: S. C. Johnson & Son, Inc., technical service information, JONCRYL 537, Vehicle 90-721.

Table 13.10. Formulation for Cellophane Ink/Nitrocellulose/Resin

Component	Percent Weight
RS Nitrocellulose, 5–6 seconds	32.5
Abitol hydroabietyl alcohol	17.5
Ethyl acetate	15.0
Ethyl alcohol	7.5
Butyl Cellosolve solvent	2.5
Toluene	25.0
(Good adhesion to Mylar or saran-coated cellophane)	

Source: Hercules, Inc., technical service report CSL-82A, COATINGS AND INKS, Formula 1.

Table 13.11. Formulation for Duplicating Fluids and Solvents

Component	Percent Weight
Ethyl alcohol	75.0–78.0
Methyl alcohol	15.0–20.0
Ektasolve EE Solvent	0.8–1.6
(Duplicating fluid 85, 95%)	

Source: Eastman Chemical Products, Inc., Publication No. M-203, DUPLICATING FLUIDS, suggested formulation.

Table 13.12. Formulation for Fluid Ink, Resin, CAB (Yellow)

Component	Percent Weight
Chrome yellow pigment	14.1
CAB-381-0.5-cellulose acetate butyrate	9.4
Uni-Rez 7024 Resin	9.4
Kodaflex DBP Plasticizer	3.3
Isobutyl acetate	12.8
Tecsol 3 Solvent	12.8
Toluene	38.2

Source: Eastman Chemical Products, Inc., Publication No. F-174B, EPOLENE WAXES AS ADDITIVES FOR INKS, Formula 1 from Table 2.

Table 13.13. Formulation for High-Solids Ink, Acrylics

Component	Percent Weight
Titanium dioxide	35.0
Joncryl 682 Acrylic Oligomer	7.5
Ammonia (28%)	1.88
Water	10.45
Isopropanol	1.50
Defoamer	0.67
Joncryl 80 Acrylic Polymer	35.00
Johnson 26 Polyethylene Wax Emulsion	3.0
Ethanol	5.0
(Color: white, high-solids ink, high gloss, good printing property)	

Source: S. C. Johnson & Son, Inc., technical bulletin, JONCRYL 682, suggested formulation.

Table 13.14. Formulation for Matt Finish Ink, Acrylic

Component	Percent Weight
Joncryl 67 Acrylic Resin	13.80
Ammonia (28%)	2.10
Morpholine	1.68
Tall oil fatty acid	1.50
Ethylene glycol monoethyl ether	0.90
Water	59.02
Organic pigment	16.00
Isopropanol	5.00
(Typically used for corrugated box board)	

Source: S. C. Johnson & Son, Inc., technical service information, JONCRYL 67 ACRYLIC RESIN, suggested formulation.

Table 13.15. Formulation for Moisture-Set Ink, Resin

Component	Pecent Weight
Organic pigment	11.0
Extender	10.0
Fumaric/maleic modified penta ester of rosin	39.0
Polyethylene wax or polycrystalline wax	2.0
Diethylene glycol	32.0
Dipropylene glycol	6.0
(This is a moisture-set ink with organic pigment)	

Source: Braznell Company, NAPIM PATTERN PRINTING INK FORMULA, Formula #106.

Table 13.16. Formulation for Newspaper Ink, Oils

Component	Pecent Weight
Multimix Flush	25.0
Magie #3 Oil	43.0
Magie #2 Oil	10.0
Clay (treated)	5.0
Petrolatum	5.0
Magie 535 Oil	12.0
(This ia a no-heat newspaper ink with versatility and economy)	

Source: BASF Wyandotte Corp., technical bulletin, THE OIL KEY, suggested formula.

Table 13.17. Formulation for News Ink, Vehicle/Oil

Component	Percent Weight
Multimix Flush	30.0
Gelled hydrocarbon vehicle	54.0
Clay (treated)	3.0
Petrolatum	3.0
Magie 535 Oil	10.0
(This is a low-heat news ink, with economy and versatility)	

Source: BASF Wyandotte Corp., technical bulletin, THE OIL INK KEY, suggested formulation.

Table 13.18. Formula for Packaging Ink, Rosin/Lacquer

Component	Percent Weight
Industrial carbon black	12.0
Methyl ethyl ketone	14.0
Toluene	20.0
Soya lecithin	1.0
N/C lacquer	17.0
Dioctyl phthalate	6.0
Limed rosin	30.0
(This is a black gravure packaging ink)	

Source: American Gilsonite Co., reprinted from AMERICAN INK MAKER, GILSONITE IN PACKAGING INKS, Formula 7 from Table II.

Table 13.19. Formulation for Packaging Ink, Rosin/Rubber

Component	Percent Weight
Industrial carbon black	12.0
Methyl ethyl ketone	25.0
Toluene	23.0
Soya lecithin	1.0
Chlorinated rubber	6.0
Dioctyl phthalate	3.0
Limed rosin	30.0

(This is a black gravure packaging ink)

Source: American Gilsonite Co., reprinted from AMERICAN INK MAKER, GILSONITE IN PACKAGING INKS, Formula 4 from Table II.

Table 13.20. Formulation for Paste Ink, Resin/Wax

Component	Percent Weight
Epolene C-10 Wax	17.0
Eastman Resin H-130	33.0
Magie 470 Oil	50.0

(This is a paste ink compound with great flexibility and toughness)

Source: Eastman Chemical Products, Inc., publication No. F-174B, EPOLENE WAXES AS ADDITIVES FOR INKS, Formula 1 from Table II.

Table 13.21. Formulation for Polyethylene Ink, Resin

Component	Percent Weight
Carboset XL-37 Resin	72.87
Benzidine yellow	6.27
Colloid 680 Defoamer	4 drops
Water	13.18
Silane A-1120 Adhesion Promoter	1.28
Ammonium stearate (33% solids)	6.40

(This is a waterborne printing ink with good adhesion to treated and untreated polyethylene; designed for flexographic printing or breadwrappers and other nonabsorbent packaging substrates)

Source: BF Goodrich Co., data sheet CR-79-7, CARBOSET RESINS, suggested formulation.

Table 13.22. Formulation for Process Ink Varnish/Oil

Component	Percent Weight
D49-2286 Flushed Color	40.0
Varnish	50.0
Tetron 60	4.0
TXIB Solvent	3.0
Magiesol 47 Oil	3.0

(This is a process blue ink)

Source: Sun Chemical Corp., technical bulletin, FLUSH COLOR PRODUCT LINE, Formulation B.

Table 13.23. Formulation for Thermoplastic Ink, Resin/CAP

Component	Percent Weight
CAP-504-0.2 Cellulose Acetate Propionate	6.10
Sucrose acetate isobutyrate (SAIB)	1.50
Kodaflex DOP Plasticizer	4.10
Uni-Rez 710 Maleic Resin	8.20
Pigment	5.10
Isopropanol (99%)	56.30
Water	18.70

(This is a thermoplastic ink with excellent adhesion to treated polypropylene, dries rapidly, and has good gloss)

Source: Eastman Chemical Products, Inc., formulator's notes No. E-4.1C, CELLULOSE ACETATE PROPIONATE INKS FOR FLEXIBLE SUBSTRATES, Formula FLPR-24.

Table 13.24. Formulation for Flexo/Gravure Acrylic Ink

Component	Percent Weight
Joncryl 142 Acrylic Polymer Emulsion	75.0
Balab 748 Defoamer	0.5
Isopropanol	5.0
Water	18.2
Ammonia (28%)	1.3

(This is a flexo/gravure low-solids, water-based system)

Source: S. C. Johnson & Son, Inc., technical service information, JONCRYL 142, Formula I.

Table 13.25. Formulation for Flexo/Gravure Ink, Acrylic/Polyethylene

Component	Percent Weight
Joncryl 87 Styrenated Acrylic Dispersion	20.0
Jonwax 22 Microcrystalline Wax Emulsion	5.0
Joncryl 67 Acrylic Resin	12.0
Ammonia (28%)	1.6
Morpholine	1.0
Isopropanol	4.0
Dibutyl phthalate	1.2
Ethylene glycol monoethyl ether	1.2
Water	39.8
Sag 471 Antifoam	0.2
Organic pigment	14.0

(This is a flexographic or gravure ink, with fast drying, good finish, water resistance, and good printability)

Source: S. C. Johnson & Son, Inc., technical service information, JONCRYL 67, suggested formulation.

Table 13.26. Formulation for Flexo/Roto Ink, Acrylic

Component	Percent Weight
Joncryl 67 Acrylic Resin	20.0
Ammonia (28%)	4.7
Water	75.3

(This is a straw-colored popular water ink varnish)

Source: S. C. Johnson & Son, Inc., technical service information, JONCRYL 67 ACRYLIC RESIN, Resin Cut A.

Table 13.27. Formulation for Flexo/Roto Ink, Acrylic/EHEC

Component	Percent Weight
Joncryl 67 Acrylic Resin	28.75
Dibutyl Phthalate	2.00
Dye	0.75
Cellosolve Solvent	1.50
Ethyl alcohol	66.10
Ethylhydroxyethylcellulose (EHEC)	0.15
Plasticizer	0.75

(This is an excellent replacement for solvent-borne systems, and has good adhesion, durability, and water resistance)

Source: S. C. Johnson & Son, Inc., technical service information, JONCRYL 67 ACRYLIC RESIN, suggested formulation.

Table 13.28. Formulation for Gravure Ink, Cellulose Nitrate/Oil

Component	Percent Weight
Inorganic pigment	32.0
Cellulose Nitrate RS Type	8.0
Epoxidized soya oil	5.0
Ethanol	30.0
Isopropyl acetate	20.0
Toluene	3.0
Polyethylene wax	2.0

(This is a gravure type C ink with inorganic pigment)

Source: Braznell Co., NAPIM PATTERN PRINTING INK FORMULAE, Formula #306.

Table 13.29. Formulation for Gravure Ink, Polyethylene/Wax

Component	Percent Weight
Polystyrene	20.0
Toluene	10.0
Isopropyl acetate	20.0
Methyl ethyl ketone	39.0
VMSP Naphtha	8.0
Refined paraffin wax	3.0

(This is a type X gravure toplacquer ink)

Source: Braznell Co., NAPIM PATTERN PRINTING INK FORMULAE, Formula #315.

Table 13.30. Formulation for Gravure Ink, Resin/Nitrocellulose

Component	Percent Weight
RS Nitrocellulose	4.62
Lewisol 28 Synthetic Resin	7.07
Dibutyl phthalate	2.53
Unitane OR-580 Titanium Dioxide	25.30
Ethanol	1.97
Isopropyl acetate	23.28
1,1,1-trichloroethane	35.23

(This is a gravure type C ink, low volatile compounds, formulated with chlorinated solvents)

Source: Hercules, Inc., technical information CSL-193D, suggested formulation.

Table 13.31. Formulation for Heatset Ink, CAP

Component	Percent Weight
CAP-482-0.5 Cellulose Acetate Propionate	10.0
Tecsol C Solvent	59.50
Ethyl acetate (99%)	25.50
Dye	5.0
(This is a heat transfer printing ink)	

Source: Eastman Chemical Products, Inc., formulator's notes No. E-4.3D, CELLULOSE ACETATE PROPIONATE IN HEAT TRANSFER PRINTING INKS, suggested formulation.

Table 13.32. Formulation for Letterpress Ink, Glycol/Resin

Component	Percent Weight
Joncryl 67 Acrylic Resin	30.0
Ethylene glycol	60.0
Diethylene glycol monobutyl ether	5.0
Ammonia (28%)	3.0
Morpholine	2.0
(This is a water-washable letterpress ink, fast drying and excellent water resistance, useful for paper napkins, etc.)	

Source: S.C. Johnson & Son, Inc., technical service information, JONCRYL 67 ACRYLIC RESIN, Formula 1904 W122.

Table 13.33. Formulation for Letterpress Ink, Oil

Component	Percent Weight
Elftex Pellets 115 Carbon Black	10.5
Gilsonite Solids	2.0
Mineral oil	87.5
(This is a black letterpress newspaper ink formulation, yields a flat, blue-toned print, and does not have strike-through or excessive ruboff)	

Source: Cabot Corp., Technical Report S-27, CARBON BLACK SELECTION FOR PRINTING INKS, suggested formulation.

Table 13.34. Formulation for Letterpress Ink, Oils/Resins/Polyethylene

Component	Percent Weight
Pigment (color)	40.0
Picco 6140 Resin	10.0
Isophthalic alkyd resin	1.5
Phenolic modified penta ester of rosin	15.0
Polyethylene wax	1.5
Hydrocarbon petroleum distillate (C_{12}–C_{16} range, IBP 470°F)	20.0
Hydrocarbon petroleum distillate (C_{12}–C_{16} range, IBP 510°F)	3.0
Petroleum distillate (C_{12}–C_{16} range, IBP 535°F)	9.0
(This is a colored heatset letterpress ink)	

Source: Braznell Co., NAPIM PATTERN PRINTING INK FORMULAE, Formula #105.

Table 13.35. Formulation for Lithographic Ink, Acrylate/Benzophenone

Component	Percent Weight
Pigment	15.0
Epoxidized oil acrylate	73.0
Benzophenone	9.0
Michlers ketone	1.0
Polyethylene wax (may be modified with microcrystalline wax)	2.0
(This is an ultraviolet curing lithographic ink)	

Source: Braznell Co., NAPIM PATTERN PRINTING INK FORMULAE, Formula 209.

Table 13.36. Formulation for Thermal Curing Lithographic Ink, Oil/Resin

Component	Percent Weight
Pigment	14.0
Castor oil (grade 3)	56.0
Maleic modified penta ester of rosin	14.0
Synthetic paraffin wax	1.5
Hexamethoxymethylmelamine	12.0
Paratoluene sulfonic acid	1.5
Glycerol-allyl ether	1.0
(This is a thermal curing, catalytic ink)	

Source: Braznell Co., NAPIM PATTERN PRINTING INK FORMULAE, Formula 210.

Table 13.37. Formulation for Lithographic Oil/Resin

Component	Percent Weight
Pigment	14.0
Naphthenic mineral oil (C_{46}–C_{50} range)	50.0
Picco 6140 Resin	15.0
Hydrocarbon petroleum distillate (C_{12}–C_{16} range, IBP 470°F)	21.0
(This is a non-heatset lithographic web offset (newspaper) ink with low pigment level)	

Source: Braznell Co., NAPIM PATTERN PRINTING INK FORMULAE, Formula 201.

Table 13.38. Formulation for Offset Ink, Oil/Varnish

Component	Percent Weight
Elftex Pellets 115 Carbon Black	18.0
Hydrocarbon resin varnish	41.0
Mineral oil	41.0
(This is black web-offset newspaper ink for porous stock)	

Source: Cabot Corp., Technical Report S-27, CARBON BLACK SELECTION FOR PRINTING INKS, suggested formulation.

Table 13.39. Formulation for Quickset Ink, Varnish

Component	Percent Weight
U49-2356 Flushed Color	32.00
Varnish	61.00
MPP-620VF Polyethylene	2.50
Fluo HT Dry Teflon Compound	0.50
Cobalt drier (6%)	0.75
Manganese drier (6%)	1.25
535 Oil	2.00
(This is an infrared heat quickset ink with good tack rise, rub, and set)	

Source: Sun Chemical Corp., technical bulletin, FLUSH COLOR PRODUCT LINE, Formulation A.

Table 13.40. Formulation for Rotogravure Ink, Acrylic

Component	Percent Weight
Moly orange	40.0
Joncryl 61LV Acrylic Resin Solution	25.0
Water	5.0
Joncryl 134 Acrylic Polymer Emulsion	30.0

(This is a fast-drying ink with resolubility, organic solvent, compatibility, low viscosity, water and grease resistance, high solids, easy washup, and no overnight settling)

Source: S. C. Johnson & Son, Inc., technical bulletin, JONCRYL 134, suggested formulation.

Table 13.41. Formulation for Rotogravure Ink, Nitrocellose/Resin

Component	Percent Weight
RS Nitrocellulose, 1/2 seconds	39.2
Dewaxed dammar	39.2
Castor oil	9.8
Dioctyl phthalate	9.8
Syloid 308 Silica	2.0

(This is a rotogravure ink, with mar resistance, good gloss and clarity)

Source: Hercules, Inc., Technical Bulletin CSL-120A, POLYETHYLENE AS A MARPROOFING AGENT, Formula 3.

Table 13.42. Formulation for Screen-Process Ink, Alkyd/Resin

Component	Percent Weight
Organic pigment	7.0
Extender	50.0
Styrenated alkyd resin	24.0
Piccotex 120 Resin	10.0
Aromatic Hydrocarbon Solvent (IBP 370°F)	6.0
Technical hydrobiety alcohol (85–90% in xylene)	2.0
Cobalt naphthenate drier (6%)	1.0

(This is an enamel type with organic pigment)

Source: Braznell Co., NAPIM PATTERN PRINTING IN FORMULAE, Formula #501.

Table 13.43. Formulation for Screen-Process Ink, Binder/Plasticizer

Component	Percent Weight
Solvent	65.0
Binder	10.0
Plasticizer	10.0
Ethylcellulose	10.0
Wetting agent	5.0
(This is a conventional air-dried process formula)	

Source: Hercules, Inc., Technical Bulletin M-340A, CELLULOSE POLYMERS IN CERAMICS, Table I.

Table 13.44. Formulation for Sheetfed Ink, Varnish/Polyethylene

Component	Percent Weight
B49-2210 Flushed Color	40.0
Infrared Quickset Varnish	49.5
Anti-offset compound	3.0
S-394 Polyethylene Wax	2.5
Teflon compound	0.5
Manganese drier (6%)	1.0
Cobalt drier (6%)	0.5
500 Oil	3.0
(This is a sheetfed quickset infrared ink with good rub and set)	

Source: Sun Chemical Corp., technical bulletin, FLUSH COLOR PRODUCT LINE, suggested formula.

Table 13.45. Formulation for Clear Varnish, Acrylic

Component	Percent Weight
Joncryl 67 Acrylic Resin	50.0
Ethanol	50.0
(This is a clear varnish for paper, etc.)	

Source: S. C. Johnson & Son, Inc., JONCRYL 67, suggested formula.

Table 13.46. Formulation for Varnish, Nitrocellulose

Component	Percent Weight
Nitrocellulose (70% nonvolatile 30/35 seconds SS)	35.0
Ethyl cellulose	35.0
Ethyl acetate	30.0
(This is a clear nitrocellulose varnish for paper, etc.)	

Source: S. C. Johnson & Son, Inc., JONCRYL 67, suggested formula.

Table 14.1. Ink Materials, Chemical Description, and Source

Raw Material	Chemical Description	Source
A49-1551 Flushed Color	Phthalo Blue (40% pigment)	Sun Chemical Corp.
Abitol Hydrobietyl Alcohol	Technical grade of hydrobietyl alcohol, derived from rosin	Hercules, Inc.
A-C 6 Polyethylene Resin	Polyethylene homopolymer resin. Softening point 222°F	Allied Chemical Corp.
Acryloid B-72 Polymer	Acrylic ester resin	Rohm & Haas Co.
Acryloid NAD-10 Polymer	Acrylic ester resin	Rohm & Haas Co.
Aerosil R-972 Hydrophobic Silica	Hydrophobic silica	Degussa Corp.
Amberol M-82 Polymer	Phenolic resin	Rohm & Haas Co.
Arochem 404 Resin	Maleic resin	Spencer-Kellogg
Aromatic Solvent SC-100	Petroleum solvent with 311°F IBP	Exxon Chemical
Aromatic Solvent SC-100	Petroleum solvent with 362°F IBP	Algan, Inc.
ASM-5029 Alglos Setmaster Varnish	Ultrafast quickset letdown varnish. Modified phenolic/T.S.O.R. in Nagie 470	Engelhard Minerals & Chemicals
B19-1750 Flushed Color	Lithol Rubine, B.S. (33% pigment)	Sun Chemical
B49-1202 Flushed Color	Phthalo Blue, G.S. (37% pigment)	Sun Chemical
B49-1752 Flushed Color	Phthalo Blue, G.S. Pigment color 49	Sun Chemical
B49-2194 Flushed Color	Carbon Black. Pigment class Black 7	Sun Chemical
B49-2210 Flushed Color	Phthalo Blue, G.S. (36% pigment)	Sun Chemical
B49-2262 Flushed Color	Phthalo Blue, G.S. (40% pigment)	Sun Chemical
B49-2316 Flushed Color	Phthalo Blue, G.S. (34% pigment)	Sun Chemical
Balab 748 Defoamer	Organic, nonsilicone proprietary defoamer (100% active)	Witco Chemical
Bartyl F Anti-skinning Agent	Proprietary composition antiskinning agent	Sindar Corp.
Beckamine 21-511 Resin	Urea-formaldehyde resin (60% solids in alcohol)	Reichhold
Bentone 38 Gelling Agent	Organo-clay thixotropic additive	NL Chemicals
Bentone 500 Rheological Additive	Organo-clay rheological additive	NL Chemicals
Bronze Powder XM18G	Gold pigment. 5.5 μm average particle size	Obron Corp.
Butyl Cellosolve Solvent	Ethylene glycol monobutyl ether acetate solvent	Union Carbide
BYK-301 Resin (50%)	Ink resin	Mallinckrodt

(continued)

Table 14.1. (Continued)

Raw Material	Chemical Description	Source
CAB-381-0.5 Cellulose Acetate Butyrate	Cellulose acetate butyrate	Eastman Chemical
CAB-482-0.5 Cellulose Acetate Propionate	Cellulose acetate propionate ester	Eastman Chemical
CAB-504-0.2 Cellulose Acetate Propionate	Cellulose acetate propionate ester	Eastman Chemical
Carbitol Solvent	Diethylene glycol monoethyl ether solvent	Union Carbide
Carboset XL-37 Resin	Acrylic polymer (35% solids)	B.F. Goodrich
Cellolyn 21 Synthetic Resin	Dibasic-acid-modified rosin ester	Hercules, Inc.
Cellosolve Solvent	Ethylene glycol monoethyl ether solvent	Union Carbide
Chlorafin 40 Chlorinated Paraffin	Chlorinated paraffin. 40% chlorine content	Hercules, Inc.
Cobalt/Manganese Drier 2.4	Cobalt/manganese tallate mixture. 2/4% ratio	Shepherd Chemical
Colloid 675 Defoamer	Proprietary composition defoamer. 100% active	Colloids, Inc.
Colloid 680 Defoamer	Proprietary composition defoamer	Colloids, Inc.
D49-2035 Flushed Color	Phthalo Blue, G.S. (37% pigment)	Sun Chemical
D49-2286 Flushed Color	Phthalo Blue, G.S. (85% pigment)	Sun Chemical
D49-2397 Flushed Color	Phthalo Blue, G.S. Color 49	Sun Chemical
Day-Glo A Pigment Series	Fluorescent pigment series	Day-Glo Color
Day-Glo AX Pigment Series	Fluorescent pigment series. Stronger than A line	Day-Glo Color
Day-Glo IRB Base Color	Fluorescent pigment series	Day-Glo Color
Day-Glo Special Heatset Base	Special heatset pigment base series	Day-Glo Color
Decotherm Varnish	Printing ink vehicle, for high gloss (78% solids)	Lawter
Diarylide Yellow 1270	Dichlorobenzidine coupled pigment. Pigment Yellow 14. Color Index No. 21095	Harshaw Chemical
Drier #1269 Paste	Metal salt of neodecanoate acid (6% cobalt paste)	Shepherd Chemical
Dyall C-124 Polyethylene Dispersion	Polyethylene dispersion	Lawter
Dyall C-306 Wax Compound	Wax compound	Lawter
Eastman Resin H-130	Ink resin	Eastman Chemical
Ektasolve EB Solvent	Ethylene glycol monobutyl ether solvent	Eastman Chemical
Ektasolve EE Solvent	Ethylene glycol monoethyl ether solvent	Eastman Chemical
Elftex 8 Carbon Black	Furnace process carbon black. 27 nm particle size	Cabot Corp.
Elftex Pellets 115 Carbon Black	Furnace process carbon black. 27 nm particle size	Cabot Corp.
Elvacite 2013 Resin	Acrylic resin	du Pont
Epolene C-10 Wax	Polyolefin wax. Softening point 104°C	Eastman Chemical

(*continued*)

Table 14.1. (Continued)

Raw Material	Chemical Description	Source
Epolene C-13 Wax	Polyolefin wax. Softening point 110°C	Eastman Chemical
Ester Gum 8D	Glycerol ester of rosin	Hercules, Inc.
Ethyl cellulose	Organosoluble ethyl ether of cellulose	Hercules, Inc.
Ethylhydroxyethyl cellulose	Organosoluble ethyl ether of cellulose	Hercules, Inc.
Exkini #2 Anti-skinning Agent	Antiskinning agent of the volatile oxime type	Tenneco Chemicals
Filtrez 525 Resin	Fumaric resin. Melt point 148°C	FRP Co.
Filtrez 526 Resin	Fumaric resin. Melt point 130°C	FRP Co.
Filtrez 530 Resin	Fumaric resin. Melt point 150°C	FRP Co.
Filtrez 593A Resin	Fumaric resin. Melt point 130°C	FRP Co.
Filtrez 5001 Varnish	Varnish for inks	FRP Co.
Filtrez 5008 Resin	Fumaric resin	FRP Co.
Filtrez 5012 Resin	Fumaric resin. Melt point 135°C	FRP Co.
Filtrez 5014 Resin	Fumaric resin. Melt point 140°C	FRP Co.
Filtrez 5400 Resin	Fumaric resin. Melt point 130°C	FRP Co.
Flexiverse Dispersion	Pigment dispersion line	Grow Polymer
Fluo HT Dry Teflon Compound	Micronized PTFE. Melt point 620°F	Micro Powders
Grocryl P-260 Polymer Emulsion	High-solids (48%), low-viscosity polymer emulsion	Grow Polymer
Grocryl 6057 Modified Acrylic	Modified acrylic copolymer (40% solids)	Grow Polymer
Groplex 6066 Vehicle	Polymer vehicle for inks	Grow Polymer
Gro-Rez 2020 Acrylic Resin Solution	Modified acrylic resin solution (30% solids)	Grow Polymer
Gro-Rez 2050 Acrylic Resin Solution	Acrylic resin dispersing vehicle solution	Grow Polymer
Gro-Rez 6064 Acrylic Resin Solution	Acrylic resin solution (24.5% solids)	Grow Polymer
Growax 35 Polyethylene Emulsion	Nonionic emulsion of 275°F melt point polyethylene	Grow Polymer
Gulf 581 Naphthenic Mineral Oil	Naphthenic mineral oil	Gulf Oil
Halex Repellant Varnish	Water/alcohol repellent varnish	Lawter
Harshaw 2737 Chrome Yellow	Chrome yellow pigment	Harshaw Chemical
Hercolyn D Resin	Hydrogenated methyl ester of rosin	Hercules, Inc.
Huber 80 Kaolin Pigment	Aluminum silicate extender pigment	J.M. Huber Corp.
Ionol CP Phenol Compound	2,6-Di-*tert*-butyl-4-methyl-phenol compound	Shell Chemical
IRB Base Color	Fluorescent pigment base color	Day-Glo Color
Joncryl 61 Acrylic Polymer Solution	Acid functional styrene/acrylic resin (34% solids)	S. C. Johnson
Joncryl 61 LV Acrylic Resin Solution	Improved acrylic resin varnish solution (34% solids)	S. C. Johnson
Joncryl 67 Acrylic Resin	Acrylic resin, versatile and hard, in flake form	S. C. Johnson
Joncryl 74F Acrylic Polymer Solution	Acrylic polymer solution (49% solids)	S. C. Johnson

(continued)

Table 14.1. (Continued)

Raw Material	Chemical Description	Source
Joncryl 77 Acrylic Polymer Solution	Acrylic polymer solution gloss vehicle (45% solids)	S. C. Johnson
Joncryl 80 Acrylic Polymer	Acrylic polymer (49% solids)	S. C. Johnson
Joncryl 87 Styrenated Acrylic Dispersion	Styrenated acrylic dispersion gloss vehicle (49% solids)	S. C. Johnson
Joncryl 89 Styrenated Acrylic Dispersion	Styrenated acrylic dispersion economical gloss (48% solids)	S. C. Johnson
Joncryl 99 Acrylic Solution Polymer	Acrylic solution polymer vehicle (37% solids)	S. C. Johnson
Joncryl 134 Acrylic Polymer Emulsion	Acrylic polymer emulsion for gravure (45% solids)	S. C. Johnson
Joncryl 138 Acrylic Polymer Dispersion	Acrylic waterborne polymer for high-gloss systems	S. C. Johnson
Joncryl 142 Acrylic Polymer Emulsion	Acrylic polymer emulsion for flexo/gravure (39% solids)	S. C. Johnson
Joncryl 537 Acrylic Emulsion Polymer	Acrylic detergent-resistant emulsion polymer (46% solids)	S. C. Johnson
Joncryl 678 Acrylic Resin	Acrylic resin in flake form	S. C. Johnson
Joncryl 682 Acrylic Oligomer	Solid grade acrylic oligomer for high-solids inks	S. C. Johnson
Joncryl 1535 Acrylic Mixing Vehicle	Acrylic mixing vehicle for metallic pigments (37% solids)	S. C. Johnson
Jonwax 22 Microcrystalline Wax Emulsion	Microcrystalline wax emulsion (35% solids)	S. C. Johnson
Jonwax 26 Polyethylene Wax Emulsion	Polyethylene wax emulsion (25% solids)	S. C. Johnson
Kodaflex DBP Plasticizer	Dibutyl phthalate plasticizer	Eastman Chemical
Kodaflex DOP Plasticizer	Dioctyl phthalate plasticizer	Eastman Chemical
Lactol Spirits Solvent	Aliphatic naphtha in the toluene evaporation range	Union Chemicals
Lewisol 28 Synthetic Resin	Maleic-modified glycerol ester of rosin	Hercules, Inc.
Lin-All P.I. Drier	Printing ink drier. 4.3% manganese metal	Mooney Chemicals
LoCal A-7-T Dispersion Vehicle	Dispersion vehicle systems	Lawter
LoCal FST Dispersion Vehicle	Dispersion vehicle system, with thixotropy	Lawter
LoCal G-33 Dispersion Vehicle	Medium heatset gel system	Lawter
Magie #2 Oil	Ink oil	Magie
Magie #3 Oil	Ink oil	Magie
Magie 415 Oil	Ink oil	Magie
Magie 470 Oil	Ink oil	Magie
Magie 500 Oil	Ink oil	Magie
Magie 535 Oil	Ink oil	Magie
Magie 590 Oil	Ink oil	Magie
Magiesol 47 Oil	Ink oil	Magie
Magruder I.R. Color Flush	I.R. color flush series	Magruder
Mineral Spirits 360	Mineral spirits solvent	Union Chemicals

(continued)

Table 14.1. (Continued)

Raw Material	Chemical Description	Source
Mogul L Carbon Black MP-22 Wax	Furnace process carbon black Micronized synthetic wax. Melt point 219°F	Cabot Micro Powders
MPP-123 Polyethylene Wax	Micronized polyethylene wax. Melt point 233°F	Micro Powders
Multimix Color Flush	Color flush series	BASF Wyandotte
Nacrylic 78-6175 Acrylic Copolymer	Solid alkali-soluble acrylic copolymer resin	National Starch
NiPar S-20 Solvent	2-Nitropropane solvent	Angus
NiPar S-30 Solvent	Mixed nitropropane isomers	Angus
Obron Bronze Pigment	Bronze pigment series	Obron
Obron XM-18 Pigment	Highest-quality bronze pigment	Obron
Obron XM-18G Pigment	Superfine ink lining	Obron
Parlon S10 Chlorinated Rubber	Chlorinated rubber viscosity grade	Hercules, Inc.
Parlon S20 Chlorinated Rubber	Chlorinated rubber viscosity grade	Hercules, Inc.
Pentalyn G Synthetic Resin	Pentaerythritol ester of rosin	Hercules, Inc.
Pentalyn K Synthetic Resin	Pentaerythritol ester of resin	Hercules, Inc.
Picco 6140 Resin	Proprietary aromatic resin	Hercules, Inc.
Piccotex 120 Resin	Thermoplastic copolymer resin. Softening point 120°C	Hercules, Inc.
Pliolite 50 Resin	High-styrene/butadiene resin	Goodyear
Poly-Em 40 Emulsion	Polyethylene emulsion	Gulf Oil
Pope BW-813 Black Flush	33% carbon black in mineral oil	Pope Chemical
Pope VWO Hydrocarbon/Mineral Oil Vehicle	Hydrocarbon/mineral oil vehicle	Pope Chemical
Raven 500 Furnace Black	Industrial furnace black. Mean particle diameter 56 nm	Columbian Chemicals
Raven 890 Carbon Black	Industrial furnace black. Mean particle diameter 30 nm	Columbian Chemicals
Regal 330R Carbon Black	Furnace process carbon black. 25 nm particle size	Cabot Corp.
Regal 400R Carbon Black	Furnace process carbon black. Medium flow	Cabot Corp.
Regal 500 Carbon Black	Furnace process carbon black. Regular color	Cabot Corp.
Resimene V-980 Resin	Ink resin	Monsanto
Rex Orange X-1939 Pigment	Coprecipitated lead pigment	Hercules, Inc.
RS Nitrocellulose, 1/2 Second	RS nitrocellulose, 1/2 second	Hercules, Inc.
RS Nitrocellulose, 5–6 Second	RS nitrocellulose, 5–6 seconds	Hercules, Inc.
S-394 Polyethylene Wax	Dry polyethylene wax	Shamrock Chemicals
Sag 471 Antifoam	Proprietary silicone antifoam	Union Carbide
Silane A-1102 Adhesion Promoter	Amino organofunctional silane	Union Carbide
Sucrose Acetate Isobutyrate (SAIB)	Sucrose acetate isobutyrate solvent	Eastman Chemical
Sunprint 996 Naphthenic Mineral Oil	Naphthenic mineral oil	Sun Petroleum
Surfynol 104-H Surfactant	Organic surfactant	Air Products
SWS-213 Silicone Compound	Silicone compound	SWS Silicones
Syloid 308 Silica	Micrometer-sized silica	Davison

(continued)

Table 14.1. (Continued)

Raw Material	Chemical Description	Source
Tecsol C Solvent	Special industrial solvent	Eastman
Tecsol 3 Solvent	Special industrial solvent	Eastman
Telura 797 Process Oil	Process oil	Exxon Chemical
Tetron 60 Heat-Set Compound	Fluorinated wax blend heatset compound (60% solids)	Lawter
Thixcin R Thixotrope	Powder form thixotrope	NL Chemicals
Ti-Pure R-902 Titanium Dioxide	Rutile titanium dioxide (99%+ assay)	Du Pont
Titanox 2090 Titanium Dioxide	Rutile titanium dioxide	NL Chemicals
Transaid 1280 Polymeric Material	Proprietary composition polymeric material	Grow Polymer
Trionol No. 7 Varnish	Quickset vehicle. #7 Litho viscosity	Lawter
TXIB Solvent	Proprietary solvent	Eastman
U49-2356 Flushed Color	Phthalo Blue, G.S. (50% pigment)	Sun Chemical
Ultrex Quickset Varnish	Gloss quickset varnish	Lawter
Uni-Rez 304 Resin	Maleic resin	Union Camp
Uni-Rez 710 Maleic Resin	Maleic resin. Softening point 143°C	Union Camp
Uni-Rez 7020 Resin	Maleic resin	Union Camp
Uni-Rez 7024 Maleic Resin	Modified maleic resin. Softening point 118°C	Union Camp
Unitane OR-580 Titanium Dioxide	Rutile titanium dioxide	American Cyanamid
V-2630 Urethane Q.S. Varnish	Urethane Q.S. varnish	Superior Varnish
Varnish 936	Ink varnish	Degen Oil
Versamid 930 Thermoplastic Polyamide Resin	Thermoplastic polyamide resin. Softening point 110°C	Henkel
WD-2507 Raw Umber	Raw umber pigment dispersion (60% pigment)	Daniel Products
WD-2509 Burnt Umber	Burnt umber pigment dispersion (40% pigment)	Daniel Products
XJ-12 Compound	Antioffset compound	Lawter
Zinc Oxide Solution #1	Zinc ammonium cross-linking agent (15% solids)	S. C. Johnson

Source: Flick (1985). Reprinted with permission of Noyes Publications.

Index

CPSIA information can be obtained at www.ICGtesting.com
Printed in the USA
LVOW08*0756150414

381775LV00005B/28/P

9 780306 455414